Real-Time and Deliberative Decision Making

NATO Science for Peace and Security Series

This Series presents the results of scientific meetings supported under the NATO Programme: Science for Peace and Security (SPS).

The NATO SPS Programme supports meetings in the following Key Priority areas: (1) Defence Against Terrorism; (2) Countering other Threats to Security and (3) NATO, Partner and Mediterranean Dialogue Country Priorities. The types of meeting supported are generally "Advanced Study Institutes" and "Advanced Research Workshops". The NATO SPS Series collects together the results of these meetings. The meetings are coorganized by scientists from NATO countries and scientists from NATO's "Partner" or "Mediterranean Dialogue" countries. The observations and recommendations made at the meetings, as well as the contents of the volumes in the Series, reflect those of participants and contributors only; they should not necessarily be regarded as reflecting NATO views or policy.

Advanced Study Institutes (ASI) are high-level tutorial courses intended to convey the latest developments in a subject to an advanced-level audience

Advanced Research Workshops (ARW) are expert meetings where an intense but informal exchange of views at the frontiers of a subject aims at identifying directions for future action

Following a transformation of the programme in 2006 the Series has been re-named and re-organised. Recent volumes on topics not related to security, which result from meetings supported under the programme earlier, may be found in the NATO Science Series.

The Series is published by IOS Press, Amsterdam, and Springer, Dordrecht, in conjunction with the NATO Public Diplomacy Division.

Sub-Series

A.	Chemistry and Biology	Springer
B.	Physics and Biophysics	Springer
C.	Environmental Security	Springer
D.	Information and Communication Security	IOS Press
E.	Human and Societal Dynamics	IOS Press

http://www.nato.int/science
http://www.springer.com
http://www.iospress.nl

Series C: Environmental Security

Real-Time and Deliberative Decision Making

Application to Emerging Stressors

Edited by

Igor Linkov

U.S. Army Engineer
Research and Development Center
Concord, Massachusetts, USA

Elizabeth Ferguson

U.S. Army Engineer
Research and Development Center
Vicksburg, Massachusetts, USA

Victor S. Magar

ENVIRON International Corporation
Chicago, Illinois, USA

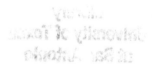

 Springer

Based on papers presented at the NATO Advanced Research Workshop on Risk,
Uncertainty and Decision Analysis for Environmental Security
and Non-chemical Stressors, Estoril,
Portugal, April 2007

Library of Congress Control Number: 2008934299.

ISBN 978-1-4020-9025-7 (PB)
ISBN 978-1-4020-9024-0 (HB)
ISBN 978-1-4020-9026-4 (e-book)

Published by Springer,
P.O. Box 17, 3300 AA Dordrecht, The Netherlands.

www.springer.com

Printed on acid-free paper

TABLE OF CONTENTS

Part 2.　Risk Assessment for Emerging Stressors

Part 3.　Multicriteria Decision Analysis: Methodology and Tools

**Part 4.   Applications of Multicriteria Decision Analysis for
         Environmental Stressors**

PREFACE

Over the past eight years, the editors of this book and their colleagues have convened a series of meetings to explore the topic of making policy decisions in the face of social and environmental management uncertainties. Each workshop has tried to demonstrate the power of risk assessment and decision analysis as tools that decision makers should use to understand complex environmental, economic, legal, social, and technological information and to make rational, informed decisions. The first workshop in this series, entitled, "Assessment and Management of Environmental Risks: Cost-efficient Methods and Applications" (Lisbon, Portugal; October 2000) [1], confirmed the role risk assessment could play as a platform for providing a scientific basis for environmentally sound and cost-efficient management policies, strategies, and solutions to various environmental problems. The second workshop, entitled, "Comparative Risk Assessment and Environmental Management" (Anzio, Italy; May 2002) [2], explored the development and application of comparative risk assessment (CRA) and other risk-based, decision-analysis tools in environmental management. The use of CRA was exceptional for facilitating decision making when various social, political, and economic activities compete for limited environmental resources. The third workshop, entitled, "The Role of Risk Assessment in Environmental Security and Emergency Preparedness in Mediterranean Region" (Eilat, Israel; April 2004) [3], focused on environmental security challenges in the Middle East and how risk assessment could resolve some of the region's pressing environmental needs. The fourth workshop, entitled, "Environmental Security In Harbors And Coastal Areas: Management Using Comparative Risk Assessment And Multi-Criteria Decision Analysis" (Thessaloniki, Greece; April 2005) [4], explored environmental security issues in ports, harbors, and coastal areas and how the use of MCDA, in conjunction with risk assessment, could evaluate environmental security threats, formulate responses to those threats, and evaluate the threat-reduction efficacy of different responses. The fifth workshop, entitled, "Management Tools for Port Security, Critical Infrastructure, and Sustainability" (Venice, Italy; March 2006) [5] merged the concepts of environmental risks and critical infrastructure vulnerabilities with the objective of harmonizing risk management and decision support methods and tools.

This book is based on discussions and papers presented at a sixth workshop, entitled "Risk, Uncertainty and Decision Analysis for Environmental Security and Non-chemical Stressors." This meeting—held in Estoril, Portugal, in April 2007—started with building a foundation to apply chemical risk assessment approaches and tools to a broad collection of non-chemical stressors, including physical (unexploded ordnance, noise, temperature, pH) and novel technologies, as well as emerging materials (nanomaterials, pharmaceuticals and pathogens), and biological agents (invasive species, biocontrol agents, biological warfare agents). More than 60 international science, risk assessment, decision-making, and security analysts from 14 countries discussed the current state-of-knowledge with regard to emerging stressors and risk management, focusing on the adequacy of available systematic, quantitative tools to guide vulnerability and threat assessments, evaluate the consequences of different events and responses, and support decision-making. This workshop, like those previous, was sponsored jointly by the Society for Risk Analysis and NATO.

The organization of the book reflects sessions and discussions during the Estoril meeting. The goals of the meeting were to review the needs, methods, and tools for real time and deliberative decision making and develop a solution-focused framework and supporting tools. The papers included in Part 1 focuses primarily on real time decision making in emergency settings. It includes reviews of military decision making, response to floods and homeland security threats as well as applications in business settings. Part 2 highlights regulatory and management challenges associated with non-chemical stressors, including nano-materials, suspended and bedded sediments, warfare agents, and physical environmental stressors. Part 3 provides multicriteria decision analysis methods and tools that are useful for supporting management decisions and focuses on methodology development. The final section encompasses a series of case studies that illustrate different applications and needs across MCDA applications.

The wide variety of content in the book reflects the workshop participants' diverse views and regional and global concerns. They also reflect the increasing complexity of societal and environmental change in response to the new global economy that requires an enhanced capacity for scientific assessment and management. Public pressure for decision transparency in government and corporations drives the need for more deliberative decision making and a framework for thinking about globalization and the prioritization of social and corporate resources that reaches beyond the realms of economics, world trade, and corporate management to include environmental protection and social goals. Risk assessment and decision analysis have been proposed to guide policies in many settings, including military, corporate, environmental, and others. Risk assessment offers a relatively objective, unbiased, and rational approach for framing and solving complex problems based on assessing the likelihood of hazard, potential population exposure, and resulting impairments. It is increasingly likely that decision analysis will supplement traditional risk assessment by providing the means for integrating heterogeneous scientific information, technical expert judgment, and seemingly intangible social values. The principles of risk assessment and decision analysis are unchanged regardless of the type of problems encountered. However, in practice, they are influenced by the temporal and spatial frame in which the decisions must be made. Some decisions, such as those made within a rapidly evolving crisis require extensive pre-planning and training in gaming scenarios to prepare decision makers so that they may react in a timely fashion. Other decisions in which the timing of the decision is less critical can occur in a more deliberative manner, including consensus building.

During the NATO workshop, we realized that much of the debate and differences in viewpoints arose from the varied experiences of the participants and differences in the nature of the decisions that they have addressed. Some scientists were influenced by experiences dealing with emergency responses to spills and catastrophic events such as the Class-4 hurricane that demolished New Orleans in 2004. Others were more familiar with slowly unfolding risks such as global climate change, nutrient enrichment, land use conversion, or the need to remediate long standing problems such as those encountered at contaminated sites. Still others were grappling with migration of exotic species or widespread exposures to pharmaceuticals and nano-particles with unknown mechanisms of action and unknown risks to people and the environment. It became apparent that risk assessment and decision analysis might be improved by considering not only the magnitude and extent of risks but also the time frame in which action must be taken to avert or minimize impacts.

The differences between rapid and deliberative decision analysis are apparent from the outset of risk assessment through the final decision analysis and remedy implementation. The differences begin with an initial recognition of the nature of the problem to be addressed and continue through the risk assessment, actions taken, and post-decision analysis of the process, including evaluation of the effectiveness of the decisions (Figure 1). In general, deliberative assessments are more likely to address a full range of conditions and outcomes. In contrast during rapid assessments, conditions and causes are often easily recognized or cursory condition and causal assessments are performed and the focus is on immediate risk and risk management. Both types of decisions are based on risk assessments that inform decision makers of their options and the potential outcomes of the choices before them. In both situations, the decision is expected to minimize harm. The assessments that inform the decisions are based on scientific knowledge that is derived from an understanding of causal mechanisms. The assessments often must consider a variety of types of information that are challenging to integrate into estimates of risk. In both situations, the assessor must transmit the synthesized information in a form that is understandable and transparent to the decision maker so that the decisions are defensible and decisive. Ultimately,

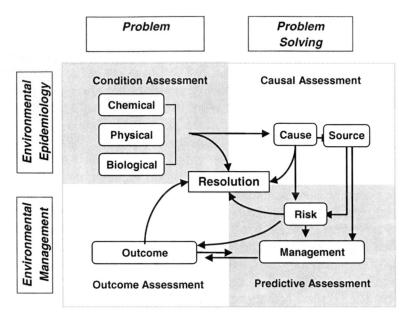

Figure 1. Environmental assessments framework [6].

only one decision can result from the assessment for a particular place and time, including a determination to defer a decision. Although risk assessments ideally are science-based, decisions about what to do are influenced by political, economic, and social drivers that may not be optimal for long term goals or for the majority of people affected by the decision.

Clearly, a continuum exists across a range of situations. Indeed, in the structuring, execution, and debriefing of planning scenarios, one goal is to achieve consensus regarding the proper course of decision-making for each scenario. This book opens a dialogue on aspects of risk assessment and decision analysis that apply to these two conditions, namely rapid versus more deliberative risk management processes.

Decision-making tools are urgently needed to support environmental management in an increasingly global economy. Addressing threats and identifying actions to mitigate those threats necessitates an understanding of the basic risk assessment paradigm and the tools of risk analysis to assess, interpret, and communicate risks. It also requires modification of the risk paradigm itself to incorporate a complex array of quantitative and qualitative information that shapes the unique political and ecological challenges of different countries and regions around the world. Establishing, maintaining, or enhancing a sense of environmental security will require (i) matching human demands with available environmental resources; (ii) recognition of environmental security threats and infrastructure vulnerabilities; and (iii) identification of the range of available options for preventing and minimizing natural disasters, technological failures, and terror actions. These three considerations will require input from different stakeholder perspectives, and a broad range of quantitative and qualitative sociopolitical, environmental, and economic information.

Assuredly, this is only the first step in a deeper exploration of the topic. We have provided this preface to draw distinctions between rapid and deliberative approaches. The subsequent contributions in this book provided by the participants of the 2007 NATO Advance Research Workshop in Estoril, Portugal give us an opportunity to consider what we might learn and how we might improve risk, uncertainty, and decision analysis by customizing methods and processes intended to deal with crisis and chronic situations.

Igor Linkov, Victor Magar, Elizabeth Ferguson, and Susan Cormier
June 2008

References

1. Linkov, I. and Palma-Oliveira, J.M. 2001. *Assessment and Management of Environmental Risks.* Kluwer, Amsterdam, 460p.
2. Linkov, I. and Ramadan, A. 2004. *Comparative Risk Assessment and Environmental Decision Making.* Kluwer, Amsterdam, 436p.
3. Morel, B. and Linkov, I. 2006. *Environmental Security: The Role of Risk Assessment.* Springer, Amsterdam, 326p.
4. Linkov, I., Kiker, G., and Wenning, R. 2007. *Environmental Security in Harbors and Coastal Areas: Management Using Comparative Risk Assessment and Multi-Criteria Decision Analysis.* Springer, Amsterdam.
5. Linkov, I., Wenning, R., and Kiker, G. 2007. *Managing Critical Infrastructure Risks: Decision Tools and Applications for Port Security.* Springer, Amsterdam.
6. Cormier, S.M. and Suter II, G.S. 2008. *Environmental Management. A Framework for Fully Integrating Environmental Assessment. published "Online First"* http://dx.doi.org/10.1007/s00267-008-9138-y.

ACKNOWLEDGEMENTS

The editors would like to acknowledge Dr. Mohammed Abdel Geleel (NATO workshop co-director) and organizing committee members (Drs. Steevens, Ramadan, Figueira, and Kiker) for their help in the organization of the event that resulted in this book. We also wish to thank the workshop participants and invited authors for their contributions to the book and peer-review of manuscripts. We are deeply grateful to Deb Oestreicher for her excellent editorial assistance and management of the production of this book. Additional technical assistance in the workshop organization was provided by Elena Belinkaia and Eugene Linkov. The workshop agenda was prepared in collaboration with the Society of Risk Analysis Decision Analysis and Risk Specialty Group. Financial support for the workshop was provided mainly by NATO. Additional support was provided by the ENVIRON International Corporation, Technical University Lisbon, Merck, Pfizer Inc., and the U.S. Army Engineer Research and Development Center.

PART 1

REAL-TIME AND DELIBERATIVE DECISION MAKING: NEEDS AND APPLICATIONS

COGNITIVE ASPECTS OF BUSINESS INNOVATION

Scientific Process and Military Experience

I. LINKOV

US Army Engineer Research and Development Center
83 Winchester Street, Suite 1
Brookline, MA, USA 02141
igor.linkov@usace.army.mil

C. SHILLING

Amarin Neuroscience Ltd.
UK

D. SLAVIN

Institute of Biomedical Engineering
Imperial College London
London, UK

E. SHAMIR

Institute for National Strategic Studies
Israel

Abstract: Increasing information richness and the changing sociopolitical environment in recent years have resulted in changes in corporate structure and organization. The growing challenges of organizational and technological complexities require the development of new organizational concepts. The effects of a combination of high complexity and high uncertainty have been recognized before in military settings. To take advantage of new technologies and manage information complexity, a theory of network-centric operations (NCO) was developed. Mission Command (MC) and NCO formulate organizational structure across functional domains (physical, informational, cognitive, and social), in a way that is also applicable in a business setting. In response to an increase in decision complexity and regulations, academia has developed risk assessment and multicriteria decision analysis (MCDA) tools for use in military and industrial settings. We believe that the combination of military science with MCDA and risk assessment has the potential to dramatically improve the credibility, efficiency and transparency of strategic and tactical decisions in industrial settings. This paper summarizes the military concepts of MC and NCO, and links them with mental modeling, risk assessment, and decision analysis tools. Application of the combined framework to the pharmaceutical industry is also discussed.

1. Introduction

The ability to make good decisions and communicate their impact is crucial to any business. Providing timely, clear direction based on the best available information is at the heart of both setting and achieving an organization's aims. Indeed, the ability to consistently make the right decision at the right time can be a significant competitive advantage. Although perhaps an obvious statement, it is important to remember that the operational implementation of a strategy requires a decision maker to guide the application of people and materials to a process, through the collection, analysis, and use of information. As information sources and volumes continue to multiply, the certainty that a decision is being based upon the right and best available information decreases—the paradox of uncertainty caused by too much information that may or may not be relevant to any given decision, resulting in an increased uncertainty as to the sound footing of any decision.

Today's competitive business environment requires cooperative internal communications and operations, and must be tempered by an understanding of the mental models of internal and external stakeholders as well as the social, cultural, and technological challenges of bringing new products and services to market. The concept of multicriteria decision analysis (MCDA) offers a framework for surfacing and balancing the various perspectives and requirements of each stakeholder, and to consider which information is most important when compromising on a course of action.

In the traditional hierarchical, "full-service" model of a business, decision makers could at least feel they had some level of control over the implementation of their decisions across the entire research, development, marketing, sales, and supply chain processes. A hierarchical structure promising long-term employment and well established career paths maintained a strong link between employee and employer, so that a company could to some extent rely on a loyal workforce as a foundation for developing its business.

Industry globalization, new business models, and a changing workforce make traditional hierarchical organizational models less efficient in executing strategic and operational plans. As more and more companies seek to focus on their core value proposition, networks of partners and suppliers make major contributions not only to manufacturing a company's new products but to the research, development, and marketing of those products. The relationships that occur in an outsourced business model introduce a greater level of complexity to the implementation of a strategy. Business development leaders, managers, and scientists are increasingly involved in operations where they must make real-time decisions in the context of a combination of the internal cultural context and those of external stakeholders (e.g., governmental agencies, industrial partners, and customers). At the heart of effective operations in new product development is an organization's ability to reconfigure quickly to exploit an opportunity, whilst retaining a robust decision-making framework that ensures overall clarity.

The rise of the dispersed collaborative model of business, now often referred to as Open Innovation, introduces greater complexity to organizational management. It requires a different way of thinking about how an organization coordinates activities to deliver and derive value from a final product or service. Relationships within such collaborations occur on many levels at the same time; between the corporate entities, principal officers, project teams, accounting departments, and lawyers. Research partners may become competitors based on the output of their research (e.g., Schlumberger in the oil and gas industry). Competing companies may be linked by a common partner that must work with each of the competitors in their own way, with very different procedures and performance expectations.

The effects of a combination of high complexity and high uncertainty have been recognized before in military settings. The breakthrough technologies the world has experienced in the last three decades have brought military organizations to some radical thinking on how to increase organizational effectiveness and remain relevant in a changing world. Military organizations are commonly perceived as conservative, hierarchical, and rigid, as well as command and control oriented.[1] In fact, although some of these attributes do exist in parts of military organizations for historical and other reasons, there is also another side of the military which is less known: an innovative and adaptive one. Military organizations are dealing with what is probably the most difficult task: wining battles and wars. Fighting wars can be a very messy and complicated thing; anything can and will happen. Clausewitz, the great war philosopher, described war as the "kingdom of uncertainty," a place which is characterized by a "clash of wills."

The organizational concepts of Mission Command (MC) and network-centric operations (NCO) that have emerged in the military have important

---

[1] "Command and control" is a military term, and is commonly used in business in a negative connotation which implies strict management rules imposed from above and micromanaging.

implications for dealing with complex and uncertain environments, not only for military organizations but also for large organizations in general. This paper links military concepts with methods and tools of real or near-real time decision making (risk assessment, mental modeling, and MCDA). The methodology we propose provides the ability to establish and maintain clarity of understanding and communication across multiple relationships, whilst preserving the flexibility and agility necessary to meet changing needs.

2. Military Concepts

Military organizations have dealt with decision and management complexity for a long time. Whilst we acknowledge that many theories and approaches to dealing with complexity have been developed by military science, we are focused on the concepts of MC and NCO because of their specific applicability to emerging industries. MC involves the assignment of a mission or task, rather than a set of instructions, to a subordinate. The subordinate then analyses the mission, having been provided with a framework of understanding or context and the support/resources needed to succeed. NCO offers a new form of organizational behavior that seeks to translate an information advantage, supported by technology, into a competitive advantage through robust networking.

2.1. MISSION COMMAND

MC, or as it has been known by its German name, Auftragstaktik,[2] is a decentralized leadership and command philosophy that demands and enables decision and action in every echelon of command where there is an intimate knowledge of the battlefield situation. MC, derived from the Auftragstaktik concept, is believed to have been initially developed by the German army in a gradual process, following the shocking defeat of the Prussians in Jena by the innovative army of Napoleon. It calls for subordinates to exploit opportunities by being empowered to use their initiative and judgment, as long as their decisions serve the higher objective communicated to them prior to the mission, which is referred to as intent. It is based on the belief in an individual's ability to act wisely and creatively in order to solve a problem without having to resort to higher authority.

MC aims to avoid the drawbacks of centralized systems, which suffer frequently from a lack of flexibility and responsiveness. It also helps avoid the

---

[2] A number of translations are often used (mission type orders, directive control). The term sed here is the most common one: "mission command" is used in American Army official doctrine papers (FMs).

usual shortcomings of decentralized systems, that is, the lack of coordination and control. Through the use of the higher intent as a coordination mechanism, it goes beyond simple decision delegation and empowers subordinates; it provides a flexible framework that allows the exploitation of opportunities while maintaining the overall purpose of a military operation.

A key element in the success of this approach is the articulation and communication of the commander's intent. This is done through a framework for meaningful reception and dissemination of information which forces the superior commander to assess information and to convert it into a plan or idea, often referred to as a concept of operation, and then translate it into orders that reflect his chosen course of action in a way that is easily communicated and executed. The executed plan is then under constant revision and alteration according to the ever-changing situation, but these changes are always done according to the higher intent. This enables flexibility and responsiveness.

MC is an approach designed to deal with complex systems, large amounts of information and an ever-changing environment. It is not easy to understand or to carry out, and its implementation might run contrary to basic existing organizational cultures. It requires above all a shared doctrine, trust which implies tolerance for learning and latitude for honest mistakes, professionalism and inclination for initiative.

MC is based on the following basic dictums regarding the nature of warfare and human behaviour:

- The complexity and chaotic nature of the battlefield—what Clausewitz called "fog of war," "friction," and "uncertainty"—are an integral part of warfare and should be taken into account.

- Commanders and managers are leaders of complex systems; their mission is to understand how complex systems work through the idea of intent and thus be able to optimize subunits to produce the best result to support the system as a whole.

- Time is a critical factor: in low tactical levels the commanders must act within a very short time frame, and decision making cycles must be quick.

- Limitation of span of control; the best commander has nevertheless a limited capacity for information processing, therefore a necessity to share the burden with a limited number of subordinates.

- Technology, regardless of its sophistication, cannot make judgement calls or generate creativity as this capacity is uniquely human. Technology can only enhance communication and more efficiently process information.

- Better motivation and commitment is gained through active participation and an individual sense of executing one's own ideas and plans.

- As long fas these dictums remain true, MC can be effectively applied to the organization.

In the post World War II years following the defeat of the German Army, MC was somewhat neglected. During the years of the Cold War, the West, facing the Soviet threat, was searching for ways to balance its relative quantitative inferiority. In its investigation to explore the fighting qualities of the Wehrmacht, it discovered MC as a central virtue that gave the Germans an edge over their rivals. More specifically, it was viewed as a major principle to enable a fast Observe Orient Decide Act (OODA) loop principle which was developed by John Boyd, emphasizing the importance of quick adjustment of decisions and executions to changing situations. MC was first officially incorporated into the US Army 1982 Doctrine, known also as the AirLand Battle, which emphasized four main tenets: agility, depth, initiative, and synchronization [1]. This doctrine was put to effective use in the first Gulf War 1991. Since then it has been adopted by all NATO members and continues to be a central command approach in all major military doctrines [2].

2.2. NETWORK-CENTRIC WARFARE

Since the early 1990s, the world has experienced what some describe as an information revolution, a shift from industrial based society to one which is information based. Network-centric warfare (NCW) is the military expression of this change. In fact, many see the Gulf War as the watershed that marks the first conflict which was significantly dominated by information age characteristics [3].

NCW refers to the "combination of emerging tactics, techniques, and technologies that a networked force employs to create decisive warfighting advantage" [4]. NCW acts as an enhancing principle to accelerate the ability to know decide and act by "linking sensors, communication systems, and weapons systems in an interconnected grid" [4]. It is based on a variety of information technologies that should allow commanders to rapidly analyze and communicate critical information to friendly combat forces and to react quicker in a hostile environment. NCW therefore offers a technical tool that further enhances the OODA loop.

However, to be able to fully exploit these advantages, new patterns of behavior and forms of organizations are required. The new focus is on access and speed of information, sharing information and collaboration, therefore a radical transition from the traditional top down hierarchical organization is required. Instead NCW would best suit flat, networked organizations [5].

The changes NCW introduces can be described through the three main domains it influences:

- Physical domain. This represents the traditional dimension of war which includes forces moving through time and space.

- Information domain. This is where information is being created, manipulated, and shared, including command and control and intent.
- Cognitive domain. This is what goes on inside the mind of each individual, or in other words, how each individual interprets the world around them. It includes moral, leadership, experience, and situational awareness.

The required attributes and new capabilities of any joint force capable of conducting network-centric operations must be carefully considered for each of these three main domains. The combined synergetic effect of these three domains stands in the core of the NCW concept and provides three distinct advantages [6]:

- Forces achieve information superiority and as a result develop better understanding of their own situation vis-à-vis their enemy's situation.
- The need to aggregate people to create mass becomes obsolete, instead, improved ability to disperse forces using speed and precision over greater geographical distances.
- Improved command and control and as a result a rapid OODA loop.

According to theory, NCW organizations should adhere to a number of principles in order to fully exploit the information advantage. Each advantage is dependent upon a few such guiding principles:

- High quality shared awareness is achieved through the application of a collaborative network of networks.
- Dynamic self-synchronization and adaptivity sustained by skipping the traditional hierarchy when change is necessary.
- Elimination of organizational boundaries and creation of new processes to achieve rapid effects.
- Rapid speed of command achieved by turning information superiority into decision superiority

Above all, the overarching principle should be the ability to empower individuals at the edge of the organization, where they have the most interaction with the environment and can quickly make a resounding impact on this environment. This involves expanding access to information and the elimination of unnecessary constraints to get it. It implies enhanced peer to peer interactions on all levels of the organization [5].

3. Conceptual Model for Decision Making in Pharmaceutical Industry

In general, we concur with Alberts and Hayes [5], who propose NCO concepts (Figure 1) may be applied to a broader set of applications, including corporate

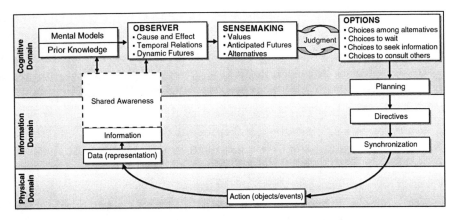

Figure 1. NCO and Effect-Based Operation Action/Reaction Cycle [7].

planning and decision making. Similar to the military, an organization's ability to reconfigure quickly to exploit an opportunity, whilst retaining a robust decision-making framework that ensures overall clarity, is at the heart of effective operations in corporate innovation. The pharmaceutical industry is a prime example of corporate complexity, and we will use it to illustrate how we think the combination of concepts and tools outlined above can be used in industry.

3.1. PHARMACEUTICAL INDUSTRY: SUMMARY OF CHALLENGES

The business of researching, developing and commercializing a new medicine is a complex and challenging undertaking fraught with uncertainty, in which scientific, technical, economic, ethical, and political issues are intertwined. During the course of the process different actors—pharmaceutical companies, academia, regulators and other government departments, contract research and manufacturing companies, pharmacies, healthcare professionals, and charitable organizations—interact at multiple levels to deliver treatments to patients. Each of these groups has a specific interest in the provision of healthcare, offering fertile ground for misunderstandings, conflict, and missed opportunities (Figure 2) .

Recent industry performance metrics put the average time to bring a medicine to the market as approximately ten years, at an average cost of $1.2 billion—the price of a 99% failure rate in the research, development and commercialization of new medicines. For many years the industry has dealt with the huge risks by extensive consolidation, driven by business economics to exploit economies of scale and scope. Scale gives an investment tolerance to cope with the risks inherent in uncertainty and to bring to bear the expertise and technology needed to deal with complexity. Scope allows companies to access diverse technology and

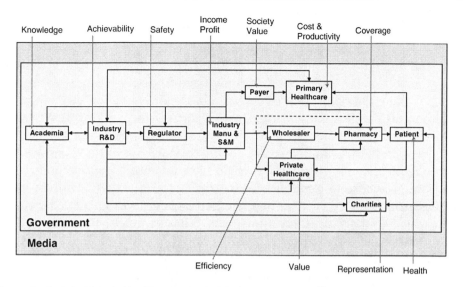

Figure 2. Stakeholders in the Pharmaceutical Industry and Different Drivers for their Actions.

intellectual capacity to apply to R&D challenges. In common with most large organizations, the challenge has been to operate at scale—enabling sharing across geographies, R&D portfolios, disciplines, and therapy areas—and has often resulted in inefficiency in decision making and communication processes.

At the same time, a number of factors have conspired to move the industry towards an increasingly extreme outsourcing model. The increased availability of cutting-edge technologies and the burgeoning biotech sector, coupled with the pressure to reduce costs in the face of falling reimbursement budgets and rising patent losses, has given rise to a dispersed "open innovation" R&D model, with some companies going so far as to do away with traditional research and look to outsource everything. Although attractive in terms of agility and cost base, this model increases the level of complexity in the interactions that deliver the body of research required by regulatory bodies to allow them to give a confident approval for a new medicine.

It is in this environment that R&D teams must discover and/or develop new medicines. The traditional corporate command and control structure may provide clear reporting lines, but can predispose an organization to follow bureaucratic, prolonged decision-making processes. Multidisciplinary teams have been used as a means of increasing organizational agility, but multiple reporting and approval responsibilities have the potential to destroy that agility. Either model (or combination thereof) has the potential to create decision conflicts. Understanding the decision and communication boundaries between multiple organizations provides an operating framework for complex collaborations to succeed.

3.2. APPLICATION OF MISSION COMMAND IN PFIZER: LESSON LEARNED

Like the human body, an organization is a complex adaptive system. Everything in it is related to everything else. Chains of causality are not linear. Picking the right point of leverage in the organization was similar to designing a treatment for a patient with a variety of symptoms. We had a treatment but had to decide upon the route of administration and dosage level. We decided to administer MC locally because we did not know all its effects. Administering it generally would have taken a long time and risked rejection. We chose to administer it to select project teams in full development because they were where the potential leverage was greatest, being the point where strategy and operations meet. They represent the main axis of value creation.

The "dosage" level we decided on was a set of two three-day workshops run by a small team which specializes in introducing MC to business. The workshops spent one day on teamwork and behavior, and one and a half days on analyzing the team's mission. The initial pilot was run with two teams whose leaders were keen to try it out.

Early indications are that applying the principles of MC in the pharmaceutical business is both safe and effective. The teams involved both responded very positively, and have reported far higher internal alignment and engagement with their projects. Clarifying their mission proved to be surprisingly valuable, resulting in what one project leader called "a real sense of clarity about what we needed to deliver and why." Internal structures have been simplified, meetings have been streamlined, and levels of accountability have increased.

People are beginning to believe that they really are empowered to take decisions and are therefore starting to take them. One of the teams achieved a filing deadline for regulatory submission, which at the beginning of the year was regarded as a forlorn hope with no more than a 10% chance of success. Another has taken a full three months out of its timeline.

The methodology also appears to be safe. It can be integrated with existing planning systems without causing disruption and does not involve costly new systems. The metrics the teams use to track their missions can be derived from a balanced scorecard. People are not abusing their freedom or running wild. One side effect of the increased focus on the main effort of getting drugs to patients was that commitments to internal projects suffered, with time allocation decisions more in favor of project teams. However, the business showed no signs of suffering as a result.

This initial treatment highlighted the need to adjust and realign the environment in which teams operate. There are implications for goal setting, performance management, budget responsibility, governance, and approval processes—indeed, our whole operating model. We can address these issues as we go, and have already started to do so. MC is increasingly setting our agenda.

As a next step we are running more teams through the workshops and have now launched an empowerment code which legitimizes the principles of MC throughout our Sandwich, UK, site. We have realized that this is not just about running some team-building workshops, but about changing our operating model and aspects of our culture. The one certainty about that is that it will take a long time. But then we are used to that. We are not certain what the operating model will look like, or how the culture will develop, but we do know what the main principles behind both of them are. The rest is uncertain. But then, we are used to that too. We are looking forward to the journey.

3.3. PHYSICAL, INFORMATION, COGNITIVE, AND SOCIAL DOMAINS IN PHARMACEUTICAL INDUSTRY

The emerging environment of complex collaborations described above increases the importance of setting criteria to drive the collection, reporting, and use of information for operational decision making. Three domains (physical, information, and cognitive) discussed in NCO literature provide a convenient way of thinking about the decision process, not only in the military but also industry. Table 1 compares and contrasts definitions of these three domains in the military and in the biotech/pharmaceutical industry.

4. Implementation Roadmap

We believe that the tools of risk assessment, MCDA, and mental modeling could be operational in transitioning the pharmaceutical industry from a rigid hierarchical structure to adaptive and efficient organizations. Risk assessment provides a quantifiable and intuitive description of actions and stimulus happening in the physical domain. Through the networked information domain, risk information can be transferred into cognitive domain. Mental modeling would allow efficient information assimilation and sense making to initiate decision-making process. Full-scale implementation of mental modeling will allow efficient communication, including cross-cultural and cross-disciplinary communication. MCDA would provide a foundation for adaptive assessment of risk and other criteria and also for influencing actions in the physical domain through selection of appropriate management alternatives. All this assessment takes place and is influenced by the social domain [7], which encompasses the sociopolitical and/or business environment where decisions take place.

TABLE 1. Attributes of Cognitive, Information, and Physical Domains in Military and Pharmaceutical Industry.

Domain	Military	Common elements	Biotech/Pharma
Physical	Theater of war Logistics Weapon systems	Physical infrastructure	Global markets Supply chain Laboratories
Information	Military intelligence, military communication networks, military information/management	Systems, databases/manuals	Competitor intelligence, corporate communication networks, corporate information/management
Cognitive	Future strategic dominance	Long-term strategic objectives	Future market position
	Clear purpose, straightforward allegiances (flag, regiment, service)		Complicated goals
	Single task/orientation		Complex allegiances (industry, company, department, site, group)
	Standardized military education		Dynamic multitasking
	Staff interoperability		Different science background
	Standardized personnel roles		Specialized expertise/difficult to replace
	Sense of history and continuity		Individualized work styles/approaches
	Societal recognition		Discontinuous careers
			Materialistic goals

4.1. RISK ASSESSMENT FOR PHYSICAL DOMAIN REPRESENTATION

For centuries, the aim of planning and war gaming within a military setting has been to understand and prepare for the potential outcomes of an action, knowing that some outcomes are more likely than others. Similarly, investment/portfolio decisions in business are increasingly complex and multivariate. Risk refers to the likelihood or probability for an adverse outcome. The concept of risk is applicable to an infinite set of decision problems in both military and corporate environment. Over the last several decades, the field of risk analysis encompassing methods for developing an understanding of the processes shaping the scope and nature of risks and uncertainties has evolved. The types of questions germane to risk analysis include:

- What are the risks?
- Why and how are the risks occurring?
- How do the risk management alternatives under consideration differ in terms of risk reduction performance?

- What is the uncertainty associated with the analysis?

Risk analysis is composed of four elements:

1. Hazard identification and characterization
2. Effects assessment
3. Risk characterization
4. Risk management

While the terminology and specific tools that are applied to risk analysis vary across disciplines (e.g., military, medicine, engineering, environmental management, economics), these four elements describe activities common to the majority of applications.

Hazard identification and characterization involves description of the nature of the events initiating and quantification of threats leading to the risks under consideration. Effects assessment involves characterization of the consequences resulting from the threat. Risk characterization integrates information about the likelihood/probability of events, or families of events, with information about consequence processes to produce a description of the likelihood for specific outcomes. Risk management concerns itself with answering questions related to evaluating what actions can be taken to reduce the risks (i.e., the probability for adverse outcomes).

4.2. MENTAL MODELING AND SENSE MAKING IN THE COGNITIVE DOMAIN

Risk descriptors of the physical domain are assessed by individuals in the cognitive domain. This sense-making step is rooted in individual cognition. Efficient sense making and further decision-making steps require understanding of the cognitive basis of sense making. We propose mental models as a tool which may be used to map cognitive drivers and corporate culture of different groups and then establish cross-group communication. Mental models are a complex web of deeply held beliefs that operate below the conscious level to affect how an individual defines a problem, reacts to issues, learns, and makes decisions about topics that come to their attention through communications. Mental models have been the focus of extensive research [8, 9]. It is well established that people's mental models vary in important but often unpredictable ways, strongly affecting their decision processes [10]. Research has demonstrated that the complexity of people's thinking makes it impossible to predict the effects of communication on people's mental models without empirical testing.

Mental models are often used to conceptualize shared cognition, which has been shown to be an essential component of team effectiveness [11].

Shared cognition focuses on peer learning and can be utilized in multiple contexts and multiple disciplines. Shared mental models may influence individual and team performance through their impact on members' ability to engage in coordinated actions. Team members with similar knowledge bases and cognitive mechanisms are more likely to interpret information the same way and to make accurate projections about each other's decisions and actions. The mental models approach to developing a sense-making process and communication entails five steps [9]:

1. Expert model (or integrated assessment). Identify the relevant aspects of a problem (in this case, specific strategies recommended for reducing PTSD impacts).

2. Lay model interviews. Characterize how members of the target audience frame and understand the problem.

3. Lay model survey. Quantify the prevalence of beliefs and misconceptions revealed in the interviews in the target population.

4. Comparative analyses of lay and expert models. Identify where members of the community need more information or guidance negotiating and implementing strategies (in this case, to reduce PTSD impacts).

5. Design and implementation of intervention. Design intervention based on these systematically identified targets, aiming to improve understanding, decision making, and negotiation, in order to reduce risk.

4.3. INFORMATION AGGREGATION AND DECISION MAKING WITHIN MCDA FRAMEWORK

Multiple streams of information originating from the physical domain and sensed through the prism of mental modeling in the cognitive domain and the external environment of the social domain need to be translated into actionable alternatives. The alternatives should be prioritized and implemented as actions in the physical domain. As with any new technology or science, developing a framework for resource prioritization and selection and making management decisions with uncertainty and incomplete information is the current challenge for industry. Risk is just one factor in making decisions in real-time situations. Making efficient management decisions requires an explicit structure for jointly considering the pros and cons of a decision, along with the associated uncertainties relevant to the selection of alternative courses of action. Integrating this heterogeneous and uncertain information demands a systematic and understandable framework to organize scarce technical information and expert judgment. Our current work for the U.S. Environmental Protection Agency and Department of Defense [12] shows that MCDA methods provide a sound approach to the management of heterogeneous information and risks. The advantages of using MCDA techniques over other less structured decision-making methods are numerous: MCDA provides a clear and transparent methodology for making decisions and also provides a formal way for combining information from disparate sources.

MCDA refers to a group of methods used to impart structure to the decision-making process to address complex challenges. Generally, these decision analysis methods consist of four steps:

1. Creating a hierarchy of criteria relevant to the decision at hand, for use in evaluating the decision alternatives

2. Weighting the relative importance of the criteria

3. Scoring how well each alternative performs on each criteria

4. Combining scores across criteria to produce an aggregate score for each alternative

Most MCDA methodologies share similar steps 1 and 3, but diverge on their processes for steps 2 and 4 [13]. A detailed analysis of the theoretical foundations of different MCDA methods and their comparative strengths and weaknesses is presented in [14].

We propose to follow a systematic MCDA framework developed by Linkov et al. [12] for alternatives generation and selection. A generalized MCDA process will be adjusted for the corporate environment. It will follow two basic themes:

1. Generating alternative management options, success criteria, and value judgments

2. Ranking the alternatives by applying value weights

The first part of the process generates and defines choices, performance levels, and preferences. The latter section methodically prunes nonfeasible alternatives by first applying screening mechanisms (e.g., significant risk, excessive cost) and then ranking, in detail, the remaining alternatives by MCDA techniques that use the various criteria levels generated by models, experimental data, or expert judgment. While it is reasonable to expect that the process may vary in specific details among applications and project types, emphasis should be given to designing an adaptive management structure that uses adaptive learning as a means for incorporating decision priorities.

The tools used within group decision making and scientific research are essential elements of the overall decision process. The applicability of the tools is symbolized in Figure 3 by solid lines (direct involvement) and dotted lines (indirect involvement). Decision analysis tools help to generate and map technical data as well as individual judgments into organized structures

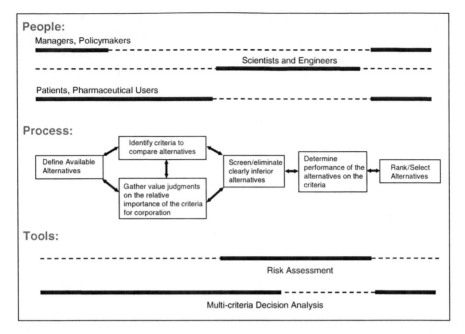

Figure 3. Example Decision Process. Dark Lines Indicate Direct Involvement/Applicability and Dotted Lines indicate less direct Involvement/Applicability.

that can be linked with other technical tools from risk analysis, modeling, monitoring, and cost estimations. Decision analysis software can also provide useful graphical techniques and visualization methods to express the gathered information in understandable formats. When changes occur in the requirements or the decision process, decision analysis tools can respond efficiently to reprocess and iterate with the new inputs. This integration of decision, scientific and engineering tools allows users to have a unique and valuable role in the decision process without attempting to apply either type of tool beyond its intended scope.

Three basic groups of stakeholders include managers and decision makers, scientists and engineers, and the affected public. These groups are symbolized in Figure 3 by dark lines for direct involvement and dotted lines for less direct involvement.

While the actual membership and function of these three groups may overlap or vary, the roles of each are essential in maximizing the utility of human input into the decision process. Each group has its own way of viewing the problem, its own method of envisioning solutions, and its own responsibility. Managers spend most of their effort defining the problem's context and the overall constraints of the decision. In addition, they may have responsibility for final alternative policy selection. Technology recipients may provide input in defining alternative pharmaceuticals, but they contribute the most input by helping formulate performance criteria and making value judgments for weighting the various success criteria. Depending on the problem and context, patients and users may have some responsibility in ranking and selecting the final pharmaceutical alternative. Scientists and engineers have the most focused role in that they provide the measurements or estimations of the desired criteria that determine the success of various materials and alternatives.

The result is a comprehensive, structured process for selecting the optimal alternative in any given situation, drawing from stakeholder preferences and value judgments as well as scientific modeling and risk analysis. This structured process would be of great benefit to decision making in management, where there is currently no structured approach for making justifiable and transparent decisions with explicit tradeoffs between social and technical factors. The MCDA framework links heterogeneous information on causes, effects, and risks for different pharmaceuticals with decision criteria and weightings elicited from decision makers, allowing visualization and quantification of the tradeoffs involved in the decision-making process. The proposed framework can also be used to prioritize research and information-gathering activities and thus can be useful for the value of information analysis.

References

1. US Army, FM 1982. Operations.
2. US Army, FM 2003. Command.
3. Toffler, A. and Toffler, H., 1993. *War and Anti War*. Boston, MA: Little Brown.
4. Director of Force Transformation, 2003. *Military Transformation: A Strategic Approach*. Office of the Secretary of Defense, Washington, DC: Pentagon.
5. Alberts, D.S. and Hayes, R.D., 2003. *Power to the Edge: Command & Control in the Information Age*. United States Department of Defense Command & Control Research Program.
6. Mandeles, M.D., 2005. *The Future of War: Organizations as Weapons*. Dulles, VA: Potomac Books.
7. Smith, E.A., 2006. Complexity, Networking, Effects-Based Approaches to Operations. CCRP Publication Series. Available at: http://www.dodccrp.org/files/Smith_Complexity. pdf.
8. Bostrom, A., Fischhoff, B., and Morgan, M.G., 1992. Characterizing mental models of hazardous processes: a methodology and an application to radon. Journal of Social Issues 48:85–100.
9. Morgan, M.G., Fischhoff, B., Bostrom, A., Atman, C., 2001. *Risk Communication: A Mental Models Approach*. New York: Cambridge University Press.
10. Fischhoff, B., Downs, J., and Bruine de Bruin, W., 1998. Adolescent vulnerability: a framework for behavioral interventions. Applied and Preventive Psychology 7:77–94.
11. Salas, E., and Cannon-Bowers, J. A. 2001. The science of training: a decade of progress. Annual Review of Psychology 52:471–499

12. Linkov, I., Satterstrom, K., Kiker, G., Batchelor, C.G., and Bridges, T. 2006. From comparative risk assessment to multi-criteria decision analysis and adaptive management: recent developments and applications. Environment International 32:1072–1093.
13. Yoe, C., 2002. *Trade-off Analysis Planning and Procedures Guidebook*. Prepared for Institute for Water Resources, U.S. Army Corps of Engineers.
14. Belton, V., and Stewart, T. 2002. *Multiple Criteria Decision Analysis: An Integrated Approach*. Boston, MA: Kluwer.

A SYNOPSIS OF IMMEDIATE AND DELIBERATE ENVIRONMENTAL ASSESSMENTS

S.M. CORMIER

U.S. Environmental Protection Agency, Office of Research and Development
National Center for Environmental Assessment
Cincinnati, OH 43268 USA
cormier.susan@epa.gov

Abstract: Environmental assessments can be classified by the urgency of the problem and therefore the amount of time allowed for the assessment before a decision is made to benefit environmental and social objectives. Deliberate (occurring in an unhurried fashion) and immediate (performed without delay) assessments have different constraints; and different value judgments or standards are used to judge their quality. Being aware of the differences and similarities can improve the quality of both deliberate and immediate environmental assessments. In particular, deliberate assessments can eventually provide knowledge or decision tools for future unanticipated emergencies.

1. Introduction

Why do some environmental assessments result in better outcomes than others? One reason is that some have a clear framework to organize planning, analysis, synthesis, and decision-making [2, 10]. Another is that circumstances place different constraints on time and resources [7]. The intention of this paper is to suggest a convenient way to organize any assessment [2] and to draw attention to the time and resource constraints by comparing the similarities and differences between immediate and deliberate assessments. The comparison itself is built upon a framework that fully integrates all types of environmental assessments and provides a clear framework to ensure good organization so that deliberate and immediate types of assessments will effectively inform decision making and achieve environmental and social objectives.

I. Linkov et al. (eds.), Real-Time and Deliberative Decision Making.
© Springer Science + Business Media B.V. 2008

2. Framework of Environmental Assessments

Environmental assessment is the process of providing scientific information to inform decisions to manage the environment [2]. They can be classified into four general types (Figure 1):

1. Condition assessments to detect chemical, physical and biological impairments

2. Causal pathway assessments to determine causes and identify their sources

3. Predictive assessments to estimate environmental, economic, and societal risks and benefits associated with different possible management actions [11]

4. Outcome assessments to evaluate the results of the decisions made using condition, causal, and predictive assessments [2]

The linkage between assessments is based on intermediate decisions that initiate another assessment or a final decision leading to the resolution of

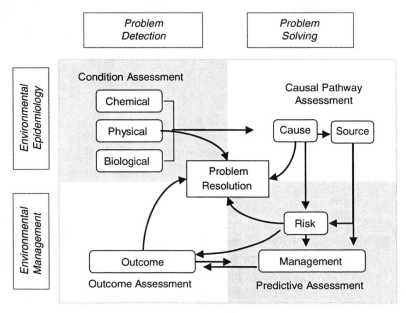

Figure 1. Flow of types (quadrants) and sub-types (oblongs) of assessments. Environmental Assessments evaluate the condition, causal pathway, prediction, and outcomes associated with problem solving or management. In general, deliberate assessments are more likely to address all of these types of assessments. Immediate assessments are more likely be a response to known causes and therefore condition and causal assessments are apt to be cursory and most of the effort focuses on risk and management options (lower two sectors) [2].

the problem [6]. By using a common structure of planning, analysis, and synthesis when describing activities within an assessment, the terminology is simplified and communication is facilitated between types of assessments and environmental programs.

Assessments can be further classified by the urgency of the problem and, therefore, the amount of time allowed before a decision is made to benefit environmental and social objectives. Although a dichotomous classification is used here, the distinction between deliberate and immediate assessments represents the extremes of a continuum of circumstances.

The differences between immediate and deliberate decision analysis are apparent from the onset of an assessment through the final decision and implementation. The differences begin with initial recognition of the nature of the problem to be addressed and they also affect the risk assessment, actions taken, and post-decision analysis of the process, including evaluation of the effectiveness of the decisions. The measures of success provide data to analyze approaches used and promote continual improvement of the process.

3. Types of Deliberate and Immediate Assessments

Deliberate assessments are undertaken when there is a long-term social commitment for implementation. For example, the goal of a deliberate assessment might involve the restoration of a river to its free-flowing condition, thus improving water quality and fish migration [12].

Immediate assessments are performed when there is imminent danger of irreversible and dire consequences, such as might occur prior to or in the aftermath of a hurricane or dam failure. Immediate assessments are sometimes called for when some action is planned that could have irreversible consequences; for example, a decision to permit mining near Yellowstone National Park [5] or to build roads into the habitat of the rare rhinoceros of Borneo [8]. Immediate data collection could be required if an ongoing effect such as an epidemic, or fish or bird kill, needs to be documented or its cause determined.

Deliberate assessments are more likely to involve long-term stakeholder interactions, data collection, uncertainty analysis, iterations, peer review, legal challenges, interventions, and reassessment. Deliberate assessments take more time, but when well planned combine all the types of assessments; that is: condition, causal pathway, predictive management, and outcome types.

Immediate assessments are more likely to depend upon past assessments, emergency action plans, scenario training, or access to experienced assessors and crisis decision analysts. They are constrained by time-critical decision points. Although the assessor and decision analysts may be cognizant of the contribution of all the types of assessments, they may be forced to bypass some assessments or take advantage of emergency action plans that attempt to provide for these needs in advance.

3.1. CONDITION ASSESSMENT

Deliberate condition assessments are performed to determine whether there is a problem and establish baselines prior to actions. They may be based on previous risk assessments or criteria; for example, comparing water quality to ambient water quality standards for metals [3] or comparing observed populations to expected assemblages [4].

Immediate condition assessment may simply document rather than assess if the condition is obviously impaired, for example, observing many beached whales. If the crisis is anticipated, baseline data may be collected for later evaluation of outcomes. If the crisis is ongoing, the condition assessment may be bypassed or samples collected and stored for later evaluation, such as taking photographs and water, soil, or tissues samples. If the crisis is past, an immediate condition assessment usually can transition to a deliberate mode of operation. In an immediate condition assessment, attention is focused on potential areas to shield and document the extent of damage [1].

3.2. CAUSAL PATHWAY ASSESSMENT

Causal pathway assessments determine the probable causes of the environmental impairments revealed by condition assessments. They consider the proximate cause, the source, and the causal pathways that connect them [2, 16].

Deliberate assessment of causal pathways can be bypassed if the cause or source is obvious; for example, a broken effluent pipe emptying waste directly into a stream. However, most deliberate assessments of causal pathways are undertaken because the cause or source is unknown. This is especially true when the condition is an identified human health or biological impairment. In these situations, a causal assessment is needed so that the management action will address the right cause. Often there are multiple causes, and these can be dealt with in many ways [16]. However, they all include a comparison of several candidate causes to identify the most probable cause(s). When there are multiple sources, they need to be identified and the amount of the causal agent allocated among them. Immediate causal assessments may be uncertain due to lack of information; while the results of deliberate assessments may have greater uncertainty compared to situations encountered with immediate ones. For example, a decline in species in a stream may be due to unmeasured, episodic, chronically low levels of a stressor, while a massive fish kill may be associated with a strong stressor such as an algal bloom or a chemical spill. An example of a deliberate causal pathway assessment is the investigation of bird kills associated with carbofuran poisonings [15].

Immediate causal pathway assessments may be bypassed if the cause and sources are obvious. When there are multiple causes or sources, the assessor identifies the most deleterious causes and sources. When the cause is unknown, exposure and effect data are collected, while the literature is searched for similar effects and potential causes. Action may need to be taken before a causal assessment is considered definitive. If the adaptive management is designed as an experiment, the attempts to manage the problem can be used to evaluate causes and sources even while management actions are underway. An example of an immediate causal pathway assessment is the investigation of an epidemic affecting humans, wildlife, or vegetation.

3.3. PREDICTIVE ASSESSMENTS

Predictive assessments estimate changes that will occur with different management actions, including the choice not to act. There are two main subtypes: risk and management assessments [11]. Risk assessments predict what will happen when a causal agent or source is altered in some way and how different management options will alter exposure to the causal agent or affect the source. Management assessments, often performed using decision analysis tools, evaluate the risk estimates in conjunction with economic, social, and political factors to predict the outcome of management actions with the intention of potentially meeting multiple goals.

3.3.1. Risk Assessment

Deliberate risk assessments may be applied locally or broadly; for example, an estimation of risks may be used to develop water quality criteria for metal toxicity to be applied nationally [13, 18]. On a local scale, a risk estimate may show that the metals are not bioavailable at that concentration, and site-specific criteria might be applied. The scope of deliberate risk assessments, because they have more time for analysis and implementation, may include a broader array of effect endpoints for consideration in addition to those that pose the greatest risks to people, property, or ecological attributes. For example, the aesthetics of scenic beauty was an important consideration in setting air quality standards for the area near the Grand Canyon [14].

Deliberate risk assessments are less likely to be limited by project length, resource distribution, or the complexity of management plans. They tend to be more limited by sustained interest of stakeholders and financial backing [7].

Immediate risk assessments focus on the impending or current crisis. Assessors adopt a triage approach with greater attention to human lives, loss of irreplaceable environmental services or resources, expensive economic scenarios, and extensive loss of property, usually in that order of priority. The speed of analysis and interpretation is improved by considering fewer options.

Immediate risk assessments are less likely to be limited by slow decision-making and tend to be limited to fewer options. That is, the options to solve the problem are focused on those that have the potential to greatly reduce deleterious effects. Moreover, the options may be limited by implementation time and by the resources accessible in the crisis area. Short-term access to skilled workers, equipment, materials, and funds may be limited, thus reducing options. For example, the only choice may be evacuation of an area; therefore, the options involve only the means to accomplish this.

Summarizing by examples, in a deliberate risk assessment, an assessor evaluates the risks and options for action regarding the planned removal of a dam. In an immediate risk assessment, the assessor evaluates the risks and options for action when a dam is in imminent danger of a breach.

3.3.2. Management Assessment

Deliberate management assessment considers the environmental decision options in light of economic, social, cultural, and other factors and values [2]. Because there is time and assessors and decision makers may need to justify decisions to stakeholders, they will more likely elect to perform surveys and gather socioeconomic data. They are more likely to balance multiple short- and long-term goals and perform decision analysis or cost-benefit analysis before choosing a management option.

Immediate management assessment is unlikely to use complex decision support systems unless assessors and decision makers are already familiar with the decision tools that are appropriate for the problem at hand. The severity of the threat overrides most other factors. While management decisions always integrate social, political, and economics costs, the information may encompass large uncertainties.

3.4. OUTCOME ASSESSMENT

Deliberate outcome assessments evaluate both the immediate impact of actions or lack of action and long-term outcomes. Cost, collateral damage, or long-term outcomes tend to be more important due to the lengthy time for implementation, visibility to society, and the nature of the types of problems. Outcome assessments that evaluate management actions that take a long time to complete are subject to second thoughts and interruption of implementation. However, these same challenges can be used to update the management plan in an adaptive management approach. Deliberate outcome assessment may require long-term commitment to monitoring. For the example, in 2006, a Superfund remediation implementation plan for 39 miles of the Fox River (Wisconsin, USA) recommended a combination of dredging, capping, and other procedures that included monitoring before and after remediation to support an outcome assessment [17]. Remediation began in 2007 and will be followed by 40 years of monitoring and outcome assessments.

Immediate outcome assessments also evaluate the immediate impact of actions or lack of action. However, long-term outcome assessments usually revert to a deliberate approach [9]. During the immediate phase, it is less likely that collateral damage from the management action or damage to less obvious yet valuable environmental entities or functions will be assessed. For example, if the threat was an imminent hurricane, a management action may have removed ships from port, and the immediate phase of environmental outcome assessment might evaluate the number and effects of boats left in port, such as damage to reefs or toxic spills. However, once the assessment shifts to a deliberate approach, attention may shift to a wider array of assessment endpoints that are environmentally or politically important.

4. Discussion and Conclusion

Environmental assessments are among the most complex analyses and syntheses that humans undertake. Organization and simplification of immediate environmental assessments can help when decisions absolutely must be made. Organization and integration of deliberate environmental assessments can help avoid indecision when decisions would benefit environmental, social, cultural, and economic objectives.

Although most assessors are familiar with the conditions that warrant different approaches to assessments, clear terminology can make it easier to communicate and integrate across types of assessments. Furthermore, deliberate environmental assessments can greatly enhance the performance of assessors under duress. Deliberate assessments can provide analytical and decision support tools that are also applicable in an emergency. Deliberate assessments can make data sets accessible for unexpected situations; for example, geographically relevant distributions of ecological, human, and physical entities. Deliberate risk assessments typically develop risk models for a wide variety of chemical and more recently physical and biological stressors. Rather than report only risk estimates, the full risk model should be easily accessible so that undesirable but inevitable tradeoffs between management objectives can be scientifically assessed in a crisis. Not only do immediate environmental assessments benefit from the products of deliberate assessments, but they also depend heavily on prior preparation. Therefore, continued development of decision support tools is needed to provide ready access to causal relationships or data and tools to quickly make scientifically informed decisions. Among these needs is the continued development of standard methods for recurring types of crises or situations that constrain time available for assessments and make these methods more widely available to smaller communities and the public.

Acknowledgements

This chapter was improved by reviews from Michael Griffith, Michael Troyer and Robert Spehar. Although this paper has been subject to U.S. EPA review and clearance, the ideas in this paper are my own and do not represent the policies or recommendations of the U.S. EPA. The paper was written on the authors' own time and was not supported financially or in-kind by any organization or agency.

References

1. Cleveland, C. (2007). Exxon Valdez oil spill. Retrieved June 3, 2007, from Encyclopedia of the earth. Available at: http://www.eoearth.org/article/Exxon_Valdez_oil_spill.
2. Cormier, S. M., & Suter, G. W. II. (2008). A framework for fully integrating environmental assessments. Environmental Management, published "Online First" http://dx.doi.org/10.1007/s00267-008-9138-y.
3. Griffith, M. B., Lazorchak, J. M., & Herlihy, A. T. (2004). Relationships among exceedences of metals criteria, the results of ambient bioassays, and community metrics in mining-impacted streams. Environmental Toxicology and Chemistry, 23(7):1786–1795.
4. Hawkins, C. P. (2006). Quantifying biological integrity by taxonomic completeness: its utility in regional and global assessments. Ecological Applications, 16(4):1277–1294.
5. Humphries, M. (1996). CRS Report to Congress. New World Gold Mine and Yellowstone National Park – NLE. (CRS Report: 96-669, p. 6).
6. Linkov, I., Satterstrom, F. K., Kiker, G., et al. (2006). Multicriteria decision analysis: a comprehensive decision approach for management of contaminated sediments. Risk Analysis, 26:61–78.
7. NRC (National Research Council). (2005). Superfund and mining megasites: lessons from the Coeur D'Alene river basin. Washington, DC: National Academies Press.
8. Padel, R. (2005). Tigers in Red Weather: a quest for the last wild tigers. London: Little, Brown.
9. Peterson, C. H., Rice, S. D., Short, J. W., et al. (2003). Long-term ecosystem response to the Exxon Valdez oil spill. Science, 302:2082–2086.
10. Suter, G. W. II, & Cormier, S. M. (2008). A theory of practice for environmental assessment. Integrated Environmental Assessment and Monitoring, 4(4).

11. Suter, G. W., II. (2007). Ecological risk assessment. Boca Raton, FL: CRC.
12. Tuckerman, S., & Zawiski, B. (2007). Case study of dam removal and TMDLs: process and results. Journal of Great Lakes Research, 33(2):103–116.
13. USEPA (US Environmental Protection Agency). (1994).Water quality standards handbook (2nd ed.). Contains Update #1. EPA/823/B-94/005a. Washington, DC: US Environmental Protection Agency.
14. USEPA (US Environmental Protection Agency). (2006). Requirements related to the Grand Canyon visibility transport commission. Federal Register 64(126):35769–35773.
15. USEPA (US Environmental Protection Agency). (2006). Interim reregistration eligibility decision: carbofuran. US Environmental Protection Agency, Prevention, Pesticides and Toxic Substances (7508P), Washington, DC. EPA-738-R-06-031. Available at: http://www.epa.gov/oppsrrd1/reregistration/RFDs/carbofuran_ired.pdf.
16. USEPA (US Environmental Protection Agency). (2007). Causal analysis, diagnosis decision information system. Last updated September 12, 2007. Available at: http://www.epa.gov/caddis.
17. WDNR (Wisconsin Department of Natural Resources), & USEPA (US Environmental Protection Agency). (2006). Final basis of design report: Lower Fox River and Green Bay Site; Brown, Otagami and Winnebago counties. Chicago, IL: US Environmental Protection Agency, Region V. Available at: http://dnr.wi.gov/org/water/wm/foxriver/documents/BODR/Final BODR Volume1.pdf.
18. Suter, G.W. II & Cormier, S.M. (2008) What is meant by risk-based environmental quality criteria. Integrated Environmental Assessment and Moniforing, 4(4).

FEDERAL DECISION MAKING FOR HOMELAND SECURITY

Mapping the Normative/Descriptive Divide

L. VALVERDE, JR.

AI Systems, Inc.
Kirkland, WA, USA
drljva@hotmail.com

S. FARROW

University of Maryland
Baltimore County, MD, USA
farrow@umbc.edu

Abstract: The events of 9/11 have dramatically shifted public and private sector priorities aimed at addressing the threat of transnational terrorism. An important issue facing public decision makers is how best to allocate scarce resources in the face of significant uncertainty concerning potential threats and hazards, together with uncertainty concerning the potential costs and benefits associated with possible prevention and mitigation strategies. Viewing this problem from the vantage point of modern economic theory, normative theories of choice provide guidance on how agents should make decisions if they wish to act in accordance with certain logical principles. Often, however, there is a discord between normative theory and how people behave in real-world decision contexts. In this paper we explore several aspects of current homeland security resource allocation practices within the federal government. We begin with an examination of two normative investment models, and we explore the linkages that exist between actual practice and the insights that economic theory lends to these problems. We then present the rudiments of a prescriptive approach to homeland security decision making and risk management that seeks to guide decision makers toward consistent, rational choices, while recognizing their real-world limitations and constraints.

1. Introduction

The events of 9/11 have brought about dramatic shifts in government and private sector investments to address the threat of transnational terrorism.

I. Linkov et al. (eds.), Real-Time and Deliberative Decision Making.
© Springer Science + Business Media B.V. 2008

An important issue facing federal agencies and public decision makers charged with managing the security of the homeland is how to best allocate scarce resources in the face of large uncertainties concerning the evolving nature of the threat, together with uncertainty concerning the potential costs and benefits associated with possible prevention and mitigation strategies. In particular, federal agencies within the homeland security domain face a number of challenges in deciding how best to allocate scarce resources in the pursuit of a broad range of strategic goals and objectives—program effectiveness and economic efficiency, to name just two. In this decision context, the allocation of resources is made difficult by:

1. The existence of multiple decision makers and stakeholders

2. The presence of multiple and often conflicting objectives

3. The prevalence of significant uncertainty surrounding key facets of the terrorism problem

In a complex, dynamic, and uncertain context like this, decision makers can avail themselves of guidance and decision aids from a variety of sources, ranging from informal, qualitative methods to the most formal, quantitative methods. In this regard, it is natural to distinguish between two types of theories: *normative* theories of choice on the one hand, which seek to provide guidance on how agents should make decisions based on logical principles; and alternatively, *descriptive* theories of choice, which seek to provide empirical explanations for actual decision-making behavior in these environments.

In this paper we explore these two decision-making perspectives, with a view towards ultimately informing a *prescriptive* view of how homeland security decision making might best be improved, given all of the attendant constraints and uncertainties. As several decades of empirical psychological research have shown, there is often a discord between normative theories of choice and observed behavior in real-world decision contexts characterized by risk and uncertainty. Our pursuit of this line of inquiry is motivated, in the first instance, by our witnessing a plurality of viewpoints and methodologies currently being applied in the homeland security domain. There are, we feel, a number of lessons to be gleaned from the current state of affairs. How issues are framed in these complex environments, how rational or cognitive decision rules are utilized, how key uncertainties are characterized and evaluated, how values are aggregated—all of these factors influence both the decision-making process itself and, ultimately, the likely ensuing outcome.

Our discussion is organized along the following lines. First, we present an illustrative pair of canonical normative investment models under uncertainty that attempt to capture and represent several salient features of the homeland security problem. In this discussion, our point of departure is a normative model for allocating security expenditures across multiple sites, given a specified security budget. A generalization of this model then allows us to capture two central and related problems in terrorism risk management; namely, how to allocate resources across *probability-* and *damage-reducing* activities. With this as background, in Section 3 we discuss current general practices within the federal government for allocating homeland security resources. This, in turn, motivates a discussion in Section 4 on the rudiments of a prescriptive framework for approaching these problems. Ultimately, the framework seeks to guide decision makers toward consistent, rational choices, while recognizing multiple limitations and constraints (e.g., cognitive, organizational, and other). We conclude with some closing remarks and a brief discussion of possible future research directions.

2. Normative Investment Models Under Uncertainty

Normative investment models under uncertainty span a wide conceptual range—from individual, utility-maximization models to market-based welfare models of rational choice. In this section, we take the rational actor model as a point of departure for highlighting several normative bases for choice in the homeland security domain.

We begin by looking, first, at a utility maximizing model for an individual decision maker who—in the context of our discussion here—considers numerous possible outcome dimensions as being important (e.g., national welfare, agency mission, government costs, political support). Given the decision maker's preferences across these dimensions, the decision maker chooses the option with the greatest expected utility. To illustrate key issues we use the simplest expected-value approach, expected-value maximization, which assumes that the decision maker values increases and decreases in risk equivalently.

The rational expected utility model provides a useful starting point for the issues under consideration here. For the purposes of our discussion, we ignore debates as to whether decision makers actually make decisions according to the classical model [8]; the position we take here is that models grounded on the maximization principle may be useful as benchmarks for evaluating the *quality* of actual decisions made in these environments. The models we consider here are intended to integrate decision, probability, and outcome information in ways that seek to inform decisions on government expenditures directed at managing homeland security.

As we discuss below, different analytical models, in effect, pose different questions. The simplest normative model involves expenditures to reduce the probability of attack at independent sites. A key result of the basic model we present is that some sites are left unprotected if, after the updating of probabilities for investment, the marginal social costs of an attack on the site are less than a threshold that is exogenously constrained by the available funds.

Presented below are two short variations based on independent sites, and consideration of both prevention and mitigation investments.

2.1. ALLOCATING DEFENSIVE EXPENDITURES ACROSS MULTIPLE, INDEPENDENT SITES

We begin with an expected cost minimization model for optimally allocating defensive expenditures across multiple, independent sites.[1] The model presented here is easily extended to allow for the treatment of complexities such as dependency between sites and other variations (see, e.g., [4]).

Whether viewed from a national perspective, or from the vantage point of a decision maker charged with infrastructure protection, we assume a unitary decision maker with two or more independent sites for which defensive resources must be allocated. The decision maker ultimately wishes to select those defensive options that minimize the expected costs associated with a terrorist attack. We begin by defining

e_i $\equiv$ Level of defensive expenditure on site i, for $i = 1, ..., n$;

Z $\equiv$ Aggregate expenditure level over all sites and vulnerability pathways;

$Pr(e_i) \equiv$ Probability of a successful terrorist attack, with $Pr'(e_i) < 0$ and $Pr''(e_i) > 0$;

$S(e_i)$ $\equiv$ Non-governmental costs of the investment expenditures, with $S'(e_i) > 0$;

$C(e_i)$ $\equiv$ Social cost, given that an attack occurs, with $C'(e_i) < 0$ and $C''(e_i) > 0$.

The government's decision problem is to choose an optimal level of expenditure, $e_i^* \geq 0$, for each site i, minimizing expected costs

$$\min \sum_{i=1}^{n} \{Pr(e_i)[e_i + C(e_i) + S(e_i)] + [1 - Pr(e_i)][e_i + S(e_i)]\},$$

subject to the constraints

$$\sum_{i=1}^{n} e_i = Z \text{ and } e_i \geq 0.$$

---

[1] Interdependencies—both positive and negative—are a central concern in evaluating homeland security investments. Positive interdependencies among sites have a possible public good component, in that expenditures at one site may have beneficial effects at other sites. Border security is an obvious example: if potential attackers are stopped at the border, the probability of an event at a number of sites is reduced. Alternatively, should an attack occur, improvements in response capabilities may mitigate or reduce damages at multiple sites.

Looking first at those sites where positive expenditures occur, we formulate the Lagrangian expression for this problem, yielding the following necessary conditions for optimization with exhaustion of the budget:

$$\text{Pr}'(e_i)C(e_i) + C'(e_i)\text{Pr}(e_i) + S'(e_i) = \lambda - 1, \tag{1}$$

The left-hand side of this equation is simply the *marginal expected social cost avoided* (MESCA) through each additional unit of expenditure, while being net of the non-governmental cost associated with each expenditure, $S(e_i)$.

All sites $i \neq j$ with positive expenditures are equated to the common shadow price of funds $(\lambda - 1)$:

$$\text{Pr}'(e_i)C(e_i) + C'(e_i)\text{Pr}(e_i) + S'(e_i) = \text{Pr}'(e_j)C(e_j) + C'(e_j)\text{Pr}(e_j) + S'(e_j),$$

such that the MESCA is equal across all sites. In this formulation it is important to note that some defensive expenditures, e_i, can be zero. Sites without expenditures are those where the MESCA is less in absolute value than the cutoff level of the shadow price of funds. Prescriptively, the model stipulates that some sites are sufficiently "small"—taking both the probability of success and the potential ensuing damages into account—that it is optimal to do nothing to protect them. Of course, all sites are characterized by some level of risk exposure, regardless of whether defensive expenditures occur. The asymmetric nature of the attacker and the intended victim(s) precludes the possibility of reducing the risk to zero.[2]

Prescriptively, then, in allocating defensive funds across independent sites, *for sites that exceed a threshold of potential impact, equate the marginal expected social cost avoided for all sites and vulnerabilities.* In this way, for any given site, there is a cutoff marginal social cost avoided where it is optimal not to expend anything on that site.

2.2. ALLOCATION OF EXPENDITURES ACROSS DAMAGE AND PROBABILITY REDUCING ACTIVITIES

A crucially important policy question in the homeland security domain is the optimal balance between actions and processes that *prevent* attacks and those that *mitigate* (partially or fully) the potential adverse consequences associated with these attacks. In practical settings, the problem may be one of deciding how best to allocate budgets between intelligence-related activities (that are, by their very nature, directed towards preventing attacks) and

---

[2]A lucid argument for this line of reasoning is provided by Posner [11].

the hardening of vulnerable physical infrastructure (aimed at minimizing the adverse effects associated with an attack). For this particular model, let

e_i ≡ Probability-reducing expenditures at site i;

h_i ≡ Damage-reducing expenditures at site i;

Z ≡ Aggregate level of probability- and damage-reducing expenditures;

$Pr(e_i)$≡ Probability of a successful attack given defensive expenditure e_i;

$C(h_i)$ ≡ Social cost given that an attack occurs.

Consistent with our earlier discussion, we assume that $Pr'(e_i) < 0$ and $Pr''(e_i) > 0$, and that $C'(h_i) < 0$ and $C''(h_i) > 0$.

As before, we assume a unitary decision maker who is charged with maintaining a finite number of sites, labeled $i = 1, 2, ..., n$. The decision maker wishes to choose an optimal level of expenditures

$$e^* = (e_1^*, e_2^*, ..., e_n^*) \geq 0 \quad \text{and} \quad h^* = (h_1^*, h_2^*, ..., h_n^*) \geq 0$$

that minimize the total expected cost

$$\min \sum_{i=1}^{n} \{Pr(e_i)[e_i + h_i + C(h_i)] + [1 - Pr(e_i)][e_i + h_i]\},$$

subject to the constraints

$$\sum_{i=1}^{n} (e_i + h_i) = Z \text{ and } e_i, h_i \geq 0.$$

As before, Lagrangian methods are used to solve this constrained optimization problem, yielding the following necessary conditions for optimality:

$$Pr'(e_i)C(h_i) = \lambda - 1 \ \forall i; \tag{2}$$

$$Pr(e_i)C'(h_i) = \lambda - 1 \ \forall i. \tag{3}$$

Equations (2) and (3) imply the equality of the marginal expected social cost at each site, with positive expenditures for each individual type of expenditure and across both types of expenditure (the latter when Eqs. (2) and (3) are set equal to each other).

It is important to note that this model does not distinguish between expenditures that are earmarked for "homeland security" and those that are directed at other types of risks or hazards. In the homeland security domain, it may, for example, be useful to distinguish between manmade hazards (like acts of terrorism) and natural hazards (like extreme weather events).

The above model can, of course, be generalized to allow for this kind of "all-hazards" conception of how best to allocate prevention and response investments.

3. What to Protect: A Descriptive View

Risk management is, in many ways, an endemic feature of public decision making in the 21st century. As a matter of course, the federal government manages a panoply of risks, ranging from employment, environment, finance, and public health to national security [1]. Managing this last component—the national security interests of the country—is, to be sure, a multifaceted task that is fraught with risk and complexity. The specter of transnational terrorism exists throughout the world, in a number of guises.[3] As a practical necessity, managing this evolving threat requires the ability to trace out the expected consequences—economic and otherwise—associated with potential acts of terrorism.

In this light, risk management in a homeland security context is seen to entail various attempts to:

1. Characterize the nature of the threat environment
2. Characterize the vulnerability of people and systems to these threats
3. Value the potential monetary and non-monetary impacts associated with these threats and vulnerabilities

In a management and planning context, decision makers utilize this information to prioritize capitol investment decisions geared at the *prevention* of undesirable events or at the *mitigation* of adverse consequences. Ultimately, the goal is to arrive at adequate levels of protection against these risks and hazards, within specified constraints.

Of course, in the wake of 9/11, all of these considerations sit in an organizational setting and context that is vastly more complex than the one that preceded it. The U.S. Department of Homeland Security (DHS) consists of 23 separate agencies with more than 183,000 employees. Given both the scale and urgency of this undertaking, the challenges that federal decision makers face are, in the first instance, *organizational*. How an organization of this size and complexity takes its congressionally legislated mandate and drives it programmatically through the entire organization is, of course, a key challenge.[4]

---

[3] For a discussion of recent trends, see, e.g., Chalk et al. [2].
[4] For one DHS insider's perspective on these organizational challenges, see Ervin [3].

At the heart of DHS's mandate is a fundamental desire to protect people and property against a broad range of potential extreme events—both manmade and natural. How, in this context, strategic intent is construed and executed rests, in large measure, on the ability to create and foster a *risk-based* culture that takes as its point of departure a coherent and rational appraisal of the threat/hazard environment, together with a flexible and adaptive organizational structure that is able to prepare for, and respond to, these threats.

Any incremental steps to this end must, in the first instance, be informed by a strategic roadmap that lays out how risk management principles should inform a broad range of homeland security decisions. Central in this regard is the ability to provide—at every level of the organization—clear and direct guidance on how risk management principles should be applied in these strategic, tactical, and operational settings. At the present time, there is little in the way of systematic guidance for how risk management principles should be applied, though some progress has been made in certain areas in recent years. In light of this situation, it is not surprising that, in the homeland security domain, there are a broad range of risk assessment models currently in use at the federal level. The diversity of models found in these environments reflects, to a large extent, the domains and mission areas from which they stem, with applications including agro-terrorism, aviation security, cargo security, port security, rail security, and critical infrastructure protection.

In the post-9/11 era, much emphasis has been placed on models that proceed from a threat, vulnerability, and criticality (TVC) mindset, for which the U.S. Government Accountability Office (GAO) provides the following characterization [13, 15]:

- *Threat Assessment:* An attempt to identify relevant threats, and to characterize their potential risk

- *Vulnerability Assessment:* The identification of weakness and susceptibility in a system

- *Criticality Assessment:* An attempt to systematically identify and evaluate an organization's assets and operations by the importance of its mission or function (and perhaps other key attributes, such as national security, public health and safety, etc.) and individuals at risk

Looking, first, at the threat assessment component, much effort currently focuses on identifying and evaluating a number of potential threats and hazards.

Specific steps in this process usually include:

1. The identification of *threat categories*, together with potential adversaries
2. The characterization of adversary *motivations*, *intentions*, and *capabilities*

3. The estimation of frequencies or likelihoods for specific threat scenarios

At the conclusion of this type of analysis, decision makers often rank threats along various dimensions; e.g., greatest likelihood or potential impact. A vulnerability assessment then takes this threat information and assesses the manner and degree to which a system's integrity and viability are compromised by specific threats. Finally, criticality assessment entails the prioritization of assets, as determined by how a particular asset compares with other valued assets, given specified threats and vulnerabilities. Often this will take the form of a prioritized list of risks (asset, threat, and vulnerability combinations) that inform resource allocation decisions. In this regard, various countermeasures can be considered in order to reduce specific vulnerabilities linked to risks that are deemed unacceptable.

The constellation of models currently in development and use represent an important first step in the government's efforts to assess and manage terrorism risk. As we discuss below, however, these models place a myopic focus on risk assessment *per se*, to the exclusion of other factors and considerations that are central to a more fully realized conception of risk management.

Current analytical approaches are characterized by several notable features. First, as mentioned above, is the focus on TVC-based approaches [10]. Second is the use of multicriteria analysis (MCA) methods [6]. Increasingly, MCA-type methods are used in homeland security applications, largely because costs and benefits are not always easily monetized. In general, these methods provide decision makers with

- A way of looking at complex problems that are characterized by a mixture of monetary and non-monetary objectives

- A set of analytical techniques for breaking complex problems into manageable pieces, allowing for data and expert judgments to be brought to bear on individual elements of the problem

- Analytically tractable ways to reassemble the pieces, and to present a coherent overall picture to decision makers

The U.S. Coast Guard's Port Security Risk Assessment Tool (PS-RAT) provides a useful case in point. This risk assessment tool is used by the Coast Guard leadership to help prioritize the allocation of scarce resources to key mission areas and activities.[5] On the threat side, the methodology is scenario-driven, with emphasis on the combination of *target* and *means of attack*.

---

[5] A detailed description and critique of the PS-RAT is provided in [15].

Relative threat frequencies are assigned for each scenario. Potential target vulnerabilities are scored based on perceived susceptibility in four potential dimensions of vulnerability:

1. *Availability*

2. *Accessibility*

3. *Organic security*

4. *Target hardness*

Consequences are similarly valued in a multi-attributed way; specifically, consequences are measured in terms of their impact on five attributes, namely:

1. *Death/injury*

2. *Economic impact*

3. *Impacts on national defense*

4. *Symbolic effect*

5. *Follow-on homeland security threat*

These attributes are combined using a simple additive value function, and a probabilistic event tree is then used to structure the information in a way that gives decision makers a snapshot view of the expected consequences associated with a given threat scenario.

4. A Prescriptive Framework for Homeland Security Decision Making

The centrality of risk management as an organizing principle around which problems of scarce resource allocation are structured and evaluated is an idea that permeates most contemporary efforts within the federal government to assess and manage the potential adverse consequences associated with extreme events—both manmade and natural [16]. To be sure, the panoply of decision-aiding and risk assessment tools currently being developed will continue to evolve and improve as new methodologies and ways of thinking are brought to bear on these complex issues. Still, as our discussion in the previous sections suggests, there is value to be gained in mapping the hinterland that exists between normative theory, on the one hand, and descriptive decision-making reality, on the other, as it relates to managing the security of the homeland. Understanding the conceptual and pragmatic terrain that defines this hinterland helps inform a *prescriptive* view of how homeland security decisions under uncertainty should be construed and evaluated. In what follows, we set out the rudiments of a prescriptive framework for decision making and risk management that encompasses a number of elements that are important in any reasoned and systematic effort to appraise and manage homeland security risks.

4.1. ELEMENTS OF THE FRAMEWORK

Our approach to risk management begins, in the first instance, with an awareness and understanding of the fact that assessing and evaluating complex risks presents decision makers with a unique set of challenges, especially in situations or contexts where the risks are ill-defined or poorly understood.[6] As we discuss in detail below, any attempt to characterize and evaluate homeland security risks leads, naturally, to a consideration of possible risk mitigation alternatives, whether at the strategic, tactical, or operational level. In the evaluation of strategic alternatives, decision makers will typically integrate and weigh knowledge and information from a variety of sources, including organizational or societal values. In evaluating potential courses of action, decision makers will also look to explore fundamental trade-offs between risk and return, short-term versus long-term gain, and so on. In the management selection process, other issues may be considered, including relevant organizational constraints and risk tolerances. And finally, any selection of risk mitigation options will entail a program for implementation and monitoring.

The prescriptive framework presented here is based on a synthesis of published literature, and is intended as an all-hazards approach, with particular emphasis on homeland security issues. The framework is designed so that the individual components of the approach do not become ends in themselves; rather, the framework entails a full cycle of activities, ranging from strategic planning all the way through to implementation and monitoring. The five elements of the framework are as follows [15]:

- *Strategic goals, objectives, and constraints*
- *Risk assessment*

---

[6]A large technical and professional literature addresses these issues. The field of risk assessment has a long history, with much attention focused on the analysis of complex systems (e.g., energy, space systems) and the evaluation of environmental problems. Various risk analysis techniques can be used in evaluating risk mitigation strategies. Fault trees, for example, can be used to focus attention and logical analysis on undesirable events. Failure modes and effects analysis is often used to analyze the effects of possible failure modes on system performance. These and other techniques are often used in probabilistic risk analyses, which seek to measure the risks inherent to a particular system's design or operation. For an overview of relevant methods and techniques, see, e.g., Haimes [5], Morgan and Henrion [9], Raiffa [12] and Viscusi [16].

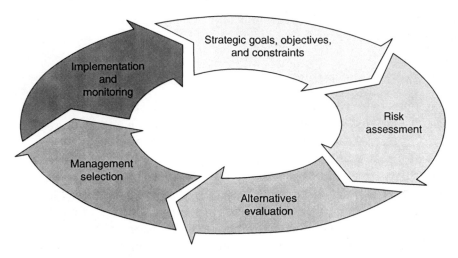

Figure 1. Elements of the Prescriptive Decision making and Risk Management Framework [15].

- *Alternatives evaluation*
- *Management selection*
- *Implementation and monitoring of risk mitigation measures*

Figure 1 illustrates the cyclical nature of the framework. Proceeding through the framework's steps is generally a linear process, though loops may feed back from later to earlier steps in the cycle. Once the process is complete, one or more iterations through various aspects of the framework are possible. The nature of the framework is such that new information can enter any element at any stage in the overall decision making and risk management process.

4.1.1. Strategic Goals, Constraints, and Objectives

The pursuit of *goals* and *objectives* lies at the very foundations of any modern conception of *strategic intent*, and this viewpoint is the conceptual starting point for our prescriptive framework. Modern management practices embed tactical and budgetary decisions in the context of a strategic plan, with clearly articulated goals and objectives that identify resource issues and external threats/hazards.

In our framework, effort is, in the first instance, directed at *structuring* strategic objectives in ways that are meaningful to decision makers, with particular attention paid to the manner in which objectives *relate to*—and *potentially conflict* with—one another. Ultimately, this focus on objectives enables decision makers to:

1. Uncover hidden objectives

2. Improve communication and facilitate involvement among stakeholders

3. Enhance the coordination of interconnected strategies and programs

4.1.1.1. Fundamental Objectives, Means Objectives, and Objectives Hierarchies. For our purposes here, it is useful to distinguish between *fundamental objectives* and *means objectives* [7]. As the name implies, fundamental objectives are those objectives that matter most to decision makers. Means objectives, on the other hand, are objectives that provide the instrumental means by which fundamental objectives are achieved.

An examination of national strategies can serve to illustrate these concepts.[7] In particular, we take the National Strategy for Homeland Security (NSHS) as a specific case in point. The overarching objective of the NSHS is, perhaps, best summarized as *maximizing homeland security*. Four fundamental objectives are seen to define this overarching objective:

- *The prevention of terrorist attacks*
- *Reducing vulnerability to attacks*
- *Minimizing damage resulting from attacks*
- *Enhancing recovery*

4.1.1.2. Linking Means and Ends Objectives. Having structured the fundamental objectives hierarchy, the next stage in our process calls for relating means objectives to the fundamental objectives in a manner that conveys the interrelationships between these entities. This linking of means and ends objectives is accomplished via a so-called *means-ends objectives network* [7]. In such a network, the goal is to provide tangible linkages between the decision makers' fundamental objectives and the instrumental means by which these objectives are realized or accomplished. In this regard, it is instructive to pose the question of how the fundamental objectives of the NSHS are achieved via means objectives. These means objectives—and their relation to the fundamental objectives of Figure 2—are depicted in the means-ends objectives network shown in Figure 3.

---

[7] For an overview of national strategies pertaining to national security and terrorism, see, e.g., [14].

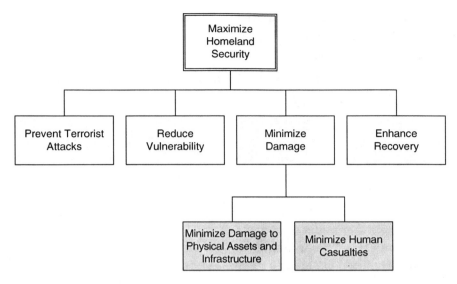

Figure 2. Fundamental Objectives Hierarchy for Homeland Security.

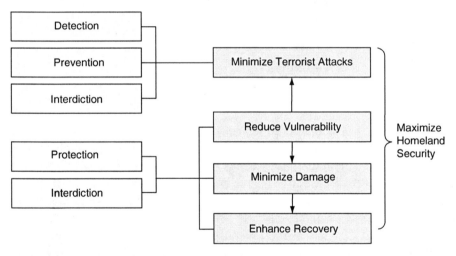

Figure 3. Relation of Key Homeland Security Mission Areas to Means-Ends Objectives.

4.1.2. Risk Assessment

Risk assessment enables decision makers to characterize and evaluate potential adverse consequences under uncertainty. In a typical risk assessment, the following questions are addressed:

- What can go wrong?
- What is the likelihood that something will go wrong?

- What are the consequences associated with these events? There may be multiple dimensions of effects, which may, in turn, be mapped into multi-attribute or benefit-cost analyses.

As a field of professional practice, risk assessment provides a powerful set of analytical tools for assessing the likelihood of events, together with their associated possible consequences. Risks can be evaluated by various methods, depending on the specific application, the available knowledge and information, and management's preferences.

4.1.2.1. Risk-Ranking Methods.   Much current risk assessment practice depends on the qualitative, relative ranking of identified risks. Such rankings may be purely qualitative (using, perhaps, ad-hoc judgments), while others may have a more formal process, using multi-attribute or multi-objective approaches. In some simple cases, direct risk ranking is possible in decision situations where the outcomes are of the same type. In most settings, though, different types or levels of outcomes occur and more complex analyses involving weights or trade-offs are required. In these latter cases, ranking risks typically follows a sequence of steps that include:

1. Identifying consequence attributes (such as exposure or consequence)
2. Defining weights and scales for the attributes
3. Scoring event-consequence scenarios on these attributes
4. Aggregating the weighted scores

From a prescriptive vantage point, the following are some questions useful for evaluating risk-ranking models:

- Is sufficient and reliable information available for the analysis?
- Are attributes that potentially include both government and nongovernment items identified by a reasoned process?
- Is the form of aggregation of the attributes justified? If weights are used in the aggregation process, what justification is given for them?
- Are the upper and lower points of a scale well defined, or at least consistent, across risks in the problem domain?
- If group facilitation or elicitation methods are used to obtain scores or weights, how are the respondents selected? What information is provided to the respondents?
- If ranges or categories (such as 'high,' 'medium,' and 'low') are used, are risks identified as being near analytical boundaries considered in more detail, given the uncertain precision of the responses?

- Is the process formally documented?

4.1.2.2. Quantitative Risk Assessment. Quantitative risk assessments give rise to a wide range of possible outputs (e.g., point estimates, probability distributions). As we discuss below, it is in this step that the discrepancy between normative models of choice such as those sketched in Section 2 and the descriptive practice outlined in Section 3 diverge most markedly. The normative models make a number of unrealistic assumptions concerning the level of precision that is attainable in a complex system such as this (e.g., that the incremental effects of alternative investment options can be distinguished, and that cost information is measurable strictly in dollar terms).

From a prescriptive vantage point, examples of useful quantitative risk assessment questions will include the following:

- Is there a formal, logical model of the risks under consideration?

- What evidence supports the functional forms for the equations that link or functionally relate variables?

- What evidence supports the distributions that are assumed for the uncertain variables?

- What quality control steps are used to assess model validity and calibration?

- Does the analysis conform to accepted practice for the quantitative methods used?

4.1.2.3. Risk Assessments Based on Threat, Vulnerability, and Consequence. As discussed earlier, *threat*, *vulnerability*, and *consequence* are a frequently used decomposition in homeland security risk assessments [10]. In most security settings, all three components are present: a specific threat, a vulnerability in the asset or system that could be exploited by a specific threat, and a damaging outcome associated with specific threat and vulnerability combinations. In the context of our prescriptive framework, questions related to threat, vulnerability, and consequence will include the following:

- Is the threat information credible? How is threat information gathered? Does it come from multiple sources? How is it combined or summarized?

- Are a broad range of threat scenarios used in the risk assessment process?

- Are the threat scenarios generic (oriented toward a general threat environment), or are they particular to specific assets and locations?

- If risk filtering techniques are used to arrive at a manageable set of threat scenarios, how are they implemented? Are 'discarded' scenarios reassessed at some later stage, perhaps in response to new or improved information?

- Are likelihoods (expressed qualitatively or quantitatively) assessed for each identified threat scenario, or are all scenarios assumed to be equally likely? What is the evidence to support the kind of likelihood chosen?

- If likelihood is characterized qualitatively, is it clearly defined?

- Are cognitive biases (such as availability or saliency) managed as part of the threat characterization process?

- How are threat assessments coupled to assessments of vulnerability and consequence?

- What attributes are used to characterize an asset's vulnerability?

- Are weights assigned to each attribute? How are the weights determined?

- How are the consequences associated with specific threats characterized? Is more than one attribute (such as 'lives lost' or 'property damage') used to characterize these outcomes? If so, are the attributes defined clearly and consistently? Are the consequences monetized or used in a benefit-cost analysis?

- If consequences depend on threat, is the threat level clearly specified as part of the consequence valuation process?

4.1.3. Alternatives Evaluation

A risk assessment is likely to identify alternative ways in which decision makers can act to alter either the likelihoods or the outcomes associated with various identified risks. Prevention or damage-reducing actions may also be generated internally or externally through a publicly informed process. The alternatives may include a full range of actions, such as procedural changes, capital investments, regulations, and other actions.

Risks can be reduced appreciably by minimizing their likelihood or by mitigating their impact. In this regard, two concepts are key. The first is that action alternatives should be fed back through the risk assessment process to determine the extent to which risks can be reduced by the alternatives being considered. The initial risk assessment establishes at least part of the structure for evaluating the benefits of alternatives. Consideration should also be given to the possibility that certain actions may simply deflect risk to other assets of the agency, other parts of the government, or to the private sector, all of which reduce the benefits of the action. The second concept is the role of costs to both government and the public; costs are a key element of alternatives evaluation. Major regulatory actions or capital investments generally require a cost- benefit or cost-effectiveness approach.

Core business and government guidance for evaluating alternatives for budgetary and regulatory purposes focuses on monetized net benefit evaluation. It is here, again, that substantial differences exist between normative best practices and current practice in many homeland security settings, due largely to the lack of accepted methods for quantifying and monetizing the full range of costs and benefits that should be considered as part of the alternatives evaluation process.

4.1.3.1. Structuring Portfolios of Risk Mitigation Strategies. The task of both identifying and structuring the risk mitigation options that will be appraised as part of the resource allocation process is an important aspect of our prescriptive framework. To this end, we are interested in characterizing and evaluating a portfolio of possible risk mitigation strategies. Moreover, we are interested in evaluating this portfolio relative to the kinds of objectives and criteria described earlier.

To this end, our first task is one of specifying the portfolio of possible risk mitigation strategies. There are numerous methods for accomplishing this task. A useful tool for this purpose is a strategy table, which provides a convenient way of summarizing a sequence of interrelated decisions. To illustrate, take the broadly defined means objectives that we described earlier. Under each of these broad categories, we can specify a set of possible risk mitigation strategies. As Figure 4 illustrates, a strategy table provides a convenient way of summarizing the overall portfolio of decision alternatives. The strategy table lists, in each vertical column, the set of risk reduction strategies identified for each means objective (e.g., 'Detection,' 'Prevention,' etc.). In this way, we are able to specify an entire portfolio of possible risk reduction strategies.

4.1.4. Management Selection

The fourth step in our prescriptive framework, *management selection*, entails choosing among possible alternative courses of action. Management's active participation is important at this stage because risk assessment tools contain various assumptions about preferences that may require value judgments and review at the management level. Management may also have values or information that analysts have not fully assessed. Once decisions have been reached, evidence that they were informed by risk-based information should be documented.

4.1.4.1. Evaluation of Risk Mitigation Strategies. As described earlier, the strategy table shown in Figure 4 represents the portfolio of all possible risk mitigation strategies that are deemed worthy of consideration. In making a

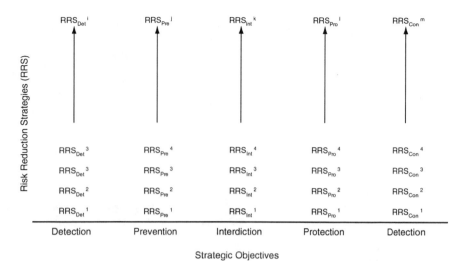

Figure 4. Strategy Table for Risk Reduction Strategies.

strategic resource allocation, our task is one of determining which combination of risk mitigation strategies provides the greatest overall value. In making this determination, decision makers will want to understand and explore key trade-offs, between, say, benefits versus costs or benefits versus risks.

To facilitate this type of analysis, it is possible to utilize objectives hierarchies like those described earlier to make the representation of such trade-offs an explicit feature of the strategic evaluation process. The objectives hierarchy shown in Figure 5 takes elements of our earlier hierarchies and marries them to an explicit consideration of benefit-cost trade-offs. Looking at the leftmost portion of the figure, we begin with the overall objective of maximizing homeland security. To the right of this fundamental objective is the key trade-off to be explored: *Benefits* and *Costs*. In this example, benefits are derived from the pursuit of the fundamental objectives described earlier (e.g., Prevention of Terrorist Attacks, Reduction of Vulnerabilities, etc.). For costs, we distinguish between monetary and non-monetary costs. At the rightmost portion of the diagram are the criteria against which the achievement of each objective is measured. For this illustrative set of criteria, it is, for example, possible to explore the trade-offs that exist between the benefits that might be derived from preventing terrorist attacks and the (social) cost associated with the potential loss of civil liberties.

4.1.5. Implementation and Monitoring

Any conceptual roadmap for how risk management principles can inform homeland security decision making must inevitably confront a number of

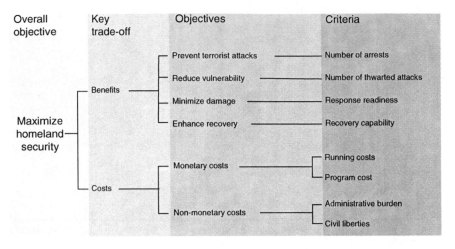

Figure 5. Hierarchical Representation of Objectives and Criteria for Benefit-Cost Trade-Offs.

issues that pertain to *implementation* and *monitoring*. Monitoring is essential to determine whether key objectives and milestones are being met, and whether policies and controls are giving rise to intended outcomes.

Risk management plans should be constructed in ways that ultimately support innovation and improvement, based on a process of continual feedback and learning. Monitoring helps ensure that the entire risk management process remains current and relevant, and that it reflects changes in the effectiveness of the actions and the risk environment in which it operates. Monitoring the risk management plan also involves assessment of the adequacy of strategic objectives and performance measures, as well as ensuring that service delivery and support functions are consistent with design specifications and implemented in accordance with the plan's timeframe.

In assessing, again from a prescriptive vantage point, the implementation of risk mitigation actions, it is useful to pose the following sorts of questions:

- Are objectives and time schedules specified for implementation actions?
- Are mitigation actions implemented as specified?
- Are mitigation actions implemented in a timely manner?
- Do mitigation actions meet cost objectives?
- Are internal controls adequate?
- Are risk communication issues considered?

In addressing monitoring and evaluation activities, critical questions will include the following:

- What types of ongoing monitoring occur as part of the overall risk management process?
- If performance measures exist, what is the outcome of performance measurement protocols and procedures?
- Has the agency previously evaluated the program or does it have a detailed plan for evaluating the program?
- Does the evaluation conform to best practices?
- Are the recommended activities reviewed periodically?
- Are risk scenarios kept up to date and is the system tested periodically?
- How often do decision makers review the entire risk management system?
- What mechanisms identify and deal with risks affected by changing circumstances or new information?
- Do barriers have a significant impact on the agency's ability to achieve its risk management goals?

4.1.5.2. Continual Feedback. Active monitoring is essential to providing feedback to decision makers for continual or periodic improvement of the risk management plan (as dictated by the situation or context), together with information as to whether the plan coordinates effectively with other relevant plans, programs, and agencies. Risk management is a dynamic process and monitoring is a check on whether resources are used effectively and efficiently. Monitoring and evaluation provide information to management and stakeholders about the status of the plan, such as if the plan is in compliance with all current applicable professional standards, and if all memorandums of understanding and mutual aid agreements are in place, and that legal liability concerns have been resolved.

5. Concluding Remarks

In this paper, we have sought to explore a number of issues pertaining to federal decision making for homeland security, looking specifically at the divide—both conceptual and pragmatic—that exists between normative theories of choice and descriptive decision-making practice as it presently exists in the homeland security domain. Our attempts to understand the nature of this divide—and its implications for decision quality, program effectiveness, and economic efficiency, among other things—has motivated a prescriptive framework that seeks, on the one hand, to make the best use of normative insights, while, on the other, candidly confronting the difficulties (cognitive and otherwise) that decision makers routinely confront in these complex and uncertain realms. If, in our approach, there is a bias, it is in strongly siding with the view that risk management is the *sine qua non* for how extreme events—both manmade and natural—must be construed and managed in the post-9/11 era.

Of course, the tragedy of Hurricane Katrina illustrates what is, perhaps, one of the most vexing challenges in the homeland security domain, namely, how best to allocate scarce resources among the vast panoply of catastrophic risks that can beset mankind in the technological society of the 21st century. Any reasoned risk management approach begins with a cold and dispassionate assessment of the true extent of the nation's vulnerability to a diverse range of threats and hazards. As we have said, at the federal level, the organizational challenges that must be confronted in these domains are significant. In this paper, we have argued for a common set of analytical tools and procedures regarding how the federal government invokes and makes use of risk management concepts and techniques. While current federal approaches to homeland security decision making is evolving towards consistency with the risk management approach articulated here, substantial gaps still exist. The challenge remains one of continued vigilance, flexibility, and resilience in anticipation of, and in response to, an ever-changing threat/hazard environment.

Acknowledgments

We are grateful to a number of colleagues for useful comments and suggestions on many aspects of the work presented here. The prescriptive framework presented in Section 4 draws heavily on work done by both authors at the U.S. GAO, where Neil Asaba, Nancy Briggs, Steve Caldwell, Nancy Kingsbury, Norm Rabkin, and numerous others provided valuable feedback and commentary. Any errors are, of course, our own.

References

1. *Risk Assessment in the Federal Government: Managing the Process,* 1983. National Academy Press, Washington, DC.
2. Chalk, P., Hoffman, B., Reville, R., and Kasupski, A., 2005. *Trends in Terrorism: Threats to the United States and the Future of the Terrorism Risk Insurance Act.* RAND Corporation, Santa Monica, CA.
3. Ervin, C. K., 2006. *Open Target: Where America Is Vulnerable to Attack.* Palgrave Macmillan, New York.
4. Farrow, S., 2007. The economics of homeland security expenditures: foundational expected cost-effectiveness approaches. *Contemporary Economic Policy* 25(1):14–26 (January 2007).
5. Haimes, Y., 2004. *Risk Modeling, Assessment, and Management.* Wiley, Hoboken, NJ.

6. Keeney, R., and Raiffa, H., 1993. *Decisions with Multiple Objectives: Preferences and Value Trade-Offs*. Cambridge University Press, Cambridge.
7. Keeney, R. L., 1992. *Value-Focused Thinking: A Path to Creative Decision-Making*. Harvard University Press, Cambridge, MA.
8. Machina, M. J., 1987. Choice under uncertainty: problems solved and unsolved. *Economic Perspectives* 1(1):121–154.
9. Morgan, M. G., and Henrion, M., 1990. *Uncertainty: A Guide to Dealing with Uncertainty in Quantitative Risk and Policy Analysis*. Cambridge University Press, New York.
10. Moteff, J., 2004. Risk management and critical infrastructure protection: Assessing, integrating, and managing threats, vulnerabilities, and consequences. Tech. Rep. RL32561, Congressional Research Service, Washington, DC.
11. Posner, R. A., 2005. *Preventing Surprise Attacks: Intelligence Reform in the Wake of 9/11*. Rowman & Littlefield, Lanham, MD.
12. Raiffa, H., 1968. *Decision Analysis*. Addison-Wesley, Reading, MA.
13. United States Government Accountability Office (GAO), 2001. *Key Elements of a Risk Management Approach*. No. GAO–02–150T. Washington, DC.
14. United States GAO, 2004. *Combating Terrorism: Evaluation of Selected Characteristics in National Strategies Related to Terrorism*. No. GAO–04–408T. Washington, DC.
15. United States GAO, 2005. *Risk Management: Further Refinements Needed to Assess Risks and Prioritize Protective Measures at Ports and Other Critical Infrastructure*. No. GAO–0691. Washington, DC.
16. Viscusi, W. K., 1998. *Rational Risk Policy*. Oxford University Press, Oxford.

GROUP INFORMATION-SEEKING BEHAVIOR IN EMERGENCY RESPONSE

An Investigation of Expert/Novice Differences

Q. GU AND D. MENDONÇA

Information Systems Department, New Jersey Institute of Technology
Newark, NJ, USA
mendonca@njit.edu

Abstract: Emergencies—whether natural or technological, random or human-induced—may bring profound changes to organizations, the built environment, and society at large. These changes create the need for reliable information about the emergency and its impacts, and thus require responding organizations to seek and process information from an evolving range of sources. By understanding how skilled versus novice response personnel search for information in emergencies, we may begin to understand how to support and train for skillful information seeking in situations characterized by risk, time constraint, and complexity. This study develops a hypothesized model of information-seeking behavior in emergency response and evaluates it using data from expert and novice groups addressing simulated emergency situations. The results suggest that experts maintain breadth in the extent of their information seeking, despite increasing time pressure. Novices, on the other hand, decrease the extent of their search under increasing time pressure. Both expert and novice groups show a decreasing effort in information seeking; moreover, effort devoted to search for common and unique information decreases over time.

1. Introduction

Emergencies—whether natural or technological, random or human-induced—may bring profound changes to organizations, the built environment, and society at large. These changes create the need for reliable information about the emergency and its impacts, and thus require responding organizations to seek information from an evolving range of sources while tracking a possibly changing set of response goals.

I. Linkov et al. (eds.), Real-Time and Deliberative Decision Making.
© Springer Science + Business Media B.V. 2008

In emergencies, as in many other situations, information needs drive decision makers' search for different types of information. Emergency situations differ from nonemergency situations in a number of ways, however [20, 22]. Time constraint forces decision makers to manage tradeoffs between the effort required to search and the anticipated value of information of various types. Indeed, in emergency situations, time spent on information seeking and other planning activities is time taken away from plan implementation. Emergencies also entail risks to life and property, adding to the need to make rapid but accurate decisions but also increasing the penalties associated with making the wrong decision or failing to make the right decision in a timely manner. Finally, emergencies may be complex, requiring coordination and shared responsibility across numerous organizations.

One approach to understanding how to train for and support skillful information seeking in emergencies is to examine differences in information-seeking behavior between novices and experts. This paper begins by reviewing the existing literature (Section 2) to develop a preliminary model of how conditions of risk, time constraint and emergency complexity may impact information-seeking behavior (Section 3). It then develops a set of hypotheses (Section 3) concerning how expert and novice information seeking may differ under these conditions and explores answers to these hypotheses by examining information-seeking behavior by experts and novices in a simulated emergency scenario (Section 4). The results are presented in Section 5 and discussed in Section 6, along with possibilities for future work in refining the proposed model.

2. Background and Related Research

Various factors may impact group information seeking during decision making. These include the degree of consensus of group opinion; whether the information is common, partially shared, or unique; public assignment of expert role; number of decision alternatives; decision deadline or time pressure; availability of a group support system; demonstrability of a fact's existence; and familiarity with the decision topic [7, 8, 18, 24, 25, 29, 30].

2.1. INFORMATION-SEEKING BEHAVIOR

Search may be characterized by its extent (i.e., how exhaustive is it) and nature (i.e., what is searched for) [6]. Information seeking is "the purposive seeking for information as a consequence of a need to satisfy some goal" [34]. Prior work [11, 16, 19] suggests that information seeking is a process driven by information needs for the fulfillment of particular tasks. Information seeking can be said to consist of setting goals, forming a search set, refining the search set, locating the desired information, and reviewing or evaluating found information. The information-seeking process exists within a context, and is influenced by such factors as environment, technology, individual characteristics, and task goals [26]. These influencing factors impact strategy selection, search efficiency, and search performance (i.e., the extent to which the search results satisfy the information needs and task goals).

Information seeking has also been characterized as dynamic and nonlinear [12, 32], "analogous to an artist's palette, in which activities remain available throughout the course of information seeking" [12]. Information seeking is not merely a step-by-step process: the loops of feedback and iterative activities happen anytime. Interaction between search processes, search outcomes, and the external context leads information seekers to adaptations that are reflected in their search patterns. Previous studies do not clearly explain how such changes happen over time, and how certain variables may impact these changes.

2.2. GRMOUP INFORMATION EXCHANGE AND USE

Prior work on information-seeking behavior has focused on how information seeking by individuals is influenced by environmental, technical, or personal characteristics [11, 16]. In a group context, decision makers from different professional domains can contribute their knowledge and cooperate to solve a task, and thus benefit from a larger pool of knowledge than might individual decision makers. The assumption is that group discussion will lead to the introduction of more relevant information. However, while availability of information is likely to be a prerequisite for high-quality group decisions [24], availability itself does not necessarily induce optimal decisions. This may be seen in how various types of information are used. From the perspective of group members, information may be common (if it is known to all group members before the discussion), partially shared (if it is known to part but not all group members), or unique (if it is held by one member before the group discussion) [7, 8, 25]. However, group members tend to discuss and think more about common information (i.e., information originally known to all group members) and less about unique information (i.e., information originally known to only one or a few members) [8, 29].

The relationship between information availability and group performance varies due to within-group processes. While information recall and information exchange lead to more information being used by groups, only when group members access, store, and utilize the information will it actually show its value in the decision-making process. These three activities are an integral part of every step in decision making, though their relative importance may change depending on the stage of the decision-making process.

2.3. TIME PRESSURE

In an emergency response situation, time is critical, since any time spent on decision planning is unavailable for decision execution. Time pressure may impact information-seeking behavior during decision making and problem solving in a number of ways [1, 9, 23]. Time pressure may impact decision makers' working rate and their confidence in judgments [1]. Under time pressure, decision makers may speed up their information processing and be more selective in choosing information to be processed. As time pressure increases, they may switch to simpler information search strategies and decision rules [33].

The impacts of time pressure on group information-seeking behavior may manifest in two ways. First, the information needs of the group will be more focused and the priorities of the information processed will change. Information seeking will be more directed towards task-related information in such situations [18]. Second, as with individuals, group members will use an "acceleration and filtration" strategy [18] by eliminating some options, accessing a smaller proportion of information, and accelerating their search by spending less time handling each item of information accessed. A hierarchy of these strategies exists in people's reactions to time pressure. Acceleration will be the first response to time pressure, and selection will most probably appear as the second reaction when acceleration is insufficient. If selection is still not sufficient, people switch information search strategies to meet their information needs within the time constraint [2].

Severe time constraint may lead decision makers to rely on information that is already on-hand. The group members' intention to enlarge the information pool would interact with their adoption of a filtration search strategy across the different stages of the decision-making process. The benefit of obtaining new information may not outweigh the risk of time delay under severely time-constrained conditions. The counterbalance of these two effects will determine which takes the dominant position in information seeking.

2.4. TASK DIFFICULTY

Task difficulty or complexity [10, 14, 31] can be defined in terms of the objective task characteristics contributing to the multiplicity of goals and ways to accomplish the goals [5]. Complex tasks are difficult by their nature, but difficult tasks may not always be complex. The point is that certain tasks can be difficult (i.e., require high effort) without necessarily being complex; in contrast, some tasks are difficult because they are complex.

Task difficulty is related directly to attributes that increase information load, information diversity, and/or rate of information change as follows [5]:

1. The presence of multiple potential ways to meet a desired goal

2. The presence of multiple desired goals to be attained

3. The presence of conflicting interdependence among ways to multiple goals

4. The presence of uncertain or probabilistic links among ways and goals

In emergency situations, task difficulty can be regarded as a function of time, risk, available resources, and changing sub-goals. Decreasing time and risks in the environment increases the rate of information change. Decreasing available resources requires additional information processing. As available time decreases, available resources—which are likely to be distributed over geographic space—also decrease, thus making certain solutions infeasible. Emergency responders must therefore devise alternative (possibly improvised) ways to solve the problem [25]. Third, multiple and possibly evolving goals increase information load. Given some criterion for efficiency (e.g., planning and executing within the decreasing available time), possible solutions need to be evaluated against it. In such cases, task difficulty grows according to the decreasing available resources and the decreasing feasible courses of action. Information processing requirements will increase substantially if the connection between potential decisions and desired outcomes cannot be established with sufficient certainty.

2.5. EXPERTISE

An expert could be a person with domain-specific knowledge or task-related experience, or both. Expertise can improve group performance by increasing each member's ability and judgment; task experience can improve group performance by facilitating problem recognition and utilization of relevant knowledge [13, 17].

The discovery of expert/novice differences has been instrumental in uncovering skills and knowledge that enable high performance. Such study has been found in a variety of areas, from individual physics problem solving [15, 28] to group decision making in complex tasks [1, 4]. Experts are expected to spend less time on a problem, to memorize more relevant information, and solve the problem faster than novices [6, 15, 28]. Moreover, expert/novice differences are also manifested as differences in confidence [28]. In time-constrained situations, experts may be more efficient in information filtering (i.e., separating relevant from irrelevant information) and exhibit more confidence about their choices. For example, a study on the decision making of air commanders in a dynamic environment under very limiting time constraints reveals that experienced commanders tend to make fewer decisions within a given time interval, and process additional information better than less-experienced commanders [1].

Differences are also expected in the information-seeking behaviors of experts and novices [27]. Experts' information-seeking behaviors are well organized according to sets of basic units while novices' are characterized by depth-first and breadth-first search, suggesting that experts utilize known facts more effectively than novices, since in the same circumstances novices may need more cues to solve a problem.

3. A Model of Group Information-Seeking Behavior in Emergency Response

Prior work on information seeking and the impact of risk, task complexity, and time pressure on the behavior of decision-making groups in emergencies is here integrated into a preliminary model (Figure 1). When decision makers at some time t are faced with a future deadline at time T, every minute spent on planning is one less minute available for plan execution.

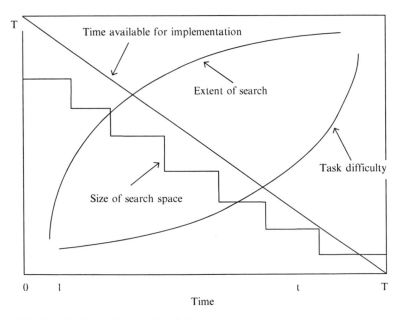

Figure 1. Model of information-seeking behavior.

Simultaneously, material and personnel resources available for responding to the event decrease, since they will typically have to be dispatched from one location to another. The number of plans (or courses of action) involving these resources decreases, thus reducing the size of the search space. As a result, a greater percentage of the resources—and therefore a greater percentage of the space—can be searched over time. In contrast, both the passage of time and the reduction in available resources contribute to increasing complexity and risk, thus making the problem of how to respond harder to solve. Consequently, response personnel are forced to "make do" with resources that are or can be made available in time. Task difficulty is inversely related to the number of available resources and the number of potential solutions.

The hypotheses that follow from this model are described below. Also included are hypotheses pertaining to the impact of expertise on information seeking, as well as hypotheses concerning the seeking of common versus unique information by groups.

3.1. H1: EXTENT OF SEARCH

The extent of search by groups could be considered from two perspectives. First, from the objective perspective, the size of the search space decreases over time because group members have fewer information sources to explore when approaching the deadline and thus are more likely to exhaust available sources. Second, from the subjective perspective, group members accelerate their search by spending less time examining each information source, leading to hypothesis H1.1:

H1.1: As time to implement decreases, the extent of search increases.

Domain knowledge and prior relevant experience can provide experts with a higher capability to deal with the emergency than novices. Under time constraint, experts are more confident in selecting the most relevant information and making decisions with a small amount of information, while novices may have to examine more information sources to enable decision making, leading to hypothesis H1.2:

H1.2: The search extent of novice groups will be greater than that for expert groups.

3.2. H2: NATURE OF SEARCH

The information-seeking process in emergency response is time critical. For two successive stages in the decision-making process (i.e., consideration set formation and final choice selection), information-seeking activities are likely to be more concentrated in the first stage than in the second one. As available time decreases, group decision makers are likely to devote more time to evaluating on-hand information and finalizing decisions, thus decreasing information-seeking activities. However, a preference for common information and the increase in time pressure may make search for common information increase but search for unique information decrease, leading to hypotheses H2.1 through H2.3:

H2.1: As time to implement decreases, search for common and unique information decreases.

H2.2: As time to implement decreases, search for common information increases.

H2.3: As time to implement decreases, search for unique information decreases.

The impact of time pressure is likely to be less for experts than for novices. Experts process additional information better than novices, and are less likely than novices to change their information-seeking strategies under time constraint, leading to hypothesis H2.4:

H2.4: As time to implement decreases, search for information (both common and unique) by groups of experts will change less than search for information by groups of novices.

4. Model Evaluation

We now turn to the design of a study used to investigate the proposed model of group information-seeking behavior in emergency response. The simulated emergency scenarios used in the study are described first, followed by the data description and the measures used in the model evaluation.

4.1. EXPERIMENT ENVIRONMENT

The data were drawn from a series of studies on group decision making in simulated emergency response scenarios [21]. Both novice and experienced groups of participants convened to work on two separate emergency response-related cases. Each group had five participants: one group coordinator (CO) acted as a facilitator and principal communicator with the decision support system and the others each represented one of four emergency services; i.e., Police Department (PD), Fire Department (FD), Medical Officer (MO), and Chemical Advisor (CA). The group's task was to allocate resources to the incident location in order to meet the goals of the emergency response. The layout of a typical experimental session is shown in Figure 2. All experimental sessions were videotaped for later transcription and analysis.

Figure 2. Layout of the experimental session.

Each group had two ways to access information during the emergency response process:

1. Track the information via the computer support system

2. Acquire information from group members via conversations

Figure 3 shows the interface of the computer support system used by the CO in Case One Phase Two. The map at the left displays the locations of resources and the incident location ("Z"). Group members obtained information for a site by clicking on its icon. A list of the equipment available at the site was displayed in the lower left. Some information was unique: each non-CO member could view only the resources at the sites controlled by that role. For example, FD could learn about sites that had firefighting equipment, but not about sites that had medical equipment. Messages were also tailored to the individual services, and could only be seen by the representatives of those services. Some information was global: all members had access to a description of the incident and all members could access information on resources (such as gymnasiums and supermarkets) that were not controlled by a particular service. Also, the CO had accessibility to information about all sites. In Figure 3, sites O, Q, L, and M are alternate resources; all other sites

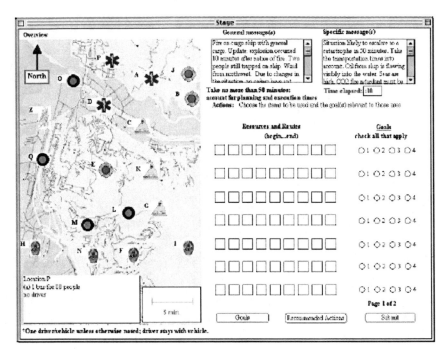

Figure 3. Computer interface for simulated emergency in Case One.

(sites A to K, N, and P) are controlled by individual services. Individuals therefore had incomplete information locally but complete information globally.

Groups were given 50 min to plan and execute courses of action to accomplish the goals of the response. In other words, every minute spent on decision planning was one less minute available for decision execution. As time passed, certain resources therefore became infeasible. Simultaneously, the situation was likely to escalate, so that problem difficulty increased due to increasing situation severity and decreased response capability. Decision support was provided to some groups when certain resources became unavailable but alternate resources could be used. The system recommended procedures that had to be assembled to form a solution. Participants elected either to accept, reject, or modify these procedures. Unsupported groups received no assistance on either case.

4.2. PARTICIPANTS

Novice participants were college students enrolled in undergraduate business or engineering programs, while expert participants were students at the U.S. National Fire Academy. Both novice and expert groups were randomly

TABLE 1. Number of observations under each condition.

		Expert	Novice
Case 1	Support	3	2
	No support	4	2
Case 2	Support	3	2
	No support	4	2

assigned to the support or no support condition in each case, with each group providing two observations via a balanced incomplete block design. The number of observations in each condition is shown in Table 1.

4.3. TASKS

Two simulated emergency cases drawn from actual accidents were used in this study. Case One concerns a cargo ship fire with an oil spill; Case Two concerns a collision between two ships with a resulting chemical emission. Each case solved by the group has two phases. In Phase One, the group is told only to plan for the activities necessary to address the emergency and is given 20 min to do so. The group then works to develop courses of action to address the emergency situation and submits these, along with the goals they wish to achieve, through the CO. Following a brief pause, Phase Two begins: the group is informed that certain resources have become unavailable but that other, nonstandard resources (specified on-screen) can be used. The time-constrained element of the experiment is also introduced: participants are told that activities have to be planned-for and completable within 50 min, at which time an event with potential for catastrophic impact is anticipated to occur. Given the nature of the Phase Two time constraint, it is essential that participants account simultaneously for planning and execution times. Phase Two (and the case) conclude once the CO has submitted the group's courses of action and corresponding goals. Each participant then fills out a questionnaire assessing their individual opinions about the course of action submitted in Phase Two. This sequence is repeated for Case Two. Participants then fill out a questionnaire assessing their professional qualifications and overall impression of the experiment. An informal debriefing session concludes the experiment, which lasts approximately 2 h.

4.4. DATA SOURCES

Data used to analyze group information-seeking behavior are stored in computer logs that contain records of which resources were examined by which group members at which time. When a group member clicked on a site to discover what resources were available, the site label and time of click were written to the log file, along with other data such as the session, group and participant role. All records are time-synchronized for analysis. Sample records from one log file are shown in Table 2. Stream indicates the category of each event in the logs. Records concerning the information-seeking process are identified with a p (for process) in the Stream column, and are here the object of analysis. (Records denoted with m mark the boundary between cases and phases; records with a d mark the point at which decisions were made.) As an example, the second record shows that participant CA in group A of session NFA1 clicked site C at 7:33:36 p.m. (148,132 ticks, where 1 tick equates to 1/60 s).

Of interest in this study is information-seeking behavior in situations requiring executing and planning at the same time. Consequently, data from Phase Two are used in the analysis.

4.5. MEASURES

Four measures are used in addressing the hypotheses, as shown in Table 3. The extent of search (M1) is measured by the proportion of search space explored (i.e., the proportion of all sites clicked by a group). The nature of search is measured by three parameters: the number of clicks on common information sites (M2), the number of clicks on unique information sites (M3), and the number of clicks on both common and unique information sites (M4). M2, M3, and M4 reflect the effort devoted to locating common, unique, and all information in the information-seeking process.

According to the measures defined above, the hypotheses proposed are summarized in Table 4.

TABLE 2. Sample records from the log file.

Session	Group	Participant	Stream	Time	Ticks	Tape_T	Event
NFA1	A	CA	m	7:33:12 p.m.	146,676	164835	"BeginC1P1"
NFA1	A	CA	p	7:33:36 p.m.	148,132	171262	"C"
NFA1	A	CA	p	7:33:44 p.m.	148,596	172035	"G"
...							
NFA1	A	CA	d	7:45:51 p.m.	192,245	292784	"Ga,1,0100"
NFA1	A	CA	m	7:45:51 p.m.	192,255	292800	"EndC1P1"

TABLE 3. Information-seeking measures.

Aspects of seeking behavior	Variable	Name	Description
Extent of search	M1	Extent	$\dfrac{\#of\ sites\ clicked\ within\ every\ \text{minute}}{\#of\ sites\ available\ within\ every\ \text{minute}}\ 100\%$
Nature of search	M2	#Common	Average number of clicks on alternative resources within every minute made by each group
	M3	#Unique	Average number of clicks on non-AR sites within every minute made by each group
	M4	#Total	Average number of clicks on all sites within every minute made by each group

TABLE 4. Summary of hypotheses.

Name	Testing hypotheses
H1.1	$M1_{t1} < M1_{t2}, t1 < t2$
H1.2	$M1_{E} < M1_{N}{}^{*}$
H2.1	$M4_{t1} > M4_{t2}, t1 < t2$
H2.2	$M2_{t1} < M2_{t2}, t1 < t2$
H2.3	$M3_{t1} > M3_{t2}, t1 < t2$
H2.4	$|M4_{t1} - M4_{t2}|_{E} < |M4_{t1} - M4_{t2}|_{N}\ t1 < t2$

* E – Expert groups, N – Novice groups.

5. Results

5.1. EXTENT OF SEARCH

All hypotheses are investigated for each case. The starting time of the session is 0 min and the ending time is 50 min. Data for evaluating the hypotheses are presented in Figures 4 through 9. In each figure, the horizontal axis represents time, which means the range of the time allowed for the task (i.e., 50 min). As Phase Two of each case progressed, the available time to implement decreased from 50 to 0 min. The groups had to consider the time remaining for execution since dispatching the available resources to the incident location takes some time. Dispatching time varies due to the distances between the resources' locations and the incident location. For example, in Case 1 (see map in Figure 3) the nearest sites to the incident location Z are O and Q, from which the resources can be delivered to Z within 5 min; the furthest site is I, from which the resources can be delivered within 23 min. Thus at the beginning of a phase (Time 0), all 17 sites in Case 1 were reachable for the group. As time passed, the reachable sites decreased. At Time 28, for example, resources at Site I could not be used in a feasible course of action since Site I was out of range. At Time 46, the remaining time to implement is only 4 min; even the resources at the nearest sites (O and Q) cannot be dispatched to the incident location Z. The size of the search space after Time 46 became 0. The change of the size of search space over time is shown in Figure 4. Decreased search-space size reduces the number of potential solutions, and further makes the task more difficult to complete.

All possible courses of action that can be taken using the available resources to meet the response goals are calculated and shown in Figure 5. In Case 1, the number of courses of action drops after the first 20 min of the task. At the beginning (Time Zero), there are 41,739 possible courses of action for implementation; at Time 20, the number of courses of action drops to 261. Case 2 is similar in this regard. The difference is that at the beginning

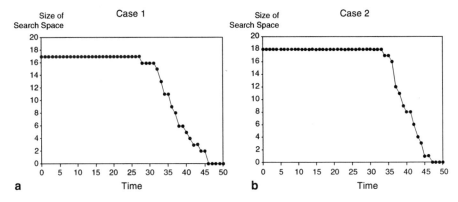

Figure 4. Size of search space over time in (a) Case 1 and (b) Case 2.

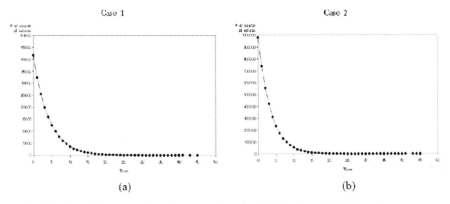

Figure 5. Number of courses of action over time in (a) Case 1 and (b) Case 2.

there are 980,128 possible courses of action, which are 23.5 times the number in Case 1, making Case 2 more complex in this sense. As discussed previously, task difficulty is inversely related to the number of potential solutions. The reductions in the number of courses of action lead to reductions of the number of potential solutions, thus increasing the task difficulty.

With the increase in task difficulty, the number of sites explored by group participants shows a decreasing trend in both Cases 1 and 2 (Figure 6). The decreasing trend is more obvious during the period when the number of courses of action drops dramatically (i.e., from Time 0 to Time 20). After that the number of sites does not vary greatly. Figure 6 also shows that novice groups explored more sites than expert groups during the first period.

The extent of search is computed according to the number of sites explored and the size of search space at every minute (Figure 7). On average

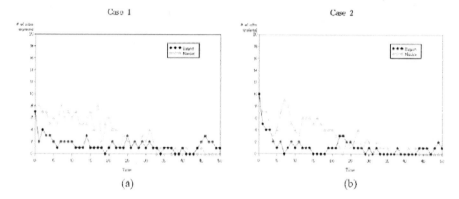

Figure 6. Number of sites explored by expert and novice groups over time in (a) Case 1 and (b) Case 2.

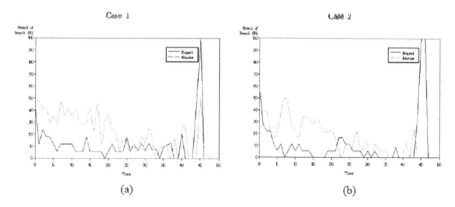

Figure 7. Extent of search between expert and novice groups over time in (a) Case 1 and (b) Case 2.

in Case 1, the extent of search by novice groups is 20.9% and the extent of search by expert groups is 11.0%. In Case 2, the extent of search by novice groups is 16% and the extent of search by expert groups is 10.7%. Novice groups have a higher extent of search than expert groups. Moreover, the higher extent is obvious in the first period of time (before Time 25). Near the end of the task expert groups show a 100% extent, which means they clicked all available sites at that time while novice groups show a 0% extent, which means they gave up the information search.

5.2. NATURE OF SEARCH

Group participants' search behavior for common, unique, and all information is shown in Figure 8.

Search for all information (both common and unique) displays an obvious decreasing pattern in both cases. Search for common information also shows a decreasing pattern in both cases. However, the search trend for unique information is not consistent: it decreases in Case 1 but persists almost at the same level in Case 2. The average number of clicks on each type of information in both cases is listed in Table 5. In Case 1, there are more

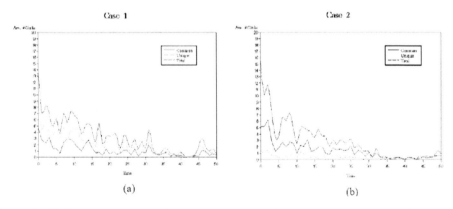

(a)   (b)

Figure 8. Clicks on common, unique, and all information sites over time in (a) Case 1 and (b) Case 2.

TABLE 5. Mean number of clicks on different types of information.

	Case 1	Case 2
Total	3.24	2.95
Common	1.11	1.37
Unique	2.13	0.36

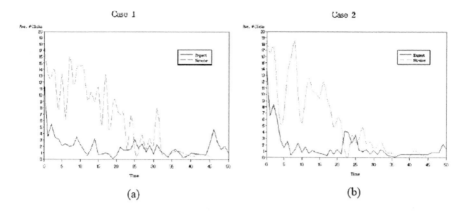

Figure 9. Clicks on information sites by expert and novice groups over time in (a) Case 1 and (b) Case 2.

clicks on unique information than on common information; the reverse is true in Case 2: there are fewer clicks on unique information than on common information.

Information search differences between expert and novice groups are shown in Figure 9. The number of clicks by novice groups is higher than for expert groups. Novice groups clicked more frequently in the first period of time and their number of clicks dropped fast as the deadline approached. Expert groups clicked quite often at the very beginning of the task, but most of the time they clicked at a relatively consistent level.

5.3. SUMMARY OF RESULTS

In summary, the extent of search displays a decreasing trend as time to implement decreases, and novice groups exhibit a higher extent of search than expert groups do. The only exception is in the last several minutes, when novice groups gave up their search and expert groups still explored all available sites, though there were only one or two sites available. As to the nature of search, the number of searches for all (both common and unique), common, and unique information decreases as time to implement decreases. Novice groups clicked much more for information acquisition than expert groups did during the first 25 min. As time passed, the number of clicks by novice groups converged with the number of clicks made by expert groups. These results are summarized in Table 6.

TABLE 6. Summary of the results.

Hypotheses		Description	Results
Extent of search	H1.1	$M1_{t1} < M1_{t2}, t1 < t2$	Rejected
	H1.2	$M1_E < M1_N$	Supported
Nature of search	H2.1	$M4_{t1} > M4_{t2}, t1 < t2$	Supported
	H2.2	$M2_{t1} < M2_{t2}, t1 < t2$	Rejected
	H2.3	$M3_{t1} > M3_{t2}, t1 < t2$	Supported
	H2.4	$|M4_{t1} - M4_{t2}|_E < |M4_{t1} - M4_{t2}|_N \; t1 < t2$	Partially supported

6. Discussion

The results suggest that the extent of search does not increase over time. On the contrary, the extent decreases as time to implement decreases, though this decreasing trend is not obvious in expert groups. As discussed in Section 3, it was assumed that the decreasing size of the search space and the acceleration strategy the groups adopted in time-constrained situations would lead to an increase in search extent. However, two other factors likely may have impacted search behavior. First, task difficulty increases over time in emergency response. The number of potential courses of action decreases over time, leading to a decrease in the number of potential solutions. Because task difficulty is inversely related to the number of potential solutions and positively related to the risks involved in the emergency response, the task will become more difficult over time. This increased difficulty leads groups to spend more time processing and evaluating on-hand information. Moreover, when the task becomes harder, both expert and novice groups tend to be more purposeful [28]: that is, the scope of their search tends to shrink in order to meet response goals. Second, considered in a broader framework of information-seeking behavior, the activities of information search, processing, and use are weighted differently in the different stages of the emergency response process. Groups' efforts will be devoted to locating information more at the beginning of the decision-making process for later filtration and final choice.

Another interesting finding is that little change has been observed in the nature and extent of search behavior of expert groups over time. Experience on prior emergency cases provides experts with skills to approach a similar emergency task. Striking changes in the extent of search only happened at the very beginning and very end of the process. An explanation for the high extent of search at the beginning of the process is that risks involved in the escalating catastrophe drive experts to learn more facts to eliminate uncertainty. An explanation to the high extent of search near the end is the extremely small size of the search space (one or two sites only). Under such

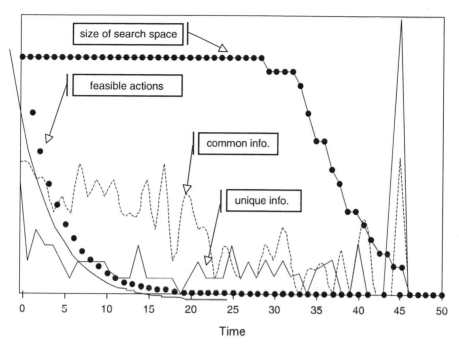

Figure 10. Revised model of information-seeking behaviors.

condition, even with severe time constraints it is easy for groups to click all sites in a very short time. Novice groups, as expected, explored a higher proportion of the available sites and sought more information than expert groups did during the emergency response process. Figure 10 shows a revised model incorporating the data from the present study.

In the simulated environment used here, the size of the search space and the number of potential courses of action decrease over time, thus increasing task difficulty. The extent of search does not show a consistent increasing trend. A number of explanations are possible. In general, the extent of search is likely to decrease in an emergency, but striking changes may be expected at the beginning and the very end, owing either to surprise at the event's sudden onset or the urgent need to complete the response before some deadline. Such changes will therefore be influenced by the risk arising from the environment and time pressure imposed on the decision groups.

7. Conclusion

Emergencies create the need for reliable information about the initiating event and its impact on society and the built environment. Response personnel may have to coordinate to seek and process information in a timely manner in order to make informed decisions about how to meet goals for the response. Understanding group information-seeking behavior is thus critical for improving group performance in emergency response situations. A computer-simulated environment was used to develop refinements to an initial model of how risk, time constraints, and expertise can influence group information-seeking behavior. The results suggest that both expert and novice groups display a decreasing trend for the extent of search as time to implement decreases, and novice groups exhibit a higher extent of search than expert groups do. Searches for both common and unique information decrease over time.

One suggestion of this research is that time pressure impacts patterns of information seeking both for expert and novice groups. Information from the same resource may not be the same as time passes; meanwhile, groups spend a decreasing amount of time and effort on information seeking. The cost of spending limited effort on unavailable information resources may be too high for emergency response groups. A second suggestion concerns the effort groups devote to seeking unique and common information. Novice groups may spend more effort locating both unique and common information than expert groups. So under the condition in which unique information is critical for decision making with time constraint, decision support systems may be of great value for decision makers in targeting search.

Future work in this area includes the consolidation of the proposed model of information-seeking behavior and the investigation of the combined impacts of time, expertise and decision support on group information-seeking behavior. It may also be advantageous to examine how learning takes place during information seeking in emergencies [30], thus contributing further to our knowledge of the factors that contribute to differences in expert/novice performance.

Acknowledgments

This material is based upon work supported by the U.S. National Science Foundation under Grant Nos. CMS-9872699 and CMS-0449582.

References

1. Ahituv, N., Igbaria, M. and Sella, A. "The effects of time pressure and completeness of information on decision making," Journal of Management Information Systems, 15(2), 153–172, Fall 1998.
2. Ben Zur, H. and Breznitz, S. J. "The effect of time pressure on risky choice behavior," Acta Psychologica, 47, 89–104, 1981.

3. Benbasat, I. and Taylor, R. N. "Behavioral aspects of information processing for the design of management information systems," IEEE Transactions on Systems, Man, and Cybernetics, 12(4), 439–450, July–August 1983.
4. Bonner, B. L., Baumann, M. R. and Dalal, R. S. "The effects of member expertise on group decision-making and performance," Organizational Behavior and Human Decision Processes, 88(2), 719–736, July 2002.
5. Campbell, D. J. "Task complexity: a review and analysis," Academy of Management Review, 13(1), 40–52, 1988.
6. Chase, W. G. and Simon, H. A. "Perception in chess," Cognitive Psychology, 1, 33–81, 1973.
7. Dennis, A. R., Hilmer, K. M., Taylor, N. J. and Polito, A. "Information exchange and use in GSS and verbal group decision making: effects of minority influence," 30th Hawaii International Conference on System Sciences, Maui, Hawaii, USA, 2, pp. 84–93, January 03–06, 1997.
8. Dennis, A. R. "Information exchange and use in group decision making: you can lead a group to information, but you can't make it think," MIS Quarterly, 20(4), 433–457, December 1996.
9. Durham, C. C., Locke, E. A., Poon, J. M. L. and McLeod, P. L. "Effects of group goals and time pressure on group efficacy, information-seeking strategy, and performance," Human Performance, 13(2), 115–138, 2000.
10. Earley, C. "The influence of information, choice and task complexity upon goal acceptance, performance, and personal goals," Journal of Applied Psychology, 70, 481–491, 1985.
11. Ellis, D. and Haugan, M. "Modeling the information seeking patterns of engineers and research scientists in an industrial environment," Journal of Documentation, 53(4), 384–403, 1997.
12. Foster, A. "A nonlinear model of information-seeking behavior," Journal of the American Society for Information Science & Technology, 55(3), 228–237, February 2004.
13. Grønhaug, K. and Haukedal, W. "Experts and novices in innovative unstructured tasks: the case of strategy formulation," Creativity and Innovation Management, 4(1), 4–13, March 1995.
14. Huber, V. "Effects of task difficulty, goal setting and strategy on performance of a heuristic task," Journal of Applied Psychology, 70, 492–504, 1985.
15. Larkin, J., McDermott, J., Simon, D. P. and Simon, H. A. "Expert and novice performance in solving physics problems," Science, 208, 1335–1342, 1980.
16. Leckie, G. J., Pettigrew, K. E. and Sylvain, C. "Modeling the information seeking of professionals: a general model derived from research on engineers, health care professionals, and lawyers," Library Quarterly, 66(2), 161–193, 1996.
17. Littlepage, G., Robison, W. and Reddington, K. "Effects of task experience and group experience on group performance, member ability, and recognition of expertise," Organizational Behavior and Human Decision Processes, 69(2), 133–147, February 1997.
18. Maule, A. J., Hockey, G. R. J. and Bdzola, L. "Effects of time-pressure on decision-making under uncertainty: changes in affective state and information processing strategy," Acta Psychologica, 104, 283–301, 2000.
19. Meho, L. I. and Tibbo, H. R. "Modeling the information-seeking behavior of social scientists: Ellis's study revisited," Journal of the American Society for Information Science & Technology, 54(6), 570–587, 2003.
20. Mendonça, D. "Decision support for improvisation in response to extreme events," Decision Support Systems 43(3) 952–967, 2006.
21. Mendonça, D., Beroggi, G.E.G. and Wallace, W.A. "Decision support for improvisation during emergency response operations," International Journal of Emergency Management, 1(1), 2001.
22. Mendonça, D. and Wallace, W.A. "A cognitive model of improvisation in emergency management," IEEE Systems, Man and Cybernetics: Part A, 37(4), 547–561, 2007.

23. Ordonez, L. and Lehman Benson, I. "Decisions under time pressure: how time constraint affects risky decision making," Organizational Behavior and Human Decision Processes, 71(2), 121–140, August 1997.
24. Parks, C. D. and Cowlin, R. A. "Acceptance of uncommon information into group discussion when that information is or is not demonstrable," Organizational Behavior and Human Decision Processes, 66(3), 307–315, June 1996.
25. Propp, K. M. "Information utilization in small group decision making: a study of the evaluative interaction model," Small Group Research, 28(3), 424–453, August 1997.
26. Ramirez, A. Jr., Walther, J. B., Burgoon, J. K. and Sunnafrank, M. "Information-seeking strategies, uncertainty, and computer-mediated communication: toward a conceptual model," Human Communication Research, 28(2), 213–28, April 2002.
27. Saito, H. and Miwa, K. "A cognitive study of information seeking processes in the www: the effects of searcher's knowledge and experience," Second International Conference on Web Information Systems Engineering, Kyoto, Japan, 1, pp. 0321–0330, 2002.
28. Simon, D. P. and Simon, H. A. "Individual differences in solving physics problems," In R. S. Siegler (Ed.), Children's Thinking: What Develops? Hillsdale, NJ: Erlbaum, 1978.
29. Stasser, G., Vaughan, S. I. and Stewart, D. D. "Pooling unshared information: the benefits of knowing how access to information is distributed among members," Organizational Behavior and Human Decision Processes, 82(1), 102–116, 2000.
30. Stasser, G., Taylor, La. A. and Hanna, C. "Information sampling in structured and unstructured discussions of three- and six-person groups," Journal of Personality and Social Psychology, 57, 67–78, 1989.
31. Taylor, M. S. "The motivational effects of task challenge: a laboratory investigation," Organizational Behavior and Human Performance, 27, 255–278, 1981.
32. Wai-yi, B. C. "An information seeking and using process model in the workplace: a constructivist approach," Asian Libraries, 7(12), 375–390, 1998.
33. Weenig, M. W. H. and Maarleveld, M. "The impact of time constraint on information search strategies in complex choice tasks," Journal of Economic Psychology, 23(6), 689–702, December 2002.
34. Wilson, T.D. "Human information behavior," Information Science, 3(2), 49–56, 2000.

THE USE OF WAR GAME SIMULATIONS FOR BUSINESS STRATEGIES

B. SHEPPARD

Simfore Ltd.
London, UK
ben.sheppard@simfore.com

D. SLAVIN

Institute of Biomedical Engineering
Imperial College London
London, UK

Abstract: War gaming, long used by military organizations to test strategies without actual combat, are now being used by nonmilitary private and public sector organizations to support the formulation of potentially high-impact decisions and plans. This chapter defines war gaming approaches, describes their application in two case studies, and identifies specific situations that they can effectively address.

1. Introduction

The concept of war gaming has its roots in military history and continues to be used extensively by armed services around the world. More recently, war gaming has been adopted and applied by businesses and non-government organizations as a tool to test and develop new strategies and procedures. The military routinely employs resources in training for operations, testing strategies, and operational plans without actual combat. These simulations are also referred to as "maneuvers" or "exercises," and underpin most collective training programs.

War gaming has also been employed to examine preparation and response measures to single or multiple chemical, biological or radiological (CBR) terrorist attacks and conventional strikes. For instance, the US TOPOFF (Top Officials) terrorism preparedness exercises mandated by the US Congress and run by the Department of Homeland Security [1].

I. Linkov et al. (eds.), Real-Time and Deliberative Decision Making.
© Springer Science + Business Media B.V. 2008

This chapter will evaluate the use of war gaming as a decision-making tool and how this provides a valuable means to examine strategies in different scenarios as an effective futures tool for the public and private sectors. The simulations discussed here are based on human interactions and not computer modeling.

The chapter is divided into the following parts:

1. Background of simulations

2. Methodology of simulations

3. Outcomes of simulations based on analysis

4. Case studies

The chapter provides an overview of how the methodology has been adapted to the business environment by the pharmaceutical industry and public health sector, and how it could be applied to other areas.

Two case studies will provide insight into where the use of war gaming has been valuable as a decision aid. The first involves a large U.S. pharmaceutical company and examines the conditions under which precision medicines (drug diagnostics combination therapy) could be attractive to the organization. This included assessing these drugs' internal and external risks and benefits, from organizational structures and decision processes to how the external environment might respond. The external environment included regulatory agencies, patient groups, and key public health bodies. The pharma company benefited from running a number of simulations to assist in their decision making processes from product development through contingency planning.

The second example is an examination of United Kingdom (UK) preparation, response, and recovery capabilities relating to a pandemic flu. Sponsored by the Bioscience Futures Forum, established by the UK government Department of Trade and Industry, the event involved six biopharmaceutical companies, the National Health Service, pharmacy bodies, and regulatory agencies (EMEA and MHRA). The simulation focused on two key themes: operational response and reputation management issues (risk communication and public relations). The outcomes helped to shape public health, government, and industry thinking to better prepare for a pandemic flu.

This paper draws upon research undertaken by this author from a three-year pharma-funded research project in 2003 at King's College London, and since commercialized into a consultancy service by Simfore and HFC. The project adapted war gaming in the defense arena into an effective risk management tool to provide government, industry, and academia a means to develop and stress-test risk assessment approaches. The tool is designed to address uncertainty and inform decision-making processes. War gaming offers a simplified and structured framework for identifying possible and probable outcomes from the interaction of qualitative variables and uncertainties and for stress-testing and identifying new strategies and approaches. The use of war gaming has become increasingly accepted in the corporate environment, with companies reporting greater demand to simulate the interactions of multiple actors in a market [2].

2. What Are Simulations and Why Are They Not Scenarios?

When discussing a simulation, we frequently are met with "Yes, we do this already. We do scenario planning." In fact, interactive simulations are the next step beyond scenarios. They can start with and are frequently adapted from scenario planning and/or financial modeling, enabling organizations and stakeholders to develop and validate novel strategies in a hypothetical but credible exercise. Simulations reveal likely outcomes, including unintended consequences, and enable the participants to challenge assumptions by allowing stakeholders' interactions to provide new insights.

To understand how simulations are adapted from the military sector, the following section provides an overview of military simulations.

2.1. WAR GAMING IN THE MILITARY

War gaming in the western military can be traced back to Prussians, whose victory over the second French Empire in the second Franco-Prussian War (1870–71) is partly credited to the senior officers receiving training from playing a war game (*kriegspiel* in German). In 1898, naval analyst and writer Fred T. Jane, who founded *Jane's Fighting Ships*, developed a series of rules depicting naval actions through the use of model ships and miniatures. Military war games evolved rapidly into more complex systems during the first half of the 20th century, which included the U.S. 'gaming' its military campaign in Asia and the Pacific Rim during the Second World War [3].

Modern armed forces run two main types of simulations: *soft gaming*, with individuals playing and interacting as teams; and *hard gaming*, or computer modeling. The present business simulation approach is adapted from soft gaming, which focuses on decision making through qualitative interactions between individuals and teams.

Hard gaming principally relies on inputting the profiles of military assets (e.g., aircraft, tanks, ships) on both the allied and enemy sides into a computer model. The computer simulation, through knowledge of military capabilities (e.g., fire power, speed, range, agility) and vulnerabilities (e.g., available countermeasures and shield strengths) calculates the attrition and casualty rates of personnel and equipment deployed in combat operations.

The objective of both approaches is to assess what force structure would best suit the desired operations. For instance, prior to the 2003 Gulf War, British forces ran simulations to assess how best to fight Iraqi forces in their approach to Basra in the event of engaging the enemy in the desert, or within the city. The advantage of hard gaming (involving computer modeling alone) is the ability to run scenarios multiple times with minimal resources. But hard gaming does not provide training or evaluate effective decision making and interactions between various groups.

While the military conducts large-scale outdoor operational maneuvers (field training exercises) involving land, air, and sea assets across thousands of square miles, to evaluate response strategies and contingencies involving a large number of personnel at once, they also conduct indoor simulations that require significantly less manpower and resources (soft gaming). One of the most common forms is the Command Post Exercise (CPX), which focuses on simulating the environments experienced by command (leadership) teams and planners without the need to physically deploy troops [4]. The CPX retains human input and is thus highly effective at simulating human imponderables and behaviors, but is easily accessible at a lower cost.

The scenario could be, for instance, a humanitarian crisis in the Balkans that requires military forces to be deployed while opposition elements are conducting offensive military activities against civilians. In the simulation there would be political interests and challenges at both the regional and international levels (e.g., the United Nations). In these types of exercises, military personnel would role-play the external political and opposition elements, while also performing their day-to-day real-world duties.

Running the exercise would be a control group responsible for umpiring the simulation, providing scenario injects (e.g., major political or military developments), and deciding what additional information the teams are allowed to receive. The control group also makes sure that the team's tasks are accomplished within the time frame allowed.

There are two clocks running. The simulation scenario environment can cover a period of days, weeks, or months. The participants are taken through the day's activities in real time. CPX simulations can last from one day to several days, or even a couple of weeks.

2.2. BUSINESS WAR GAMES

Unlike the armed forces, the business environment has a wide variety of war game options offering various degrees of complexity and value. Some options do not necessarily involve interactive simulations. For instance, Shell scenario planning provides alternative views of the future. This first came to prominence in the 1970s.

Business war game simulations, which are the focus here, are typically played over one or two days to simulate a period of weeks, months, or years in a series of sessions. The simulations can either be used to explore new strategies or as a training tool. Typical uses include:

- Ethical preparation to understand social stakeholder opinions.
- Evaluate the understanding of a strategic plan to accelerate strategy implementation.
- Assess reactions and possible responses.
- Explore intended and unintended consequences.
- Practice/rehearse communication.
- Evaluate behavior of competitors—blue and red teaming.
- Use time compression so teams can see the longer term implications of decisions.

The output value derives from three distinct phases: simulation development, execution, and analysis. Simulations address uncertainty and inform decision making, and can test assumption robustness under various conditions. These simulations are played by human subjects rather than involving computer modeling; however, they may include databases and computer models as part of the event. For instance, teams may model their financial strategies or clinical trial options using existing tools from their day-to-day activities. In such cases, the computer model is then customized with a user-friendly interface that is flexible in the simulation, for instance, populated with profiles of products that are being examined in the exercise.

One of the most powerful aspects of simulations is the lessons learned by participants as a result of their experience. Unlike other styles of workshops, these simulations are not about instructing participants. But through their experience in the simulation, participants encounter learning opportunities by living through the scenario and witnessing how their decisions and the consequences of their actions and the actions of those around them could impact their future.

There are two broad categories where simulations can be used: research and training.

2.2.1. Research Simulations

These typically allow a client to develop and stress-test the robustness of current or alternative business assumptions. Their value derives from evaluating concepts in a safe environment before implementing them in the real world. Participants are encouraged to be less risk-averse than they might be in the real world when exploring and developing new strategies. The simulation tool that has been developed also creates a collaborative space that brings together leading industry peers or other stakeholders to develop strategies and create opportunities. A key feature is that it accelerates the decision making and negotiating time.

2.2.2. Training Simulations

Training simulation offers a powerful experiential tool to immerse groups and individuals into testing their decision-making processes or learning new procedures and routines. This can include new day-to-day decision-making processes that might be implemented by an organization or new standard operating procedures. Where individuals and groups have been used to one set of routines, a simulation would enable individuals to fully explore their potential value and challenges to implementation, and test their adoption in a safe environment. A second main use of training simulation is crisis management or contingency planning. Organizations can test their emergency response public relations and risk communication procedures following an adverse event (e.g., a major product recall following contamination, or pressure on a company to withdraw or revise the labeling of a high-profile drug following reports of severe side effects). In both cases one would be training and testing the organizational structure in what information and tacit knowledge from individuals is available within and outside a company to make informed decisions in the context of uncertainty. A simulation could test an organization's public presentation of issues with invited external consultants role-playing stakeholders like the media, consumers, and the regulatory authority.

3. Methodology

Developing and running a simulation is a three-stage process:
1. Building the customized model
2. Running the simulation
3. Reporting the key findings and recommendations

3.1. BUILDING THE CUSTOMISED MODEL

A key aspect to this approach is customizing the model to the client's needs. There is no one-size-fits-all model. Key questions that need to be addressed include capturing the key objectives, identifying the timeline to be examined, and determining the key variables that need to be populated in the simulation. As it is not possible to have all the internal and external variables running at once, the simulation designers have to prioritize which ones should be factored in the model. Finally, the client's teams and line functions are identified.

During this process, the simulation design team identifies the individuals within and outside the client's organization who should be invited. Parallel to this is the development of the simulation scenario. To ensure that the simulation moves smoothly through the time period being examined, the simulation design team needs to build in advance the scenario to be examined. This includes a case study (for instance a mock product profile), scenario injects in the form of mock newspaper stories and company announcements, and one or two major external shocks. The latter could be developed in conjunction with the client to meet the needs of and stress-test the decisions being made in the simulation. Throughout simulation development, those who will take part in the player teams should not be aware of what the unfolding scenario will entail.

Although the scenario material and injects are developed in advance, there is a fine balancing technique involved to make sure that the simulation model is not overburdened with too much information and interaction nor does it have so little that the output is superficial. Getting the right balance also extends to compiling the briefing for all the participants. For instance, participants may not have that much time to read through all the material. Therefore, when building the scenario and related material, the designers have to be aware of the capabilities and time participants have available to prepare and be engaged in the simulation.

There is no one set way of getting the right balance. Developing a successful simulation requires the experience to know how much information should be included. This will partly depend on the topic at hand; for instance, the degree of familiarity the player teams will have with the issue being examined and the case study at hand.

3.2. RUNNING THE SIMULATION

The one- or two-day simulations establish all links and partnerships via player interactions. For the pharmaceutical area, these include physicians, pharmacists, payers, wholesalers, economists, and commercial interests. These can be role-played in the exercise by consultants and client employees with expertise in these areas.

At the beginning of the simulation all participants are in one room for the scenario briefing, and then move to their separate team rooms to work on their set tasks and interact with other groups During this time participants receive scenario injects of mock news stories and press releases. After a set time period, participants reconvene in one room for the report back session to present their recommendations and agreements they may have reached. This concludes one time frame move. Each move covers part of a period of weeks, months or years which forms part of the overall scenario being examined. There are several moves in one simulation. During the work stages where participants are given set tasks and objectives to fulfill, communication between teams and those representing the external environment is conducted by email and face-to-face contact. Decisions and deliberations are captured and later analyzed.

While the simulation has a prepared scenario with set aims and objectives for each of the moves (time frame segments), it is important that the simulation is not too structured to constrain freedom for the variables to interact. At the same time, there should not be so much freedom that the set tasks and objectives cannot be accomplished. As with balancing the variables in the model, there is no set way of doing this other than by experience in running simulations.

A simulation is effectively a time and space entity that you can expand and contract in segments as you see fit to meet the purpose. The only real restriction is the actual real time one has to run the simulation. Like a piece of plasticine, you can mold and move it into the shape and length you wish within the constraints of the amount of plasticine you have. The amount of plasticine in this case represents the real time and resources you have to run an event. But it is also important to keep in mind the main objectives.

3.3. REPORTING THE KEY FINDINGS AND RECOMMENDATIONS

Following a simulation, the key aims and objectives are extrapolated from an analysis of the material generated in the simulation. This includes team presentations to meet the key tasks set throughout the simulation, and report notes of meetings and interactions that have taken place between the different groups. Typically a simulation provides an extensive amount of material to assess, which is captured through specific tools and approaches.

With all the material and data at hand, the process begins to reverse-engineer the key decision points and events. Analyzing the results leads to two main outputs. The first is a timeline diagram capturing the key decision points and outcomes. The second is a series of key findings and recommendations. At this stage, the simulation output entails proprietary elements in analyzing and presenting the data. The timeline graph of the key interactions includes junctures where certain decisions were made, and their consequences. From this, a series of alternative scenarios and outcomes can be extrapolated from which the end user can see the upside and downside of various options from both internal and external perspectives.

4. Case Studies

Below are outlines of two simulation case studies that highlight how the simulations were compiled and the resulting key findings and recommendations. Given the client confidentiality of the simulations, only selected lessons are included.

4.1. PANDEMIC FLU

4.1.1. Background

While the UK Government has run pandemic flu simulations (including a Whitehall exercise run in early February 2007 called "Winter Willow," and Health Protection Agency simulations), this was an opportunity for stakeholders to collectively challenge their thinking and behavior in response to a pandemic. The simulation was run in June 2006 for the UK government body, the Bioscience Futures Forum [5]. It tested the impact of different levels and types of stakeholder engagement across public and private sector organizations and how they can best come together to coordinate and communicate complex operational policies and procedures to engage with the public. Eight biopharmaceutical companies took part as one main pharmaceutical company. Public health representatives and stakeholders included a London Primary Care Trust, the Health Protection Agency, physicians, pharmacists, and wholesale distributors. Regulatory participation included the UK's MHRA and the European body EMEA. The Department of Health observed the simulation.

4.1.2. Scenario Structure

The scenario covered a ten-month time period from July 2006–May 2007 covering one wave of a pandemic. The time period was divided into the following four sections:

- Pre-pandemic: first UK human H5N1 bird flu case from a Norfolk poultry farm (July–December 2006)
- Wave 1: pandemic starts in Sumatra, Indonesia (December 2006–January 2007)
- Wave 1: pandemic reaches the UK (January–April 2007)
- Inter-pandemic: preparation for a second wave (April–May 2007)

4.1.3. Key Findings and Recommendations

The simulation identified a number of ways to better utilize resources and capabilities, including repurposing existing assets. Many of the recommendations could be implemented through better coordination and alignment among key public and private sector stakeholders.

The first recommendation was to create a list of "essential drugs" the delivery of which needs to be maintained to ensure critical healthcare delivery. The list included a number of existing drugs that could be used to treat secondary infections from pandemic flu.

Death from pandemic flu is usually due to respiratory failure or other complications from secondary effects. Young adults can have an exaggerated immune response ("cytokine storm") that can lead to extensive damage in the lungs and cause multiorgan failure. There are a number of existing drugs that can treat this response, of which adequate supplies need to be maintained. As a vaccine based on the pandemic strain is unlikely to be available in the UK during the first wave based on what vaccines had received regulatory clearance in 2006 and there will be limited supplies of antivirals, reliance on existing therapies to treat secondary infections becomes of greater importance.

The second recommendation noted that alternative forms of pandemic flu vaccines with higher production yields could reduce the overall fatality rate. While the current egg-inactivated vaccine can provide around 300 million doses globally (insufficient to meet the needs of a global pandemic), alternative vaccines like the cold-adapted egg-based vaccine codeveloped in the UK and U.S. can increase the number of doses by several times and have a shorter development period. Another strategy is providing a lower dose of a pandemic flu vaccine to individuals to provide some protection to a larger number rather than more complete protection to a smaller number of people. While some individuals may still not survive, the overall fatality rate would be lower. This decision would require ethical considerations of whether what is better for society outweighs individual treatment needs.

The simulation also identified inconsistencies in national and local public health contingency plans that must be resolved and distribution channels for disseminating antivirals and vaccines that must be strengthened. Pharmaceutical wholesalers and pharmacists called for their supply and delivery channels to be fully integrated and consulted as part of the UK government's response plan.

To address these and other issues, the simulation recommended the establishment of a biopharmaceutical working group to ensure close collaboration and communication within the industry in order to share knowledge and expertise and communicate credibly to the wider public. Finally, effective pharmacovigilance measures to monitor the safety and efficacy of vaccines, within an acceptable safety framework, should be established to accelerate vaccine approval during a pandemic, particularly for novel drugs.

4.2. PRECISION MEDICINE SIMULATION FOR A MAJOR PHARMA COMPANY

4.2.1. Background

In 2004, the first simulation took place as part of a major pharma company's research project to pilot the methodology. The simulation examined under what conditions precision medicines (pharmacogenomics) could be attractive to that company. Precision medicines are compounds that—when combined with a diagnostic device—can identify and treat subsets of a population (e.g., responders, nonresponders, and those who may experience severe adverse effects). The concept behind this approach is that if one could identify these population subsets through a biomarker (for instance, a genetic test), then patients could be provided with the most suitable treatment from the outset. At the time of the simulation, the company wanted to investigate whether precision medicines would be of value to the organization. The traditional business model for this company and other pharmaceutical organizations has been the blockbuster model, which entails developing a product for the mass market without specifying population subsets.

The key aims and objectives of the simulation were to:

- Provide insight into the environmental challenges and opportunities of precision medicines for the pharma company.

- Illustrate the realities of drug development and external conditions.

- Determine whether simulations can capture and manipulate the main pharmaceutical variables.

- Explore to what degree the simulation can define new deliverables.

- Ascertain to what extent the results provide operational utility to further understand whether precision medicines can be attractive to the company.

The project team represented a hypothetical drug development team of a dozen company employees. Although the simulation was originally intended to look at pharmacogenomics, post-simulation analysis revealed some fundamental lessons for how the pharma company's drug development teams and governance bodies should function. This demonstrated that war game simulations have the advantage of identifying opportunities and challenges far beyond other methods like brainstorming.

4.2.2. Scenario Structure

Two key groups participated in the simulation: the facilitators, who ran the exercise and represented the external and internal stakeholders; and the drug development team, who actively played the simulation. The external variables were represented by individuals with in-depth expertise about the roles they were playing. Each team had its own room. In addition, the pharma company had a governance body to which the drug development team reported their strategy to get a go/no-go decision.

Figure 1 shows the simulation timeline examined by the pilot simulation. This timeline extends from 2007 to 2017. It was split into four moves, from Phase III of development to the fifth year of the product's launch in 2017. The last move followed a time jump, from 2011 when the drug development team submitted their proposal, to the regulatory bodies FDA and EMEA for new drug approval.

4.2.3. Key Findings and Recommendations

The post simulation analysis revealed the following.

1. The simulation design produced a workable futures simulation that could record decision paths, capture data, and identify problems and opportunities. The exercise demonstrated that it is possible to simulate and manipulate the key internal and external drivers of the pharma company's business environment over two days and develop a credible output.

2. The simulation identified the stage points at which additional knowledge of the diagnostics industry and stakeholders was essential to make informed development decisions concerning the co-development of a diagnostic device with a compound. Through analyzing the information, it became clear at what junctures during a product's drug development teams would need to have Dx information. This would enable improved decision making and informed thinking about the options available to the teams.

3. Compressing the development timeline to two days allowed the simulation to discover unexpected issues and identified solutions to implement. While the simulation's main aim was to identify the conditions under which precision medicines would be of value to a major pharma

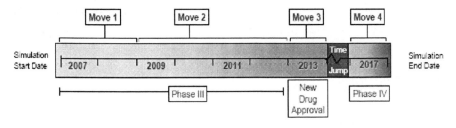

Figure 1. Timeline for the precision medicine simulation.

company, a byproduct of the exercise was to identify new issues regarding decision-making processes. This is a key advantage of simulations over more traditional management consultancy approaches, which would focus on a particular set of issues, but would not enable variables to interact freely to identify other key issues.

4. Internal governance bodies could be revised to deal with the uncertainties and identify opportunities of the changing pharma environment.

5. Potential efficiencies were identified by empowering drug development teams to influence the positioning of similar products to optimize the portfolio.

6. Product development could be more efficient and opportunities identified through reconfiguring drug development and worldwide teams.

7. Adopting the format the drug development teams experienced in the simulation could make for a more streamlined decision-making process.

5. Conclusion

Simulations can be used as a research tool to investigate alternative business strategies, and identify lessons that could streamline and improve product development and decision-making processes. In the pharmaceutical area, it is possible to use simulations to illustrate the realities of drug development and external conditions, and to provide insight into alternative drug development beyond the self-evident. The key output is a process map of the decisions made to identify the junctures where an organization can proactively influence and engage with its external environment and seize the initiative over the timeline examined. Subsequent simulations can always be conducted under alternative environmental conditions to test the robustness of strategies. The series of simulations that have been run provided operational utility to further understand the challenges and opportunities facing the pharmaceutical industry in developing innovative research and development strategies.

While the simulation model has been developed initially for the pharmaceutical and public health areas, the general concepts and approaches can be applied to other sectors, such as telecommunications, petroleum, the defense industry, and finance.

The following list summarizes key areas that war game simulations can address within each sector:

- Business optimization
- Evaluating the robustness of existing and proposed business models

- Exploring new stakeholder engagement solutions
- Increasing value from products and services from R&D to launch
- Identifying and assessing new organizational structures
- Technology optimization pathways
- Evaluating regulatory filing strategies
- Organizational optimization for team decision making

Key to running simulations is knowing the right time to apply the technique. While simulations can be applied to a broad variety of issues for training and research purposes, there may be occasions where a facilitated meeting or more traditional management consultancy approach is needed. It is up to those with experience of running business simulations to know when it is appropriate to recommend this tool, and how the simulation should be constructed.

It has been shown that simulations provide a highly innovative futures tool that provides the public and private sectors with a valuable means to test and develop new strategies in a safe environment prior to implementation. While futures and scenario tools tend to rely more on workshops and brainstorming activities, simulations have the benefit of identifying known and unknown elements, together with identifying critical but unrecognized aspects of a problem that could prove critical to the successful implementation of a new strategy. The freedom of the variables to interact in a safe environment, replicating the operating environment as much as possible, has a powerful effect.

References

1. Christian W. Erickson and Bethany A Barratt, 'Prudence or Panic? Preparedness Exercises, Counterterror Mobilization, and Media Coverage - Dark Winter, TOPOFF 1 and 2', Journal of Homeland Security and Emergency Management, Vol. 1, No. 4, 2004, p. 3
2. Shall we play a game? Economist, May 31, 2007.
3. Wilson, A., War gaming, Penguin, Harmondsworth, 1970..
4. Correspondence with Mark Bouch and Sykes Fairbairn, August 2007.
5. Interview: Testing the UK's response to a global flu outbreak. RUSI Jane's Homeland Security Monitor, April 2007, pp. 18–19.

AN INTEGRATED APPROACH FOR FLOOD RISK MANAGEMENT

J. GANOULIS

*UNESCO Chair and Network INWEB: International Network
of Water/Environment Centres for the Balkans
Department of Civil Engineering
Aristotle University of Thessaloniki
54124 Thessaloniki, Greece
iganouli@civil.auth.gr*

Abstract: In most developed countries, floods are the result of extreme hydrological events in combination with human activities and land use in both rural and urbanized areas. This paper reviews lessons learned from a European Union (EU) cooperative project, Cooperation Along a Big River: the Case of the Volga River (CABRI-VOLGA), and suggests a methodology for developing integrated flood management plans, taking into account technical, environmental, economic, and social objectives.

1. Introduction

Floods are very well known extreme events, which have been described and reported since early historical times, when human activities and human settlements were relatively limited. The floods of the Nile River in ancient Egypt and similar flood events in ancient Greece and the Roman Empire are related mainly to extreme hydrological events such as heavy precipitation and sea level rise in coastal areas. However, recent floods in Europe and other developed regions in the world have demonstrated the additional role played by different human factors and anthropogenic activities such as extensive urbanization, and river floodplain occupation and land use, in producing catastrophic floods [8, 12].

Recent catastrophic flood events both in Europe and the United States (Elbe River, 2002; Danube River, 1999; Rhine River, 1995; Mississippi River, 1993) have shown that human activities and traditional river engineering works may result in an increase in the frequency of small and medium floods and, most importantly, in negative economic consequences such as loss of property, destruction of livelihood, and loss of human life. Possible climate change could increase both the intensity and frequency of catastrophic floods.

I. Linkov et al. (eds.), Real-Time and Deliberative Decision Making.
© Springer Science + Business Media B.V. 2008

An extreme flood event in Central Europe in August 2002 caused heavy damages and loss of human life in Austria, the Czech Republic, and Southeast Germany. The total cost of these flood losses is estimated to be €15–16 billion. One hundred people lost their lives and about 100,000 were displaced. The flood peak of the Elbe River in Dresden is classified as at least a 500-year-recurrence event. In Austria, it is estimated that the flood peak along the Danube River corresponds to a 70–100-year-recurrence event, while in some tributary basins, floods with a return period of about 1,000 years and above can be assumed.

Human activities on the river basins have increased flood risk by:

- Aggravating flood events. Urbanization, agriculture, and water drainage have diminished the retention capacity of the vegetation, soil, and ground, amplifying flood scales. In addition, structural flood defenses can increase flood level and speed.

- *Aggravating flood consequences.* With increasing human presence in floodplains, floods have higher destructive potential.

In the past, the most common solution to flood risk exposure was river containment via construction of levees, embankments, canals, and dams. The global efficiency of a flood defense system based only on structural devices has proven unsatisfactory. Apart from the residual risk of failure, and increases in downstream water levels, these flood-protection devices seriously interfere with natural river flow and prevent alluvial deposits in floodplains. Moreover, as their protection is effective only against low- or medium-intensity events, flood-protection devices may give people a false feeling of security (such devices may be ineffective during a rare or extreme event). Under the illusion that there is no flood risk, people are unwilling to adopt necessary preventive measures and thus increase their vulnerability to losses in case of flooding. Integrated water basin management aims at strengthening man's ability to cope with water-related problems and to govern wisely in water-related issues. This is vital if increased water security is to be achieved, extreme poverty eradicated, and environmental sustainability ensured.

In the past few years, many initiatives have been taken to improve flood prevention and remediation at the river basin scale; and new paradigms and new tools have been developed within the frame of integrated river basin management. The majority of these studies and plans have been developed for specific areas, each with their own social, economic, geographic, and hydraulic characteristics [5].

This paper gives an overview of flood risk management plans, with special focus on existing practices, initiatives, and research results in the domain of human security and vulnerability in large European river basins. It is based on documents produced by national and international organizations, as well as cross-border European river basin boards.

After defining frequently used topics and terms—such as hazards, risks, vulnerability, and human security—lessons learned from the European cooperative project Cooperation Along a Big River: the Case of the Volga River (CABRI-Volga) are reported [2]. Then, a multiple criteria-based approach is suggested to account for technical, environmental, economic, and social objectives in developing integrated flood management plans [7, 9, 10].

2. Terminology and Useful Definitions

The terms *flood, flood risk, vulnerability, human security,* and *environmental protection* (as related to floods) are used by different specialists without consensus about their meaning. A new EU directive, 2007/60, on the assessment and management of flood risks [4], which complements the water framework directive, 2000/60 [3], adopted the engineering definitions of these terms:

- *Flood* means the temporary covering by water of land that is normally dry. This includes floods from rivers, mountain torrents, Mediterranean ephemeral water courses, and floods from the sea in coastal areas.

- *Flood risk* means the combination of the probability of a flood event and of the potential adverse consequences for human health, the environment, cultural heritage, and economic activity due to a flood event.

Although not formally included in the directive, the following definitions may be also useful:

- The term *vulnerability* (as applied to a given environmental area or to humans and ecosystems) denotes the sensitivity of the system to risk. For example, when the aim is to reduce the flood vulnerability of the river basin, measures should be taken to increase the ability of the system to sustain floods; i.e., reduce the system's sensitivity to floods. Vulnerability may be considered a performance index of the system that can be estimated by measuring the possible degree of system damage or severity of consequences due to an incident such as a flood [6].

- The term human security (as applied to large river basins) focuses on reducing risks to people and the environment from hydrological extremes such as floods. The concept of human security has broadened from its traditional context (local and worldwide civil and military security) to embrace the idea that every human being should be able to benefit from sustainable socioeconomic development. From among different natural resources, water has been recognized as the key environmental resource for social security, economic growth, and prosperity. Human security can therefore be seen to be related to environmental preservation (water, ecosystems, and biodiversity) and to socioeconomic stability and sustainable development [11].

- *Hazard* is the potential to cause harm. According to the EU flood directive, *hazard maps* should be drawn in those areas for which potential significant flood risks exist. These maps should indicate possibly inundated areas, water depths, and possible flow direction and velocities.

- In addition, *flood risk maps* should show the potential adverse consequences associated with flood scenarios.

3. The CABRI-Volga EU Project

CABRI-Volga was an international cooperative project aiming to facilitate cooperation and coordinate research in environmental risk management in large river basins in the EU, Russia, and the New Independent States (NIS). It was initiated by the European Commission 6th Framework Programme (EC-FP6) and was successfully developed in the period 2005–2007. It was selected by DG Research as "an excellent example of the positive impacts EU research can achieve." Along with 40 other selected projects, CABRI-Volga is featured in a catalog of success stories of the EC FP's six research programs.

It focuses on the Volga basin, for which environmental risk management is fundamental to protecting the environment, improving socioeconomic conditions, and promoting agricultural and industrial economies as well as the health of the Caspian Sea. The mission statement of the CABRI project was to establish an international cooperation and consensus process contributing to environmental risk reduction, sustainable development, and enhancing human security in the Volga basin through improving institutional coordination, developing cooperation and partnerships between multiple stakeholders, and strengthening research potential [2].

CABRI-Volga was developed by activities coordinated by the following Expert Groups (EG):

- EG1: River and Environmental Rehabilitation
- EG2: Human Security and Vulnerability
- EG3: Natural Resources and Their Sustainable Use
- EG4: Connecting Goods and People
- EG5: Institutional Coordination and Cooperation

The UNESCO Chair/Network INWEB, Aristotle University of Thessaloniki, coordinated the EG2 group and focused on flood risk assessment and management to achieve human and environmental security at the basin scale.

3.1. MAIN CONCLUSIONS AND RECOMMENDATIONS OF EG2

Recommendations that may be of general value and could be applied to different river basins are the following:

- Assess the water-related multi-hazard risk/vulnerabilities, including not only technical but also social, economic, and environmental/ecological analysis at the local level.

- Assess and produce scientific advice on a possible package of mitigation and adaptation responses to water-related natural disasters in the basin, which is based on a combination of structural and nonstructural measures within each stage of disaster risk reduction (i.e., mitigation, preparedness, emergency response, and rehabilitation).

- Assess the hydrological regimes, water quality indicators, ecological parameters, and social and economic impacts of artificial reservoirs and the Volga cascade.

- Assess the risks associated with hydro-technical facilities and related local ecological and socioeconomic impacts.

- Prepare an inventory of knowledge and best practices in river basins across Europe for flood-related disaster risk reduction.

- Perform a comparative analysis and evaluation of existing international and national methodologies for integrated assessment of hydraulic facilities and their adaptation to specific needs.

For future research and development, the following recommendations were made:

- Future research should favor proactive approaches to water-related disaster risk reduction, including a combination of mitigation, prevention, emergency, and rehabilitation responses.

- Reference should be made to a number of studies of project partners from Europe and Russia containing assessments of good practices and major problems during the recent flood events, which are based on the above approach.

- An integrated approach is recommended, which envisages a combination of structural and nonstructural solutions. Indeed, there is a growing understanding that structural measures alone are not able to eliminate flood risks. Coordination within a river basin—of responses from many sectors including construction, land-use planning, forestry, agriculture, and environmental protection—is essential for enhancing the security of livelihoods against floods.

- The CABRI-Volga experts also recommend that further inventories and further comparative analyses of knowledge and best practices in river basins across Europe in water-related disaster risk reduction be performed.

- CABRI-Volga experts noted the poor multidisciplinary assessment of risks of water-related disasters in the Volga Basin. Development, testing, and application of indicators that allow the assessment of vulnerability and coping capacity of a society to floods are important for effective disaster risk reduction.

- Such indicators are useful to enhance "knowledge for action" and should be taken into account by decision makers.

- Special emphasis should be placed on a rationale for development of vulnerability indicators reflecting the situation at the local- and community-based levels, because they help to identify vulnerabilities and capacities of households and local communities to manage and overcome disasters, including floods.

- Experts also noted the importance of new behavioral research on members of the local public in the Volga Basin. Local public participation in flood risk reduction needs to be enhanced and should be regarded as a way of life and a crucial element in the integrated flood risk reduction approach.

- Currently, there are significant loopholes in knowledge and data relating to human and ecological security in existing systems of hydro-technical facilities and the Volga cascade of reservoirs. There is a common expert opinion that thorough assessments are urgently needed, which could be part of international cooperation initiatives.

- CABRI-Volga experts recognized that the state of some facilities pose serious threats to the safety of communities living in the Volga and its tributaries.

- Innovative solutions need to be found to ensure both the safety of the population and integrated water management in the basin. There is also a need for further multidisciplinary evaluations of the social and economic impacts of the Volga's artificial reservoirs.

4. Integrated Flood Management Plans

When considering how the risk of floods can be reduced and how the consequences of floods can be alleviated, two different attitudes can prevail (Table 1): the first is to consider the flood as a random natural disaster and

TABLE 1. Alternative actions for flood control.

FLOOD CONTROL ALTERNATIVES	*Alleviation* • Flood mitigation • Vulnerability reduction	Emergency
Prevention	***Structural measures*** (levees, diversions, channel regulation) ***Nonstructural activities*** (open space preservation, planning, zoning ...) **Property protection** (insurance, relocation, acquisition ...) **Public information** (Flood maps, outreach ...)	• Emergency plans • Warnings
Coping post-facto	• Technical assistance • Rehabilitation	• Evacuation • Technical assistance • Rehabilitation

only respond on an ad-hoc basis through emergency programs. The alternative, favored by the CABRI-Volga project, is to recognize that floods are recurring phenomena and adopt a proactive and strategic approach including a combination of mitigation measures with emergency response and rehabilitation along with incorporation of disaster risk reduction into sustainable development strategies. In this way, the hazard is "internalized" and as a result vulnerabilities can be reduced and coping capacities enhanced.

Additional recommendations made following CABRI's EG2 meeting include:

- Flood management and protection of people and property should take into account the fact that major cities are often better protected than small settlements and rural communities. Therefore special emphasis should be given to the problems and vulnerabilities of rural communities and small- and medium-sized cities [5].

- The raising of awareness is an important issue, particularly for those people living in areas prone to floods.

- It is recommended that structural and nonstructural measures be integrated and considered at the same time, instead of sequentially.

- A key element in integrated river basin management and the reduction of potential damages and losses is the strategy based on allocating more space to the river bed through effective national and local planning.

- Floods cannot be avoided; however, human intervention—especially land use patterns and engineering works—is a key factor affecting the impact and magnitude of medium- and small-scale flood events. Particular attention should be paid to deforestation, changes in the hydromorphological situation of a river, the conversion of open space in a settlement area, and the construction of infrastructure such as roads and highways.

- Furthermore, it was mentioned that a recent study in Switzerland came to the conclusion that increasing investments in systems of flood protection lead to higher economic losses after catastrophic floods. There will always be a risk element when catastrophic floods occur, and incorrect perceptions of risk and reliability may create problems, especially for people living in floodplains, who have high exposure to such hazardous events.

- Increasing extreme weather events and rapid temperature changes resulting from climate change, leading to snowmelt, can be dangerous for dams, dikes, and engineering structures used for flood control. The possibility of dam failure cannot be neglected.

- Improved monitoring of flood events, impacts, and vulnerabilities is important to increase human security. It has been shown that poor people generally face a higher risk of mortality and relatively higher economic losses from natural hazards.

- The quality of data and reconstruction of the monitoring systems should be focused on, particularly after the decline in standards in the 1990s due to the general economic crisis in the post-communist countries.

- Additionally, building codes and guidelines for flood proofing constructions and structural measures (e.g., giant levees) are important elements that can increase human security in terms of natural hazards, includeing floods [1].

4.1. INSTITUTIONAL CONSIDERATIONS

A commission for emergency management should decide how institutions should respond in cases of emergency and how disaster risk can be reduced. This commission should encompass local and regional authorities of the respective river basin. It should be linked to important agencies and enterprises. Together with engineers and emergency response agencies, the commission should prepare a planning document every year for the spring floods in the region.

A special safety brigade should be responsible for rescue operations and emergency management during the event. The emergency plan for flooding should focus on aspects of evacuation, potential coping capacities, and places of evacuation.

Specific plans should also be formulated regarding the dissemination of information to radio and TV stations. The recommendations below resulted from experience in the Volga River basin:

- Information exchange and close cooperation between institutions as well as the active participation of the public in developing strategies for integrated flood management are essential.

- A lack of appropriate cooperation such as a lack of information sharing between national states along the same transboundary river is a major problem for human security.

- The different steps in the disaster phase (prevention and coping) and level of regulation, such as normal regulation and emergency regulation, should also be considered. A crucial question is: who is able to act appropriately in the different phases of disasters and what can s/he do? The coordination of different functions and institutions is essential. Institutional solutions cannot be generalized.

- Moreover, it is recommended that the historical dimension for risk assessment related to certain processes or events be included; for example, in the Netherlands, water management and water-related risks have been key issues for several decades.

- A serious problem regarding human security and vulnerability reduction involves false alarms that delay services giving out early warning information. Missing or late warnings may cause fatalities and increase damage. Using local information services (radio, newspapers, or TV), with which people are already familiar, is the most effective way to spread information about flood warnings.

- A crucial issue is the organization and promotion of a quick and effective response. The floods in New Orleans highlighted the need to also take into account the multi-ethnic aspect of different social groups and their social structure.[1] This leads to the recommendation that cultural, social, and linguistic aspects should be considered.

Public participation and socioeconomic issues:

- Public participation is especially well developed in countries like the Netherlands, where lifestyles and risk perceptions have also been addressed in integrated flood risk and flood vulnerability reduction.

---

[1] Elderly people are very vulnerable (e.g., casualties in old people's homes in New Orleans). the protection of societies where elderly citizens are in the majority is a challenge today.

- Besides early warning and people's level of awareness, the general status of maintenance of infrastructure is also a key element affecting vulnerability. Therefore, one can conclude that disasters are often due to a combination of different causes.

- More attention has to be given to secondary damage and secondary effects. Often only primary effects and damages are considered.

- Holistic and integrative risk and vulnerability assessment also has to be based on ex-ante and ex-post analysis. The analysis of past events is not adequate for the estimation of present and future vulnerabilities. In this context, scenario-based assessment strategies are important.

5. Conclusions

All the technical, economic or institutional tools necessary for integrated flood management already exist or are under development in Europe and other parts of the world. In relatively big rivers, floods may take place in transboundary river basins. Cooperation between countries and the backing of national institutions is necessary to enable the pooling of experiences and initiatives and the mobilization of transnational institutional structures (for example—in Europe—Joint Research Centre (JRC), European Space Agency (ESA), and the International River Commissions).

A review of the flood plans for the largest river basins in Europe (Danube, Rhine, Meuse), shows a relative homogeneity of principles and goals, moderated by the particularities of each catchment. The concepts of risk assessment and river restoration, and the importance of preventive actions, are widely understood and applied. A major breakthrough is being achieved with the use of catchment modeling for risk assessment, anticipation of consequences of land use change, flood forecasting, and economic analysis.

However, even though decision makers on a national and European level have already adopted most integrated flood strategy concepts, these have not yet been encompassed in the workings of institutions, nor is there yet a satisfactory level of public participation in areas prone to floods. This means that there is still a gap between theory and practice or between science and decision making.

References

1. Brilly, M. (2001) The integrated approach to flash flood management. In: E. Gruntfest and J. Handmer, eds., *Coping with Flash Floods*, NATO ASI Series 2, Environmental Security, Kluwer, Dordrecht, Vol. 77, pp. 103–113.

2. CABRI-VOLGA. (2008) Available at: http://www.cabri-volga.org/about.html.
3. Directive. (2000) 2000/60/EC of the European Parliament and of the Council of 23 October 2000 establishing a framework for community action in the field of water policy, *Off. J. Eur. Commun.* L 327, 22.12.2000.
4. Directive. (2007) 2007/60/EC of the European Parliament and of the Council of 23 October 2007 on the assessment and management of flood risks, *Off. J. Eur. Commun.* L 288/27, 6.11.2007.
5. Ganoulis, J. (2003) Risk-based floodplain management: a case study from Greece, *Int. J. River Basin Manage.* 1(1):41–47.
6. Ganoulis, J. (2008) *Risk Analysis of Water Pollution*, Wiley-VCH, Weinheim/Oxford/New York.
7. Ganoulis, J., Duckstein, L., Literathy, P., and Bogardi, I., eds. (1996) *Transboundary Water Resources Management: Institutional and Engineering Approaches*, NATO ASI SERIES, Partnership Sub-Series 2, Environment, Vol.7, Springer, Heidelberg, Germany.
8. Gardiner, J., Strarosolszky, O., and Yevjevich, V., eds. (1995) *Defence from Floods and Floodplain Management*, NATO ASI Series E, Vol. 299, Kluwer, Dordrecht.
9. Goicoechea, A., Hansen, D. R., and Duckstein, L. (1982) *Multiobjective Decision Analysis with Engineering and Business Applications*, Wiley, New York.
10. Pomerol, J. C., and Romero, S. B. (2000) *Multicriterion Decision in Management: Principles and Practice*, Kluwer, The Netherlands.
11. Renaud, F. (2005) Human and environmental security in the context of the CABRI-Volga project, Proceedings of the Great Rivers Forum, Nizhny Novgorod, Russia.
12. Rossi, G., Harmoncioglu, N., and Yevjevich, V., eds. (1994) *Coping with Floods*, NATO ASI Series E, Vol. 257, Kluwer, Dordrecht.

RATIONALE AND DEVELOPMENT OF A SCALE TO COMMUNICATE ENVIRONMENTAL AND OTHER COMMUNITY RISKS

M.M. PLUM

Idaho National Laboratory
Idaho Falls, ID, USA
plummm@inel.gov

Abstract: Efforts to secure any community from environmental and other community risks must be able to convincingly argue that:

1. The stressors impact the community's well being.

2. The community is ultimately responsible in the mitigation of these stressors.

3. Limited resources must be committed to manage this risk to an acceptable level.

Assuming any community has a fixed quantity of risk management funding, environmental security must compete for resources traditionally allocated to other well-recognized risks of war, terrorism, and natural disasters. Even if the quantity of funding is flexible, it must convincingly argue for the reallocation of scarce resources from the activities of consumption to investment. Economists refer to this discussion as "Guns or Butter."

Similarly, the Department of Homeland Security must allocate resources to many risks. Calculated from probable events, probable outcomes, and probable life and economic losses, a risk scale developed within one DHS-funded program uses the rationale of many successful threat and risk scales. It was suggested that this scale could be used to measure and rank the risk of all international events of terror, disaster, and calamity for the allocation of risk management efforts. This paper examines a few notable and successful scales of risk, the rationale for these and the development of the Security Assurance Index, and the recently proposed Global Risk Index and its application to environmental security.

1. Copyright Statement

2. Background

As established by the Homeland Security Presidential Directive (HSPD-7), U.S. federal departments and agencies are directed to identify and prioritize critical infrastructure and key resources within the U.S. for protection from terrorist attack. Identifying 12 critical sectors and five key assets, HSPD-7 allocates this responsibility to the following departments and agencies:

- Agriculture & Food – Department of Agriculture
- Water – Environmental Protection Agency
- Public Health – Department of Health & Human Services
- Defense Industrial Base – Department of Defense
- Energy – Department of Energy
- Transport – Department of Homeland Security
- Telecommunications – Department of Homeland Security
- Postal & Shipping – Department of Homeland Security
- Banking & Finance– Department of Treasury
- Emergency Services – Department of Homeland Security
- Chemical & Hazardous Materials Industry – Department of Energy
- National Monuments & Icons – Department of Interior
- Nuclear Plants – Nuclear Regulatory Commission
- Dams – Department of Homeland Security
- Government Facilities – Department of Homeland Security
- Commercial Assets – Department of Homeland Security & the National Institute of Standards and Technology

In support of this national effort, the Idaho National Laboratory (INL) supports many Department of Energy (DOE), Department of Defense (DOD), and Department of Homeland Security (DHS) programs with engineering, science, and management expertise. One INL program is the DHS-funded Control System Security Program (CSSP) and the associated Control System Security Center (CSSC). Established to coordinate the efforts of control systems owners, operators, and vendors as well as the federal, state, local, and tribal governments, the CSSP's mission is to reduce the likelihood of success and severity of impact of cyber attacks on critical infrastructure. The primary objectives of the CSSP are:

- Enhance incident response capabilities
- Assess vulnerabilities and risk
- Enhance industry practices
- Enhance security awareness
- Recommend R&D needs

A significant problem for this program is the communication of risk from cyber vulnerabilities. Whether this risk is within the industry sector, between sectors, or between the sector and the general economy, parochial interests and information isolation usually result in underestimating or overestimating risk. However, even if risk is accurately evaluated, it is not effectively communicated due to personal perspectives of how great or insignificant this risk is on a national scale. For this reason, an INL economist and systems engineer developed a scale to effectively communicate risk.

From this beginning, it was suggested that a similar rationale for a risk scale could be applied to communicate risk at an international or global extent.

3. Problem Statement

Currently, the risk of cyber attacks is a difficult problem to manage due to its many complex aspects, including but not limited to:

- Constantly changing technology and their associated vulnerabilities
- Constantly changing threat actors
- Difficulty in predicting attack vectors
- Difficulty in predicting consequences and their impacts quantifiably
- Extremely low probability and extremely high consequence of a cyber attack

However, even if we can obtain the resources to manage these issues, communicating the calculated risk to individuals, communities, and agencies is difficult, ineffective, and inefficient. This is the result of many intermingled problems:

- Human behavior chooses to ignore risk if it was determined from the outside
- Human behavior chooses to ignore risk if it negatively impacts the community

- Human motivation chooses to overestimate risk if it results in increased funding
- Humans sometimes do understand large scale and large numbers

From our experience, the risk communication process was a review process where decision makers required a full understanding of the detail and methods of the assessment process. Often, calculations and methods were meticulously reviewed; and their derivations and assumptions questioned. Not only did this process require significant resources from the risk management process, it also resulted in a modified risk assessment process that would favor a community's desire. From the perspective of the risk assessors, these modifications were almost always motivated by outside decision makers who understood the risk management process and modified the assessment to a desired outcome. Bottom line: true risk reduction was not being achieved expeditiously, effectively, or efficiently. It became obvious to many that there needed to be a more effective and simple method to communicate cyber risk.

4. The Existing DHS Scale and Its Application to Cyber Security

After the establishment of DHS, Presidential Directive 3 required a: comprehensive and effective means to disseminate information regarding the risk of terrorist acts to Federal, State, and local authorities and to the American people.

What was presented is the Homeland Security Advisory System (HSAS), which consists of five color-coded levels to communicate the department's calculation of risk from and potential gravity of terrorist attack (Figure 1). Simple and visual, the colors are typical of those associated with threat or danger: deep reds are associated with fire or blood, yellows and orange with caution, and green or blue with calm and peace.

Though simple and similar to many other risk scales, this scale has received criticism from many quarters. The primary criticism is the lack of any criteria to define the threat level or a methodology for its calculation; thus, it is almost impossible to deduce the threat and determine possible actions in response. Without resolution, these issues create an environment of distrust. Furthermore, since the nation has been at an Elevated Alert since 2002, many have learned to mistrust this scale and its use as they have become numb to a threat that appears to be diminishing. Lastly, since blue and green levels have never been used, many argue that this is a three-level scale in practice. Bottom line: many critics argue that this scale has done more to create an environment of apathy, ignorance, and even suspicion due to its lack of transparency. The

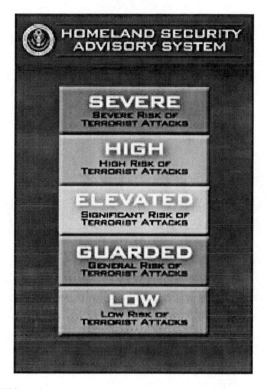

Figure 1. Existing DHS scale.

only exception may be airport security, where increased levels always result in more intensive luggage and personal-effects searches.

For these reasons, relying on the HSAS for cyber security risk communication was seen to be difficult at best.

5. Development of CSSP Risk Scale

Given the history of the HSAS, the CSSP decided to develop its own scale for internal use, and if successful in communicating risk within the program, hoped it would be adopted informally between and within DHS and other federal agencies. From the outset, the goals of the CSSP scale were to communicate the assessed risk effectively, to require little interpretation, be easy to use, and avoid parochial interests through a transparent evaluation process. If these goals could be achieved, it was argued that the scale would become trusted and eventually codified through use.

The development strategy was simple. First, we assumed the HSAS scale would not go away; however, we believed that DHS would not resolve many of the problems as this would require a transparent methodology. Second, there appeared to be a large number of other threat and risk scales from which success and failure could be determined; we did not have to start this development from scratch. Lastly, when the attributes to success were identified, we would incorporate these when possible. In the end, our test for success was always:

- Easy to use
- Easy to communicate
- Easy to understand

6. Review of Existing Risk Scales

One will notice that risk scales are a relatively new phenomenon. Neither Columbus nor the generals of World War II benefitted from their use (though they would have). A lot of this is due to timing. First, using risk scales for communication is relatively new due to the assessment and communication technologies that enable them. However, risk scales have also been associated with the development of risk-averse and modern societies, which are primarily driven by the power of the individual. Whereas past societies were largely controlled by a king or queen whose primary responsibility was to oversee the protection of the kingdom, the path of a modern society is dominated by the well being of the individual and his or her influence on society. Thus, whether one is mitigating the risk of a flu outbreak, hurricane, earthquake, or terrorist attack, modern societies are motivated by the actions of the individual, which then determines the impact to the community. Fortunately, sufficiently robust communication tools are available to enable this *laissez-faire* in societal risk management.

It is important to note that many scales are available for inquiry. There are scales of Snowfall Impact, Volcanic Explosivity Index, Drought Indices, and the Beaufort Wave Index as well as traditional scales for noise and light. In summary, all of these have the identical objectives: to employ a quantifiable system of measure that is then communicated with ease and clarity to the audience. In the following review, many scales are presented. Some are common while others are less common yet used extensively in certain circles. In almost all cases, they are noted for their simple assessment, their ease of use and communication, and their effective communication.

6.1. FUJITA SCALE

The Fujita Scale (F-Scale) measures the damage from a tornado that is the result of wind intensity (Figure 2).

F-Scale Number	Intensity Phrase	Wind Speed	Type of Damage Done
F0	Gale tornado	40-72 mph	Some damage to chimneys; breaks branches off trees; pushes over shallow-rooted trees; damages sign boards.
F1	Moderate tornado	73-112 mph	Peels surface off roofs; mobile homes pushed off foundations or overturned; moving autos pushed off the roads; attached garages may be destroyed.
F2	Significant tornado	113-157 mph	Roofs torn off frame houses; mobile homes demolished; boxcars pushed over; large trees snapped or uprooted; light object missiles generated.
F3	Severe tornado	158-206 mph	Roof and walls torn off well constructed houses; trains overturned; most trees in forest are uprooted, medium object missiles generated
F4	Devastating tornado	207-260 mph	Well-constructed houses leveled; structures with weak foundations blown off some distance; cars thrown and large missiles generated.
F5	Incredible tornado	261-318 mph	Reinforced concrete structures badly damaged, frame houses carried considerable distances; automobiles generated as missiles and thrown over 100 meters
F6	Inconceivable tornado	319-379 mph	Damage could not be differentiated from F4 or F5 winds. Possibly identified if cars and refrigerators were carried 1000s of meters or if ground swirls are found

Figure 2. F-scale.

Due to the nature of tornadoes, they are often rated after the fact, though forecasts and warnings will rate an approaching tornado to indicate the nature of the tornado and the risk of injury and death from this threat. Originally created by Tetsuya Fujita in 1971 as a 13-level scale, this design was driven by the objective to smooth the Beaufort Scale (which ranks violent storms) and the Mach Scale (which measures relative speed of wind as compared to the speed of sound in air). In practice, the Fujita scale contains only six levels (F0–F5); Fujita himself reserved an F6 ranking for a tornado of "inconceivable" magnitude and probable damage. Significant about this scale are:

1. Its simple design to warn of predicted tornado threat.

2. It incorporates only six color-coded levels.

3. Its popular adoption for the communication of tornado risk.

The simplicity of the scale corresponds easily to the number of digits on one hand. Additionally, the corresponding colors of cyan-blue to orange-red are typical of how people visualize threat or risk, where cool colors suggest serenity and red suggests blood and death. Lastly, the continued use of the F-Scale by the general public demonstrates its acceptance even though this scale was modified in 2007 as an EF-Scale (Enhanced Fujita) to reflect the current state of the science and art of predicting storm strength.

6.2. RICHTER MAGNITUDE SCALE

Also known as the Richter Scale, this scale assigns a single number to the maximum amount of seismic energy released by an earthquake using a base-10 logarithmic scale (Figures 3 and 4).

Originally developed by Charles Richter in partnership with Beno Guttenburg for an academic study of California earthquakes in 1935, it has been adopted worldwide due to its ease of assessment. Earthquakes of 10 or greater are conceivable; however, practice and physical maximums tend to keep earthquakes within the R-0 to R-8 range. Some people suggest the eventual occurrence of an R-9 earthquake—one of biblical proportions. However, earthquakes of R-10 or higher are thought to be impossible due to the stress or strain that rock is able to accept without failing. Additionally, one should be aware that a Richter Scale rating is applied after the fact; that is, earthquakes are rated after the released energy is measured. Richter ratings are not used for warning except in discussions of a scenario.

It is of interest that the Richter scale is one of the oldest threat scales in existence; possibly, for this reason it was never color-coded. Also of interest is the fact that earthquake frequency has a logarithmic relationship, similar to the logarithmic energy designed into the scale. Lastly, the logarithmic relationship between each classification has similar logarithmic effects on the number of people killed and injured.

Earthquake Description	Richter Scale Number	Human Impact	Frequency
Micro	Less than 2.0	People cannot feel these	~ 8,000 /day
Very minor	2.0-2.9	People cannot feel these, but inexpensive seismological tools can record them	~ 1,000 / day
Minor	3.0-3.9	People often feel these, but they rarely cause damage	~ 135 / day
Light	4.0-4.9	People will notice objects shaking with noise, yet little or no damage	~ 17 / day
Moderate	5.0-5.9	Poorly constructed buildings can be damaged to significantly damaged, well built and designed buildings are not, injury and possible death noted	~ 2 / day
Strong	6.0-6.9	Damage and possible death may be noted over an area of 100 square miles with damage to well constructed buildings and possible failures	~ 0.3 /day
Major	7.0-7.9	Major to significant damage is noted over an even larger area with damage to significant damage to very well designed and constructed buildings	~ 0.05 / day
Great	8.0 or greater	Substantial damage to all structures and failure to all poorly designed or constructed buildings over several hundred square miles; expect tremendous injury and loss of life	~ 0.003 / day

Figure 3. Richter scale.

Earthquake Description	Richter Scale Number	Human Impact	Frequency
Micro	Less than 2.0	People cannot feel these	~ 8,000 /day
Very minor	2.0-2.9	People cannot feel these, but inexpensive seismological tools can record them	~ 1,000 / day
Minor	3.0-3.9	People often feel these, but they rarely cause damage	~ 135 / day
Light	4.0-4.9	People will notice objects shaking with noise, yet little or no damage	~ 17 / day
Moderate	5.0-5.9	Poorly constructed buildings can be damaged to significantly damaged, well built and designed buildings are not, injury and possible death noted	~ 2 / day
Strong	6.0-6.9	Damage and possible death may be noted over an area of 100 square miles with damage to well constructed buildings and possible failures	~ 0.3 /day
Major	7.0-7.9	Major to significant damage is noted over an even larger area with damage to significant damage to very well designed and constructed buildings	~ 0.05 / day
Great	8.0 or greater	Substantial damage to all structures and failure to all poorly designed or constructed buildings over several hundred square miles; expect tremendous injury and loss of life	~ 0.003 / day

Figure 4. Richter scale.

6.3. SAFFIR-SIMPSON SCALE

The Saffir-Simpson Scale has received an unlikely amount of use recently due to the number of deadly and damaging hurricanes in the U.S (Figure 5). Developed in 1969 by a civil engineer and the Director of the National Hurricane Center (NHC) for the purpose of warning the public of possible storm danger, this system warns the public about sustained winds, likely flooding, and probable damage when a hurricane landfalls. Due to the lengthy name of the scale, hurricanes are simply described by *category*, thus a *Category 2* or *Category 3* storm.

Per *Wikipedia*:

the initial scale was developed by Saffir while on commission from the United Nations to study low-cost housing in hurricane-prone areas. While performing the study, Saffir realized there was no simple scale for describing the likely effects of a hurricane. Knowing the utility of the Richter Magnitude Scale in describing earthquakes, he devised a 1–5 scale based on wind speed that showed expected damage to structures. Saffir gave the scale to the NHC where Simpson added in the effects of storm surge and flooding.

Of interest about this scale are:

1. The number of categories

2. Its use in every day communication

3. Its color coding

Category	Sustained Wind	Storm Surge	Central Pressure	Potential Damage
1	74-95 mph 119-153 km/h	1.2-1.5 m 4-5 ft	28.94 inHg 980 mbar	No real damage to building structures. Damage primarily to unanchored mobile homes, shrubbery, and trees. Also, some coastal flooding and minor pier damage.
2	96–110 mph 154-177 km/h	1.8–2.4 m 6-8 ft	28.50–28.91 inHg 965–979 mbar	Some roofing material, door, and window damage. Considerable damage to vegetation, mobile homes, piers and small craft in unprotected boats may break their moorings.
3	111–130 mph 178-209 km/h	2.7–3.7 m 9-12 ft	27.91–28.47 inHg 945–964 mbar	Some structural damage to small residences and buildings, some curtainwall failures. Mobile homes are destroyed. Coastal flooding destroys small structures, floating debris damages large structures. Local terrain flooded.
4	131–155 mph 210–249 km/h	4.0–5.5 m 13-18 ft	27.17–27.88 inHg 920–944 mbar	More extensive curtainwall failures with some complete roof structure failure on small residences. Major erosion of beach areas. Terrain may be flooded well inland.
5	≥156 mph ≥250 km/h	≥5.5 m ≥19 ft	<27.17 inHg <920 mbar	Complete roof failure on many residences and industrial buildings. Some complete building failures. Flooding causes major damage to lower floors of all structures near the shoreline. Massive evacuation may be required.

Figure 5. Saffir-Simpson scale.

Similar to the Fujita and Richter Scales, the Saffir-Simpson Scale has five categories which are easy to remember and rank on one hand. Note that a Category 6 has been discussed to signify a hurricane storm of biblical proportions; and though not recognized, many believe that such a ranking would garner the attention of the media and convey the threat to the public better than a Category 5 ranking, which is at times ignored. Also of interest is how the scale is communicated. The scale is rarely referenced as the Saffir-Simpson; rather, the hurricane and its warning are simply described with a category number. In fact, the word category has almost become synonymous with hurricane due to its common use during the summer storm season.

6.4. TORINO SCALE

The Torino Scale is scarcely mentioned in general conversation, which is fortunate because it measures the risk of near-earth objects such as comets or asteroids striking the earth (Figures 6 and 7). Intended as a tool for astronomers, governments, and the public to assess the seriousness of collision predictions, this scale combines probability statistics and kinetic damage potential into a single risk value.

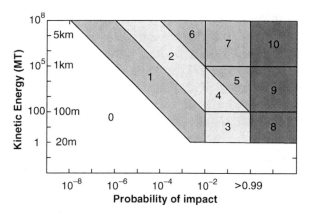

Figure 6. Torino scale.

NO HAZARD (white)	
0	The **likelihood of a collision is zero**, or is so low as to be effectively zero. Also applies to small objects such as meteors and bodies that burn up in the atmosphere as well as infrequent meteorite falls that rarely cause damage
NORMAL (green)	
1	Near Earth pass is predicted with no unusual level of danger. Current calculations show the **chance of collision is extremely unlikely** with no cause for public attention or public concern
MERITING ATTENTION BY ASTRONOMERS (yellow)	
2	An object making a somewhat close but not highly unusual pass near the Earth. While meriting attention, there is no cause for public attention or public concern as an **actual collision is very unlikely**
3	A close encounter, meriting attention. Current calculations give a **1% or greater chance of collision capable of localized destruction**. Most likely, telescopic observations will lead to re-assignment to Level 0. Attention is merited if the encounter is less than a decade away
4	A close encounter, meriting attention. Current calculations give a **1% or greater chance of collision capable of regional devastation**. Most likely, new telescopic observations will lead to re-assignment to Level 0. Attention is merited if the encounter is less than a decade away
THREATENING (orange)	
5	**A close encounter posing a serious but uncertain threat of regional devastation**. Critical attention by astronomers is needed whether or not a collision will occur. If the encounter is less than a decade away, governmental contingency planning may be warranted
6	**A close encounter by a large object posing a serious but uncertain threat of a global catastrophe.** Critical attention by astronomers is needed to determine whether or not a collision will occur. If the encounter is less than three decades away, governmental contingency planning may be warranted
7	**A very close encounter by a large object posing an unprecendented threat of a global castastrophe.** For such a threat, international contingency planning is warranted, to determine conclusively whether or not a collision will occur
8	Collision is certain with localized destruction on land or shoreline if ocean, sea, or lake landing. Events occur on average between once per 50 years and once per several 1000 years
9	Collision is certain and capable of causing unprecendented regional devastation Such events occur on average between once per 10,000 years and once per 100,000 years
10	Collision is certain, capable of causing global climatic catastrophe that may threaten the future of civilization as we know it, whether impacting land or ocean. Such events occur on average once per 100,000 years or less

Figure 7. Key to the Torino scale.

Of particular interest to our efforts, this scale is notable in communicating a wide array of possible scenarios, simplified by the number of categories and their color coding. Similar to the Fujita, Richter, and Saffir-Simpson scales, the Torino Scale has five probability categories within a matrix of three categories for object size and implied kinetic energy. As a result, there are ten risk categories, which are then allocated into five color categories. Also of interest is the scale's global perspective in communicating the threat of worldwide catastrophe. Lastly, the color coding of cool to hot colors is similar if not identical to other threat scales.

6.5. UV INDEX

The UV Index is an international standard measurement of the strength of ultraviolet (UV) radiation from the sun at a particular place on a particular day (Figure 8). The primary purpose of this scale is to communicate the threat of a sunburn or sunstroke to the general public. Originally developed in Canada in 1992, this index has been adopted by many countries due to its effectiveness in communicating the risk from sunburn. The UV index is a calculated prediction of how strong the actual UV intensity will be at the sun's highest point in the day, which typically occurs during the 4-h period surrounding solar noon. This prediction is made by calculating the impact of latitude and local altitude, weather, and pollution to determine UV intensity at the earth's surface. The calculations are weighted in favor of UV wavelengths to which the human skin is most sensitive. The date, the most severe UV warning has been an UV Index rating of 17.

Figure 8. UV index.

Of particular interest:

1. The number of color codes
2. The speed of its adoption
3. Its quantitative approach to a relatively innocuous subject

First, it is important to notice that though UV Index ratings of 17 are possible, these ratings are communicated within five risk classifications, again, probably due to ability of the common individual to identify and remember levels of threat. Additionally, the colors are similar if not identical to other previously discussed scales while being representative of an outcome from unprotected sun exposure. Also noteworthy is the scale's rate of adoption; it has been adopted in most developed countries where skin cancer is a serious health issue and is often broadcast along with weather predictions. Lastly, it is important to note that although every person reacts to UV intensity differently, this quantitative scale allows the user to interpret his or her risk based on experience in its use.

6.6. DEFCON

Applicable only to the U.S. military, DEFCON measures defense readiness (Figure 9). Similar to all other scales discussed, this scale provides a

Description	Recommended Mitigation
DEFCON 5	This is the condition used to designate normal peacetime military readiness. An upgrade in military preparedness is typically made by the Joint Chiefs of Staff and announced by the United States Secretary of Defense.
DEFCON 4	This refers to normal, increased intelligence and the heightening of national security measures.
DEFCON 3	This refers to an increase to force readiness above normal. Radio call signs used by American forces change to currently-classified call signs.
DEFCON 2	This refers to a further increase in force readiness just below maximum readiness. Declared during the Cuban Missile Crisis, although limited to Strategic Air Command only. It is not certain how many times this level of readiness has been reached.
DEFCON 1	This refers to maximum readiness. It is not certain whether this has ever been used, but it is reserved for imminent or ongoing attack on US military forces or US territory by a foreign military power.

Figure 9. DEFCON scale.

quantitative measurement of the risk of attack from a hostile armed force. Of interest is the use of five levels on an inverted scale to the degree of risk. Also notable is the lack of color coding.

6.7. HOWE-DEVEREUX FAMINE SCALE

The Howe-Devereux Famine Scale is one of the newer scales for threat or risk (Figures 10 and 11). Introduced in 2004 by Paul Howe and Stephen Devereux at the University of Sussex, this scale measures the risk of famine according

Intensity	Phrase	Crude Mortality Rate (CMR)	Livelihood
0	Food Secure	CMR < 0.2/10,000/day and/or Wasting < 2.3%	Cohesive social system; food prices stable; Coping strategies not utilized
1	Food Insecure	0.2 <= CMR <0.5/10,000/day and/or 2.3% <= Wasting < 10%	Cohesive social system; Food prices unstable; Seasonal shortages; Reversible coping strategies taken
2	Food Crisis	0.5 <= CMR < 1/10,000/day 10% <= Wasting < 20%, and/or prevalence of oedema	Social system stressed but largely cohesive; Dramatic rise in food and basic items prices; Adaptive mechanisms fail; Increase in failed coping strategies
3	Famine	1 <= CMR < 5/10,000/day 20% <= Wasting < 40% and/or prevalence of oedema	Clear signs of social breakdown; markets begin to collapse, survival strategies initiated; migration begins, weaker family members abandoned
4	Severe Famine	5 <= CMR <15/10,000/day Wasting >= 40% and/or prevalence of oedema	Widespread social breakdown; markets close; survival strategies widespread; affected population identifies food scarcity as the major societal problem
5	Extreme Famine	CMR >= 15/10,000/day	Complete social breakdown; widespread mortality; affected population identifies food scarcity as the major societal problem

Figure 10. Howe-Devereux famine scale—intensity rating.

Magnitude	Phrase	Mortality range
A	Minor famine	0-999
B	Moderate famine	1,000-9,999
C	Major famine	10,000-99,999
D	Great famine	100,000-999,999
E	Catastrophic famine	1,000,000 and over

Figure 11. Howe-Devereux famine scale—magnitude rating.

to its intensity and magnitude. This scale replaced prior famine scales due to a more quantitative approach in measuring social and human conditions. Thus, while many organizations will continue to have their own qualitative interpretation as to the specific indicators, this scale requires quantitative data to assess a famine's magnitude (individual impact of nutrition provisions and death rates) and intensity (its impact on social functioning).

6.8. INTERNATIONAL SCALE OF RIVER DIFFICULTY

This scale is an international standard to rate and convey the dangers and potential risks of a river or a single rapid (Figure 12). In summary, this scale provides a rating to reflect the river's technical difficulty, skill level requirements for safe passage, and the associated risks of failure. Six nonlinear and nonfixed class ratings are provided with the option of plus (+) and a minus (−) to denote added ease or difficulty.

Similar to the scales discussed previously, this scale is simplified into six categories of simple threat descriptions. At this time, there are no color codes to visually signal danger or risk, though certain books on river running as well as local signing have incorporated a nonstandard color code similar to the previously discussed scales where red is associated with the most dangerous level, orange and yellow with less danger, and green and blue with the least danger.

		Description of Risk	Probability of Injury or Death
Class I	Easy	Waves small; passages clear; no serious obstacles	Risk to swimmers is slight
Class II	Medium	Rapids of moderate difficulty with passages clear. Requires experience plus suitable outfit and boat.	Swimmers are seldom injured and group assistance is seldom needed
Class III	Difficult	Waves numerous, high, irregular; rocks; eddies; rapids with clear passages, requires expertise in maneuvering; scouting usually needed. Requires good operator & boat.	Injuries are rare and group assistance not necessary but avoids long swims
Class IV	Very Difficult	Long rapids; waves high, irregular; dangerous rocks; boiling eddies; best passages difficult to scout; scouting mandatory first time; powerful and precise maneuvering required. Demands expert boatman & excellent boat	Swimmers are in moderate to high danger, probable injury, rescue difficult
Class V	Extremely Difficult	Exceedingly difficult, long and violent rapids, following each other almost without interruption; riverbed extremely obstructed; big drops; violent current; very steep gradient; close. Requires the best person & boat. All possible precautions must be taken.	Swims are definitely at risk of injury, death very possible, rescue very difficult
Class VI	Not runnable	Luck rules the day for any level of expertise.	Definite risk to life in and out of watercraft, rescue almost impossible

Figure 12. International scale of river difficulty.

It is notable that this scale was devised by the river running community to describe rivers and rapids with ease. All river runners know to scout a river or rapid of Class IV or higher, regardless of their level of technical ability.

6.9. APGAR SCALE

The APGAR Scale measures the relative health and immediate risk of newborns immediately after birth (Figure 13). What is most notable about this scale is its impact on society in general; the scale has been described as "the most important medical practice or technology in reducing infant mortality."

Developed in 1952 by anesthesiologist Virginia Apgar, it is best known by its mnemonic reference:

- Activity
- Pulse
- Grimace
- Appearance
- Respiration

Designed to be administered by any hospital staff person twice after birth of the newborn (at 1 min and at 5 min), this evaluation provides an immediate assessment of risk to the infant. Like other successful scales, it is quantitative in nature, evaluating newborn health attributes on a scale of 0–10.

An APGAR 0 requires immediate response whereas an APGAR 10 requires no attention from medical staff. Newborns with an APGAR 7 or

	APGAR Score		
APGAR Sign	**2**	**1**	**0**
Heart Rate	Normal (above 100 beats per minute)	Below 100 beats per minute	Absent
Breathing	Normal rate and effort	Slow or irregular breathing	Absent (no breathing)
Grimace or reflex irritability	Pulls away, sneezes, or coughs with stimulation	Facial movement only (grimace) with stimulation	Absent (no response to stimulation)
Activity or muscle tone	Active, spontaneous movement	Arms and legs flex ed with little movement	No movement, "floppy" tone
Appearance by skin coloration	Normal color all over (hands and feet are pink)	Normal color (but hands and feet are bluish)	Bluish-gray or pale all over

Figure 13. APGAR scale.

above 1 min after birth are generally considered in good health; however, a lower score doesn't necessarily mean that your baby is unhealthy or abnormal. APGAR scores between 4 and 6 may simply demonstrate the need for some immediate, low-technology care such as suctioning of the airways or oxygen to help him or her breathe. At 5 min after birth, the APGAR score is reassessed and recalculated and if the baby's score hasn't improved to 7 or greater, doctors and nurses will continue medical care as required and closely monitor the newborn.

What is significant in developing trust of this scale and its ranking are the ease of calculating an APGAR score and its timeliness for mitigating newborn health issues. For this reason, it is very transparent. Also, it is interesting that this score is determined with five variables that are easy to remember with a mnemonic device. Response is simply a test of whether the newborn exceeds a minimum score rating of 6.

7. Security Assurance Index

The Security Assurance Index (SAI) was developed at INL for the CSSP using many of the attributes that we felt contributed to the successful risk scales (Figure 14). This included ease of use, communication, and understanding as well as a transparent, understandable, and quantifiable risk methodology. It is believed that these attributes, plus successful use, build trust—trust from the user that the index accurately reflects risk and trust from the risk manager that the users will respond appropriately. Thus, trust is an important outcome of this risk communication process.

expected life event	life loss assurance level	low-end	median	high-end
Total loss of US civilization	SAL 10	300,000, 000		
significant loss of US civilization	SAL 9	30,000,000	95,000,000	300,000,000
Loss of regional civilization	SAL 8	3,000,000	9,500,000	30,000,000
loss of metropolitan area	SAL 7	300,000	950,000	3,000,000
loss of city	SAL 6	30,000	95,000	300,000
loss of town	SAL 5	3,000	9,500	30,000
loss of community	SAL 4	300	950	3,000
loss of neighborhood	SAL 3	30	95	300
loss of family of related group	SAL 2	3	9.5	30
loss of individual	SAL 1	0.3	0.9	3

Figure 14. Security assurance index.

With respect to the CSSP, it is important to understand the program's mission, and what is considered to be at risk. First, DHS and CSSP have identical missions: the protection of U.S. infrastructure and population from terrorists' threat. The only difference is that the CSSP manages risk specific to control systems. Second, DHS clearly states that 300,000,000 people, a $12 trillion annual economy, and $120 trillion in capital investment are at risk. This is not to say that DHS does not recognize other equally important and intangible measures such as social, political, religious, economic, and psychological freedoms as well as cultural confidence, national influence, and morale. However, it simply recognizes that people and material wealth enables many other aspects of social welfare.

Given these maximum risk conditions, a scale was drafted of reasonable dimensions using the logarithmic logic of some of the scales we reviewed. As for those scales that did not incorporate this logic, this scaling often reflected event probability, probable death, and economic loss (Figure 15). Thus, a rating of 3 or 4 often suggests loss of 1,000 people and 10,000, respectively, without mitigation efforts.

Interestingly enough, such a concept in scale development was proposed by Gustav Fechner, a 19th century, German psychologist who advanced the theory that the intensity of a human sensation increases in arithmetical progression based on a geometric increase in stimulus. Describing human reaction to stimulus, Fechner's famous equation (the first to describe human psychology mathematically) is the basis of many scales that relate to human comfort and discomfort:

$$S = c \log R$$

expected life event	economic loss assurance level	low-end	median	high-end
total loss of US civilization	SAL 10	$3 Q		
significant loss of US civilization	SAL 9	$300T	$1Q	$3Q
loss of regional civilization	SAL 8	$30T	$100T	$300T
loss of metropolitan area	SAL 7	$3T	$10T	$30T
loss of city	SAL 6	$312B	$1T	$3T
loss of town	SAL 5	$31.2B	$100B	$312B
loss of community	SAL 4	$3.1B	$10B	$31.2B
loss of neighborhood	SAL 3	$312M	$1B	$3.1B
loss of family of related group	SAL 2	$31M	100M	$312M
loss of individual	SAL 1	$3	10M	$31M

Figure 15. Economic loss index.

Where:

S = sensation

R = numerically estimated stimulus

c = a constant that must be separately determined by experiment for each sensibility

This concept has been used to develop many human response scales to noise (decibel), light (lumens), and vibration (Richter). Interestingly, it appears that we have taken the liberty of applying this same concept to human sensation and response to threat, danger, risk to life and limb, and property loss. What this means is that humans, in general, will notice a perceptible difference (a doubling) in risk only if it increases by an order-of-magnitude. Thus, the risk of one to nine deaths is relatively the same; it is not until ten people are at risk that a human will notice the increase. It is from this observation that the SAI is developed.

Lastly, the SAI methodology scale was made to be transparent and reproducible to gain acceptance, trust, and usefulness for application to other risk scenarios. Accordingly, risk is calculated from the basic equation of risk under deliberate and targeted threat (versus the risk of statistically random threat):

$$Risk = Threat * Vulnerability * Consequence$$

Where

Threat = the probability of threat actor's capability to deliver an attack successfully

Vulnerability = 1 minus the probability of a target to protect itself from the attack threat

Consequence = the likey outcome or distribution of outcome of the specific attack

CSSP recognized the tremendous difficulty in obtaining data and then calculating risk; however, an evaluation process was established to calculate threat and vulnerability as a probability and consequence as a life and limb or economic statistic based on life loss and injury or loss of human and durable capital investment, inventory, S-T market disruption, and environmental loss. Thus, based on a maximum loss condition, the following SAI was developed:

First, one must note that an SAI level is referred to as a Security Assurance Level (SAL) since an index begs for a level or ranking. Also note that although life and economic losses of less than one and $1 million are calculable, they fall below the threshold level for risk management by DHS.

In practice, the greatest level attained on either scale would warrant the highest ranking.

As a test of reasonableness, two recent and well known events were used to demonstrate and understand the scale's usefulness. The 1992 World Trade Center (WTC) bombing resulted in 13 people killed, more than 1,000 injured, and $600 million in economic losses. In this event, a 600-lb car bomb was used within the WTC parking garage in hopes of destroying the structure from within. Though insufficient for the task, a bomb of this size could have delivered considerable loss of life (which it did not) and considerable facility loss (which it did). Prior to this event, when security was lax, one would have rated this scenario as a SAL 3 risk event. On the other hand, 9-11 presents a scenario of significantly higher risk. Although the final economic losses have not yet been tallied, we know that almost 3,000 people lost their lives, tens of thousands were injured, and there was a loss of more than $80 billion in structures and business to the local economy (and even more if one includes national economic disruptions). Prior to this event, the risk may have rated low since no one believed in such a strategy of execution; however, there would have been no argument that if a group of people possessed these characteristics, these buildings and the nation would have been vulnerable to a SAL 5 or SAL 6 event.

The SAI has proven to be just as valuable as hoped. The evaluation process is transparent as are the results; scenarios are given rankings based on the highest probable loss (economic or human life), organized according to the SAL levels, and if needed, are prioritized for a more intensive risk evaluation. Furthermore, this process has been adaptable for evaluating many risk scenarios, with little time, few resources, and for the purpose of ranking risk. This process has contributed significantly to ranking issues within and between industry groups; thus, scenarios are often compared, as in the example above.

Please note that a color schematic has yet to be assigned to any SAL level. This is in part due to the nature of the group, which is small, and the clear understanding of risk associated with each level. However, it may also be due to the sensitivity associated with any loss of human life. For example, it would not be politically or programmatically acceptable to associate the loss of one or even ten lives with a blue or green color. A warning color of at least yellow would have to be used to signal concern and action. It has been suggested that grouping these 10 SAL levels into five color groups would be easy for people to remember, similar to other scales ... or possibly grading the 10 levels from yellow to magenta. Either way, such a ranking would help provide the necessary urgency for mitigating the possibility of such highly ranked scenarios.

8. Development of a Global Risk Index

Presentation of the SAI to the risk management community resulted in many suggestions that a scale of similar design could be applied to global-related risks as a Global Risk Index (GRI). A risk scale based on loss of life was presented at Risk Analysis 2006 (Malta 2006) (Figure 16). (A risk scale based on economic loss was not presented due to the inappropriateness of mapping material and financial losses across the many social, cultural, and economic environments of the world.)

Similar to the SAI development, objectives such as a transparent evaluation process, and ease of use, communication, and understanding continue to be paramount to developing trust in communicating risk. As to the logic of the scale, the logarithmic scale was again selected due the gravity and orders of magnitude it demonstrates in conveying the urgency of a risk scenario. Of note, similar to the Howe-Devereux Scale of Famine, a measure of global risk may find insignificant meaning in small numbers; that is, the possibility of death on a magnitude of 1,000 people or less. From a world health perspective where events of this size happen daily (such as famines, floods, earthquakes, and disease), they seem to be of a national interest and often allow only national response. Events that have a potential for crossing over into the hundreds of thousands to millions garner world interest, attention, and possibly response. Recent tsunamis, earthquakes, famines, and even global warming have resulted in responses of differing degrees due to the magnitude of resources required to react to and mitigate these events. However, for purposes of demonstration and consistency, the categories of 1, 2, and 3 are retained.

Lastly, two notes of importance. The first note is that the original Malta presentation did not include colors to associate risk with the magnitude or urgency of a situation. This was probably an oversight on my part because it would have had a much more powerful impact in communicating risk. The second note is that if this were to be adopted as an index of world risk, there should be little discussion as to the description of risk. Thus, I would use a word such as "category" to describe world risk.

	GLOBAL RISK INDEX Category	lower bound	MEDIAN	Upper bound
total loss of civilizations	Category 10	312,000,000	1,000,000,000	3,120,000,000
significant loss of large civilizations	Category 9	31,200,000	100,000,000	312,000,000
loss of regional civilization	Category 8	3,120,000	10,000,000	31,200,000
loss of metropolitan area	Category 7	312,000	1,000,000	3,120,000
loss of city	Category 6	31,200	100,000	312,000
loss of town	Category 5	3,120	10,000	31,200
loss of community	Category 4	312	1,000	3,120
loss of neighborhood	Category 3	31	100	312
loss of family of related group	Category 2	3	10	31
loss of individual	Category 1	0	1	3

Figure 16. Global risk index, in terms of numeric impact.

Additionally, a scale of the same rationale was presented as a percentage of the population (Figure 17). This has the advantage of being applicable to any nation-state context; thus, if Malta were at risk of an event that could result in 200,000 lives lost (50% of its population), this "category 10" event could present as compelling a need for mitigation as a loss of 500 million people in China. Because of this, I prefer the scale of percentage to just numbers. Again, its use would be tailored to whether national or global concerns are at stake.

As a test of reasonableness, two worldwide flu pandemics were discussed. The Spanish Flu Outbreak of 1918 resulted in an estimated loss of 40–100 million human lives. Given the contemporary expert estimate of world population, this was a 2–5% reduction. Significant by any measure, this pandemic would have been labeled a Category 9 event. However, due to the state of the world's information, health, and pharmaceutical response at that time, it was not possible to predict a reasonable outcome for this scenario.

The current H5N1 flu strain presents a totally different scenario. The science concerning pathogens, medicine, and pharmaceutical products has progressed to a sufficiently advanced state that the outcome of the H5N1 flu strain can be predicted with some accuracy. Given this information, risk can be adequately evaluated to devote resources to mitigating the possible outcomes of scenarios concerning this pathogen. Currently, health organizations around world predict probable population reductions of between 15–150 million given the current state of the flu strain. Assuming this range of life loss (0.2–2.3% of the current world population), this pandemic would rank as a "Category 8" world event; still significant, but not as significant as the Spanish flu outbreak.

The response to this presentation was mixed, if not binomially distributed. It was obvious that more than half of the conference attendees

| | GLOBAL RISK INDEX | | | |
| | | lower | | Upper |
	category	bound	MEDIAN	bound
total loss of civilizations	Category 10	10.00%	31.60%	100.00%
significant loss of large civilizations	Category 9	1.00%	3.16%	10.00%
loss of regional civilization	Category 8	0.10%	0.32%	1.00%
loss of metropolitan area	Category 7	0.01%	0.03%	0.10%
loss of city	Category 6	0.001%	0.003%	0.010%
loss of town	Category 5	0.0001%	0.0003%	0.0010%
loss of community	Category 4	0.00001%	0.00003%	0.00010%
loss of neighborhood	Category 3	0.000001%	0.000003%	0.000010%
loss of family of related group	Category 2	0.0000000%	0.0000003%	0.0000010%
loss of individual	Category 1	0.00000000%	0.00000000%	0.00000010%

Figure 17. Global risk index, in terms of percentage impact.

found such a scale useful and beneficial in ranking risk scenarios. The other attendees were emotional about the issues of life, death, and life valuation techniques. Of particular concern was the depiction of the loss of one or even ten lives as being less significant than that 1,000 or even 10,000 lives. For this reason, caution must be taken anytime loss of life is predicted with acceptable levels.

9. Development of an Environmental Security Index

From the information presented, an index for environmental security would be larger in scope than any of the existing risk scales; however, it would be smaller in scope than the proposed world risk index. From these observations, there should be no doubt that such a scale would prove effective in communicating environmental security risk within a sphere of influence provided that the five identified factors are present:

- Easy to use
- Easy to communicate
- Easy to understand
- Transparent
- Trusted

I would propose an index with characteristics similar to the GRI. They are logical, easy to understand and use, easy to communicate, and transparent in derivation. Most importantly, a risk scale that could be used to communicate the risk of any scenario would significantly contribute to communicating the risks of any event. Thus, the risk of H5N1 would be able to be compared against the risk of global warming or the risks of nanomaterials, long-lived chemicals, or nuclear isotopes. The only recommendation is that a name or acronym that is easy to remember and say would be highly desirable. I would argue that "Category" be reserved for referencing risks that are global in significance.

Without a doubt, risk scales have found a place in modern societies. They are commonly used to communicate risks efficiently and effectively while being easy to replicate through a transparent evaluation process. Often, even if citizens know little of how the risk was determined, they seek paths of mitigation based on this communication. I believe that eventually a global risk index will become reality. This may be through the combination of other accepted risk scales such as the Howe-Devereux or Torino Scale. Or it may be through the effort of international groups who are working on developing a common language and response to world risk events.

Note of Appreciation

I want to thank George Beitel—a friend and retired INL Physicist and Systems Engineer—who encouraged, funded, and provided review in my efforts to develop an SAI to communicate risk for the CSSP. This effort in risk communication would not have been possible without his vision of a possible solution.

PART 2

RISK ASSESSMENT FOR EMERGING STRESSORS

NANOMATERIAL RISK ASSESSMENT AND RISK MANAGEMENT

Review of Regulatory Frameworks

I. LINKOV

U.S. Army Engineer Research and Development Center
83 Winchester Street, Suite 1
Brookline, MA 02446, USA
Igor.Linkov@usace.army.mil

F.K. SATTERSTROM

Harvard University School of Engineering and Applied Sciences
Cambridge, MA, USA

Abstract: Managing emerging risks, such as those posed by nanotechnology, is a challenge that requires carefully balancing largely unknown benefits and risks. Here we review current nanomaterial risk management frameworks and related documents, with a focus on identifying and assessing gaps in their coverage. We do so using a *regulatory pyramid,* with self-regulation at the pyramid base and prescriptive legislation at its apex. We find that appropriate regulatory tools, especially at the bottom of the regulatory pyramid, are largely lacking. In addition, we recommend that regulatory agencies employ an adaptive, tiered framework to manage nanotechnology risk. The framework should utilize multiple tools at different levels of the pyramid, with specific tools chosen on a case-by-case basis.

1. Background

Managing emerging risks poses a challenge to regulatory agencies because decisions must be made based on extremely limited information in the face of significant public scrutiny. Regulatory agencies worldwide have successfully implemented health and safety procedures to address environmental and occupational exposure concerns for traditional industrial materials. Newly emerging risks in the realm of nanomaterials may differ from past stressors, but they involve many similar issues, including public pressure, the necessity of making regulatory decisions, and a significant level of uncertainty regarding material properties and impacts throughout

I. Linkov et al. (eds.), Real-Time and Deliberative Decision Making.
© Springer Science+Business Media B.V. 2008

product life cycles. For many emerging risks, regulatory agencies may need to modify their traditional risk management paradigm, explore innovative hazard identification and risk characterization methods and tools, communicate risks to the public, and integrate risk management with larger societal considerations during the decision-making process.

As with many new technologies, developing a framework for making risk management decisions with regard to nanotechnology is a challenge. Around the world, regulatory agencies, trade organizations, nonprofit organizations, academics, and members of industry are proposing nanomaterial risk management models and frameworks. This chapter reviews current risk management frameworks and related documents for nanotechnology. Many of the regulatory frameworks are designed to address a specific issue, industry, or single class of nanomaterials, and thus may not be directly relevant for every aspect of nanomaterial management. Even though the current knowledge base is limited, this review and evaluation allows identification of gaps in existing frameworks that may be important to managers and other stakeholders.

Thirteen frameworks and related documents were selected for in-depth review. Data were summarized according to criteria associated with each of our four categories, and narratives were developed that describe which documents pertain to which criteria. Preliminary identification of gaps—those criteria that are relatively unaddressed by the reviewed documents—and suggested approaches for formal gap prioritization are given after the review. Taken together, this information could provide the basis for selecting an instrument of choice for regulating nanomaterial risks.

2. Approach

We reviewed documents from a range of countries and purposes. We reviewed comprehensive state-of-the-science regulation framework documents, such as USEPA's "Nanotechnology White Paper" [48], the Royal Society's "Nanoscience and nanotechnologies" report [38], and the International Risk Governance Council's "Nanotechnology Risk Governance" white paper [20]. We also reviewed documents for voluntary programs, such as the Environmental Defense-DuPont "Nano Risk Framework" report [15] and the Voluntary Reporting Scheme for nanomaterial information of the UK's Department for Environment, Food and Rural Affairs [45]. J. Clarence Davies's "Managing the Effects of Nanotechnology" [11] focuses on the regulation of nanomaterials, and the position statement "Ethics and Nanotechnology: A Basis for Action" from the *Québec Commission de l'éthique de la science et de la technologie* [35] gives an ethics-focused view of nanotechnology. A list of documents reviewed and the focus of each is provided below (Table 1).

Our summary and assimilation of current approaches focused on specific criteria identified as important for a nanomaterial regulation framework. We developed the list of criteria based on Health Canada's framework for nanotechnology products, using its categories as the basis for our review. Our categories are: (1) Science and Research Aspects; (2) Legal and Regulatory Aspects;

TABLE 1. List of Documents Reviewed. Description of Document Focus is often taken Directly from the Document Foreword.

Document	Focus	Citation
USEPA White Paper	Comprehensive framework intended to set forth current scientific knowledge and its gaps related to possible environmental benefits of nanotechnology as well as potential risks from environmental exposure to nanomaterials	[49]
FDA	Report intended to help assess questions regarding the adequacy and application of the FDA's regulatory authority to nanomaterials, and to provide findings and recommendations to the FDA Commissioner	[49]
Woodrow Wilson Center	Paper intended to describe the possibilities for government action to deal with the adverse effects of nanotechnology, and to provide evidence relevant for determining what needs to be done to manage nanotechnology	[11]
ED DuPont	Comprehensive framework for the responsible development, production, use, and end-of-life disposal of nanomaterials, intended for use by companies and other organizations	[15]
Québec Commission	Comprehensive discussion of the scientific, legal and ethical implications of nanotechnology, intended to help uphold the protection of health and the environment, as well as respect for many values such as dignity, liberty, integrity, justice, transparency, and democracy	[35]
Royal Society	Comprehensive framework intended to summarize current scientific knowledge and applications of nanotechnology, and to identify possible health and safety, environmental, ethical, and societal implications or uncertainties	[38]
DEFRA	Trial Voluntary Reporting Scheme to collect data from organizations in the nanotechnology industry to help the UK develop appropriate controls for risks to the environment and human health from nanomaterials	[44]

(continued)

TABLE 1. (continued)

Document	Focus	Citation
Responsible NanoCode	Paper intended to highlight key issues that emerged from a business workshop on nanotechnology, including development of a responsible nanotechnology code	[36]
EC SCENIHR	Technical document intended to assess the appropriateness of current risk assessment methodologies for the risk assessment of nanomaterials, and to provide suggestions for improvements to the methodologies	[14]
EC Action Plan	Plan intended to help Europe build on its strengths and advances to ensure that nanotechnology research is carried out with maximum impact and responsibility, and that the resulting knowledge is applied in products that are useful, safe, and profitable	[13]
IRGC Policy Brief	Brief intended to assist policy makers in developing the processes and regulations to enable the development and public acceptance of nanotechnology	[21]
IRGC White Paper 1	Comprehensive framework intended to advance the development of an integrated, holistic, and structured approach for the investigation of risk issues and the governance processes and structures pertaining to them	[19]
IRGC White Paper 2	Comprehensive framework which applies general IRGC risk governance framework to the field of nanotechnology	[20]

(3) Social Engagement and Partnerships; and (4) Leadership and Governance. Within each category, we modified Health Canada's specific criteria to fit our categories. For example, our "Science and Research Aspects" are adapted from the US Nanotechnology Environmental and Health Implications Working Group research needs categories [51], and our Legal and Regulatory Aspects are adapted from Davies [12]. The categories and criteria used in the review are shown below; there are four criteria per category.

- Category 1: Science and Research Aspects

  1. Development of methods for detection/characterization/data collection

  2. Assessment of environmental fate and transport/impacts

  3. Assessment of toxicology/human health impacts

  4. Assessment of health and environmental exposure

- Category 2: Legal and Regulatory Aspects
  1. Voluntary regulatory and best-practices measures
  2. Information-based regulatory tools (e.g., labeling)
  3. Economics-based regulatory tools (e.g., tax or fee for safety testing)
  4. Liability-based regulatory tools (e.g., penalty for pollution)
- Category 3: Social Engagement and Partnerships
  1. Promotion of education and distribution of information/use of risk communication tools
  2. Use of stakeholder engagement tools
  3. Development of partnerships with academia, industry, public organizations, provinces, and international regulators
  4. Emphasis on ethical conduct
- Category 4: Leadership and Governance
  1. Transparency in nanotechnology-related decisions
  2. Consideration of the benefits of nanotechnology
  3. Adaptive modification of existing or development of new legislation
  4. Consideration of precautionary principle

3. Results

3.1. SCIENCE AND RESEARCH ASPECTS

We have divided the review by category, and within each category we discuss the documents that relate to each criterion. We begin by discussing science and research—a topic covered, of course, by every document reviewed.

3.1.1. Development of Methods for Detection/Characterization/Data Collection

Various frameworks discuss the scientific and research aspects of nanomaterial regulation, including methods for detection and characterization of nanomaterials. In the U.S., the EPA Nanotechnology White Paper [48] comprehensively describes the aspects of nanotechnology relevant to USEPA, as well as the many gaps in current scientific knowledge that will need to be filled before the Agency can reliably regulate nanomaterials. An entire chapter is dedicated to the risk assessment of nanomaterials, and it discusses at length the current scientific knowledge of detection and characterization methods (for example, dynamic light scattering to obtain particle size distributions, mass spectrometry to obtain chemical composition, and electron microscopy to obtain images).

The Nanotechnology Report by the US Food and Drug Administration Nanotechnology Task Force [49] focuses on how the FDA will need to change in order to be better prepared to regulate products that contain nanomaterials. The report describes the agency's science needs, such as the development of methods for identifying FDA-regulated products that contain nanomaterials. The report also describes the agency's regulatory needs, including a discussion of the need for more guidance as to when a nanomaterial becomes a dietary ingredient that requires regulation.

Like the USEPA White Paper, the Royal Society report "Nanoscience and nanotechnologies: opportunities and uncertainties" [38] is comprehensive, containing a thorough view of nanotechnology, including knowledge gaps and regulatory issues, and scientific issues such as detection methods. Detection is also discussed in the ED-DuPont framework for nanomaterial management, which includes base sets of data that describe basic characteristics to be taken into account during the risk management process [15].

The European Commission's Scientific Committee on Emerging and Newly-Identified Health Risks (SCENIHR) offers a guidance document [14] that has a technical focus and is an excellent resource for the details of risk assessment. The EC SCENIRH covers measurement methods for nanoparticles, and, like other documents, it cautions that current detection methods need to be improved.

Detection and characterization are, of course, important steps in nanomaterial risk assessment, and these are included in many of the other documents reviewed, including all three IRGC publications [19–21].

3.1.2. Assessment of Environmental Fate and Transport/Impacts

The assessments of environmental fate, transport, and possible environmental impacts are other important steps in risk assessment. The USEPA White Paper's chapter on risk assessment [48] includes a discussion of environmental fate and transport of nanomaterials and their possible ecological effects; in many cases, current knowledge is quite uncertain, and the report describes areas that will require further research. Likewise, the Royal Society report [38] discusses risks to the environment, including environmental fate and transport, and these concerns are included in the ED-DuPont framework's base sets (such as bioaccumulation potential) as well [15].

The EC SCENIHR guidance document also includes consideration of the environment, including ecotoxicology issues [14]. The SCENIHR, among others, believes that the appropriateness of existing methodologies for evaluating environmental effects is not clear. It recommends that environmental exposure models be validated, and that additional research be conducted into the fate, transport, and effects of nanomaterials in the environment.

The International Risk Governance Council addresses environmental concerns in its nanotechnology Policy Brief [21]. The IRGC divides the development of nanotechnology into two *frames*: Frame 1 includes *passive* nanostructures (those with a "steady function" that is constant over time) and Frame 2 includes *active* nanostructures (those with an "evolving function" that can change during operation). The IRGC Policy Brief discusses the possible risks of both passive and active nanostructures, ranging from health and environmental risks to ethical and social concerns, and notes that more information will be needed to assess the environmental and human health impacts of nanomaterials.

3.1.3. Assessment of Toxicology/Human Health Impacts

Consideration of the possible toxic effects of nanomaterials on human health was the only one of the 16 review criteria to be discussed in some form by every document reviewed. The USEPA White Paper's chapter on risk assessment [48] includes a lengthy discussion of human health effects, as does the Royal Society report [38]. Assessment of possible toxic effects is also included in the ED-DuPont base sets [15].

The EC SCENIHR thoroughly describes the risk assessment process for nanomaterials, including toxicity assessment, and it believes that current methodologies are generally likely to be able to identify human health hazards of nanoparticles [14]. It describes the relevant physicochemical properties for hazard characterization, the steps of health effects assessment, and toxicology concerns such as absorption, distribution, metabolism, and excretion.

The IRGC Policy Brief on nanotechnology [21] includes a discussion of human health concerns, and health concerns are the primary focus of the general IRGC risk governance framework, described in the IRGC white paper on "Risk Governance: Towards an Integrative Approach" [19]. The general governance framework consists of three main phases: pre-assessment, appraisal, and management. First, the pre-assessment phase includes risk framing, early warning and monitoring, prescreening, and selection of assumptions and conventions for the subsequent risk assessment. Second, the risk appraisal phase includes both risk assessment and "concern assessment." Risk assessment pertains to the scientific aspects of the risk, including hazard identification, exposure estimation, and risk estimation, with a focus on human health. Concern assessment, meanwhile, deals with the social aspects of the risk, such as the public's concerns and perceptions of the risk, as well as possible socioeconomic impacts. Finally, the risk management phase includes the actions taken to mitigate the risk. This phase includes six steps: generation of management options, technical evaluation of options, subjective evaluation of options, option selection, implementation, and—lastly—monitoring and review. The decision should take possible benefits and tradeoffs into account, and the framework is cyclical to allow for adaptation of the risk governance process based on new information gained during monitoring and review. It intends to be a holistic framework for the governance of risk, with a focus on human health.

Most documents note that current information on the toxic effects of nanomaterials is greatly lacking, and the FDA in particular notes that it will need further toxicology studies and greater in-house expertise to develop a knowledge base suitable for reviewing nanomaterials [49]. Because of this uncertainty, the *Québec Commission de l'Éthique de la Science et de la Technologie* recommends thorough toxicology studies be undertaken of the long-term use of any product that will be released to the public [35].

3.1.4. Assessment of Health and Environmental Exposure

Assessment of exposure to possibly hazardous materials is, of course, another important step in health risk assessment. The USEPA White Paper's chapter on risk assessment includes discussion of human exposures [48], as does the Royal Society report [38]. Exposure assessment is also included in the ED-DuPont framework [15].

The EC SCENIHR covers steps for exposure assessment and exposure control measures [14]. The document cautions that mass concentration may not be the best metric for measurement of exposure, since numbers of (solid) nanoparticles, given differing surface area-to-volume ratios, may also be important.

Exposure assessment is part of IRGC general risk governance approach [19], and is also in the IRGC's white paper on "Nanotechnology Risk Governance" [20], which applies the IRGC risk governance framework to nanotechnology. This white paper includes nanotechnology-related ideas and concepts from the IRGC Nanotechnology Policy Brief [21], differentiating between Frame 1 (passive) and Frame 2 (active) nanostructures; like the risk governance framework, it adopts an adaptive structure that includes pre-assessment, risk assessment, concern assessment, risk management, risk communication, and stakeholder participation. The paper identifies scientific needs for risk assessment such as better tools for measuring exposure and—like the Policy Brief—notes that more attention to exposure monitoring is needed.

3.2. LEGAL AND REGULATORY ASPECTS

3.2.1. *Voluntary Regulatory and Best-Practices Measures*

Many of the documents reviewed recommend that industry voluntarily adopt best-practices measures. The Environmental Defense-DuPont Nano Risk Framework mentioned in the scientific sections above [15] is a good example of this: the document describes a voluntary, adaptive framework for the risk management of nanomaterials within a company. The framework includes an initial step in which risk managers describe the material to be managed and its application. The managers then consider the properties, possible health and environmental hazards, and possible exposures to the material. When assessing risks, the framework takes a lifecycle approach in which all phases of the material's production, use, and disposal are considered. The managers then consider different risk management options, make a decision, and take an action. The action's performance is monitored, adapted if necessary, and the process then iterates. The framework is intended to provide best-practices guidance for companies and other organizations.

A voluntary code of best-practices conduct for businesses in the nanotechnology industry is called for by the Responsible NanoCode workshop report, which describes a November 2006 meeting between the Royal Society, Insight Investment, and the Nanotechnology Industries Association [36]. The workshop report discusses uncertainties faced by businesses in the technical, social, and commercial arenas. The report stresses that the risks and uncertainties are all interconnected, and the workshop participants agreed that they need a new approach to responding to these risks. The next steps recommended after the workshop include the development and implementation of a voluntary code of responsible conduct for the nanotechnology industry.

In another voluntary effort, the United Kingdom Department for Environment, Food and Rural Affairs enacted a Voluntary Reporting Scheme for engineered nanoscale materials in September 2006 [44]. The program requests submission of data related to the material and its production and use (including composition, manufacturing process, size and shape, intended use, exposure pathways, and benefits), its health- and environment-related properties (including physicochemical properties, toxicology, ecotoxicology, environmental fate), as well as measurement techniques and current risk management practices. DEFRA is not asking companies to generate new data for submission; it is simply asking that companies which generate data during the course of their normal business submit the data to the agency so that it may gain a better knowledge base for the regulation of nanomaterials. The program is a two-year trial, and it has received nine submissions (seven from industry and two from academia) as of December 2007 [45].

Davies [11], in the Woodrow Wilson Center document "Managing the Effects of Nanotechnology," focuses on the regulation of nanomaterials. One of the options he considers is voluntary self-regulation; he believes voluntary measures must include incentives for companies to participate, and he notes that companies that do not volunteer might be those most in need of regulation. Davies concludes that nanotechnology risk management will likely require new laws, and he imagines a product-focused, rather than environment-focused, law in which the manufacturer must provide reliable evidence to support the proposition that its nanomaterial-containing product is safe.

Many of the other documents reviewed also discuss voluntary programs, including the USEPA White Paper [48], which describes a voluntary nanomaterial stewardship program undertaken by USEPA's Office of Pollution Prevention and Toxics (OPPT). The OPPT held several public meetings to discuss the program. Other documents such as the Québec Commission report [35] advocate the development of a best-practices guide.

3.2.2. Information-Based Regulatory Tools

The Québec Commission [35] believes that labeling is important for enabling freedom of choice, but it also believes that labeling will not be useful for nanomaterials until they are better understood. Labeling is an option that Davies considers [11], but he does not believe that labeling specific products would necessarily change consumer behavior.

Other documents also discuss labeling, including the Royal Society report [38], which recommends that products' ingredients lists should declare the presence of any added nanomaterials. The IRGC also considers labeling to be a useful tool for communicating possible risk to consumers.

3.2.3. Economics-Based Regulatory Tools

Davies [11] includes economics-based regulatory tools as part of his four possible incentives for promoting uses of nanomaterials that benefit the environment or improve public health. He suggests: (1) research funding to facilitate the identification of helpful and harmful applications of nanomaterials; (2) tax breaks and tax penalties to promote government-defined environmentally beneficial behaviors while penalizing pollution; (3) acquisition programs in which federal and local governments, as significant and large consumers, are required to purchase or underwrite products deemed environmentally beneficial; and (4) regulatory advantages that accelerate the review and approval process for environmentally beneficial new products.

Regulation is also discussed by the Royal Society [38], which includes a case study of the regulation of nanomaterial-containing cosmetics, and the

FDA [49], which discusses its pre-market review process. These documents do not go into depth on specific types of regulatory tools.

3.2.4. Liability-Based Regulatory Tools

The documents in our review cover liability-based regulatory tools only to the extent that they discuss existing regulations. USEPA [48] has regulations in place (e.g., under the Toxic Substances Control Act, Superfund, and the Clean Water Act) to control toxic substances and contaminated sites and to manage the effects of hazardous substances. All other developed nations have similar laws and regulations. Nanomaterials that meet the criteria of these acts would be subject to the regulations imposed on these substances. Similarly, the Royal Society holds that regulations currently in place are broad enough to have authority over harmful nanomaterials [38].

Davies [11], in contrast, recommends that nanomaterials be treated as if all are entirely new substances that fall under the regulation of the Toxic Substances Control Act (TSCA). Davies warns that existing legal measures do not necessarily apply to nanomaterials, which by virtue of their size may be exempt from regulation (because they would not reach 10,000 kg of production per year) or may display properties that are inconsistent with similar but larger materials. Also, Davies points out the seeming contradiction that the default position of TSCA is to not regulate substances with unknown health and environmental effects unless there is "unreasonable risk," yet these are the substances whose risk is not known.

3.3. SOCIAL ENGAGEMENT AND PARTNERSHIPS

3.3.1. Promotion of Education and Distribution of Information/Use of Risk Communication Tools

Many documents discussing the regulation of nanomaterials consider public information and risk communication to be vital parts of the process. The IRGC general risk governance approach recommends the use of risk communication tools at each step of its framework [19]. This is intended to enable citizens to become involved in the process, the decision, and its implications. The IRGC Policy Brief on nanotechnology also advocates public education [21], and the IRGC nanotechnology governance framework emphasizes risk communication and recommends that the public be provided with information [20]. In addition to discussing health and environmental concerns in its risk assessment framework, it also considers educational gap risks, such as when technical knowledge is not shared with regulatory agencies, civil

society, and the public, leading to skewed perceptions of health and environmental risks.

The IRGC nanotechnology governance gives specific examples of risk
communication tools and information to be communicated. Information to
be communicated could relate to the benefits and harmful effects of nanotechnology, updates on scientific research, information on the methods used
to test nanotechnology products and assess potential health or ecological
impacts, and debate on the ethical acceptability of certain nanotechnology applications. Risk communication tools include product labeling; press
releases and consumer hot lines; risk communication training courses and
exercises for scientists; and integrated risk communication programs for
scientists, regulators, industrial developers, representatives of NGOs, the
media, and other interested parties.

Other documents also recommend the use of risk communication tools.
Davies [11] holds that the public needs to be included for nonmaterial management to be successful. The Royal Society report discusses stakeholder and
public dialogue, including the importance of working with the public with
regard to nanotechnology-related issues and promoting a wider public dialogue about the field [38]. The FDA recommends communication with the
public about the presence of nanomaterials in FDA-regulated products [49],
and the ED-DuPont framework [15] and several other documents also state
that public involvement is important.

3.3.2. Use of Stakeholder Engagement Tools

Like risk communication tools, stakeholder engagement tools are advo-
cated by many frameworks. The IRGC general risk governance approach
recommends the use of stakeholder engagement tools at each step [19].
This is intended to learn about citizens' opinions; the document contains
a discussion of risk perception and the factors that affect it, including
availability bias, anchoring effect, and uncertainty. The IRGC Policy Brief
and IRGC nanotechnology governance framework also advocate government interaction with stakeholders and opinion research to improve both
risk management and public acceptance of genuinely benign technologies
[20, 21].

Other documents, such as the USEPA White Paper, recommend stakeholder engagement as well, and call for public meetings and interactions with
stakeholders [48]. The Royal Society report includes research into public
knowledge of nanotechnology in Britain, workshop findings, and the incorporation of public values into decisions [38]. Davies [11] maintains that the
public needs to be listened to, and he wants greater public participation in
the regulatory process. Responsible NanoCode workshop participants also

believe that they should develop a forum for discussion of responsible work in the nanotechnology sector [36].

Many social recommendations are also made in the European Commission report "Nanosciences and nanotechnologies: An action plan for Europe 2005–2009," which gives an outline of the actions and infrastructure required for European Union (EU) countries to succeed in the nanotechnology industry [13]. It lists actions that the Commission will take and that it calls on the EU member states to perform. For example, the report recommends that the EU invest more money in the nanotechnology industry, construct new research infrastructure, and increase funding for the training of scientists in nanotechnology. Notably, it recommends that governments provide multilingual information about nanotechnology to the public and pursue a dialogue with stakeholders about nanotechnology. It calls for an increase of nanotechnology awareness at universities and in industry, and for programs that encourage university students to pursue nanotechnology research. It also calls for the international exchange of best-practice guidelines, the development of common standards for nanotechnology, and the development or adaptation of existing regulations for nanomaterials.

3.3.3. Development of Partnerships with Academia, Industry, Public Organizations, Provinces, and International Regulators

Many documents recommend collaboration. The Eurpean Community (EC) Action Plan, for example, calls for the development of partnerships and collaborative efforts across the EU [13]. The report recommends that the EU states increase collaborative research and coordinate research programs, support networking and integration of resources, promote networking of people, promote international collaboration, and increase industrial involvement in collaborative efforts.

The USEPA White Paper recommends that USEPA collaborate with other countries, and that its own researchers collaborate more actively among themselves [48], while the FDA recommends that it pursue collaborative relationships with other federal agencies and other stakeholders [49]. The IRGC policy brief and general governance approach recommends collaboration with and among stakeholders, and the IRGC nanotechnology governance framework identifies better collaboration between institutions and better coordination among stakeholders as institutional and social needs [19–21]. Davies [11] also encourages greater institutional coordination in the nanomaterial regulation process.

Several other documents echo the message of collaboration. The Québec Commission [35] recommends wide collaboration in nanotechnology regulation, and the Royal Society recommends that scientists collaborate, as well as regulators [38]. Responsible NanoCode workship participants decided that their code of conduct should be developed in cooperation with a wide range of stakeholders [36]. DEFRA's solicitation of voluntary information, meanwhile, is essentially a collaboration with willing stakeholders [44].

3.3.4. Emphasis on Ethical Conduct

Ethics also play an important role in nanotechnology regulation, as elaborated by the *Québec Commission de l'Éthique de la Science et de la Technologie* [35]. It begins its position statement "Ethics and Nanotechnology: A Basis for Action" by discussing the state of nanotechnology science, possible risks, and regulatory tools, but its main focus is ethical issues. For example, the Commission believes that companies must protect human dignity by not treating workers simply as means of production, but rather as people whose exposure to harmful materials must be minimized, especially when possible effects are not known. When nanomaterials are used in biomedical applications, the Commission believes that researchers must consider ethical issues such as confidentiality of personal information and respect for free and informed consent. When nanotechnology is used in surveillance, biometric controls, or substance detection in the name of security, the Commission warns that they must not be used in a way that impinges upon civil liberties. The Commission also discusses other ethical issues, such as the purpose and secrecy of military applications, the legitimacy and transparency of the government decision-making process, the fair worldwide distribution of nanotechnology benefits and risks, and whether nanotechnology can fundamentally alter human identity (through performance enhancement) or human relationship with nature (by modifying the environment).

The Royal Society report includes a focus on the social and ethical implications of nanotechnology alongside its discussion of science issues [38]. For example, the Royal Society notes that nanomaterials in devices capable of collecting personal information must not be used to compromise people's civil liberties. The report also considers the possibility that nanotechnology may primarily benefit the well-to-do social classes, and that this might exacerbate the problems of class division. The Royal Society takes these issues seriously, and it recommends that all scientists working in the field consider the social and ethical consequences of nanotechnology as part of their training.

The IRGC policy brief also expresses concerns about whether the advantages of nanotechnology will favor one country over another, or whether certain countries will lower safety requirements in order to gain a competitive technological advantage [21]. Other ethical concerns include whether human identity will be compromised by nanotechnology, as well as what might happen if hybrid "nanobio" devices escape human control. In the IRGC general risk governance approach, the ethical acceptability of the process and its outcome is also emphasized [19]. Ethical acceptability is emphasized in the IRGC nanotechnology governance document as well; it also considers political and security risks, such as uneven distributions of risks and benefits in the international community [20]. The EC Action Plan includes consideration of broader social impacts and recommends an ethical analysis of nanomedicine and a study of nanotechnology's likely impact on society [13]. For its voluntary reporting scheme, DEFRA specifically discourages the generation of new information that would require animal testing [44].

3.4. LEADERSHIP AND GOVERNANCE

3.4.1. Transparency in Nanotechnology-Related Decisions

Many guidance documents recommend transparency in the regulatory process [35, 38, 48, 49]. The ED-DuPont framework recommends that decisions are documented to increase the transparency of the process [15]. Because of collaborative and inclusive nature of the IRGC general risk governance approach, each step of the process is intended to be transparent to the public, and transparency is emphasized [19]. Transparency is emphasized in the IRGC nanotechnology governance document as well [20].

3.4.2. Consideration of the Benefits of Nanotechnology

While assessing the possible risks of nanomaterials, many frameworks appropriately weigh the risks against a given nanomaterial's possible benefits. The USEPA White Paper has a separate chapter to consider the environmental benefits of nanotechnology, including zero-valent iron for the remediation of chlorinated hydrocarbons in groundwater, nanosensors for the detection of pollutants, and nanotechnologies that support—or could support—sustainability [48], however defined.

The Royal Society report also considers the beneficial applications of nanotechnology, as do ED-DuPont, the IRGC, and other organizations [15, 21, 38]. The Québec Commission discusses possible applications and benefits of nanotechnology in relation to their ethical employment, and the EC Action Plan is predicated on building infrastructure to take advantage of nanotechnology's benefits [13].

3.4.3. Adaptive Modifications of Existing or Development of New Legislation

Two of the framework documents discussed contain important adaptive elements. The ED-DuPont framework, for example, is essentially an adaptive management procedure [15]. Its iterative framework allows for the incorporation of new information into the management process, so that the regulation evolves to incorporate best practices and recently acquired scientific knowledge. The IRGC general risk governance approach is also iterative, enabling adaptive learning to take place [19]. The IRGC nanotechnology governance approach, meanwhile, recognizes that existing legislation might need to be adapted [20].

Davies [11] maintains that if existing laws are to be applied to nanotechnology, they will need to be strengthened or adapted for their new purpose, because each suffers from certain shortcomings. For example, the Toxic Substances Control Act has broad coverage, but it would be complicated to apply because it covers substances "of a particular molecular identity," and Davies notes that the physicochemical properties of a nanomaterial may change with its size or form, even if its molecular identity does not change. The possible need to modify laws to accommodate nanotechnology regulation is also mentioned by the USEPA White Paper, the Royal Society, and others [38, 48].

3.4.4. Consideration of Precautionary Principle

Many of the documents reviewed discuss the precautionary principle. The Québec Commission holds that use of the precautionary principle is essential to nanotechnology regulation in the face of uncertainty, claiming that use of this principle will ensure that no harm is caused [35]. Given the significant uncertainty in the field of nanotechnology risk assessment, the Royal Society takes a similar stance, saying that environmental releases of nanoparticles should be avoided until more is known about their effects [38]. The ED-DuPont framework says it espouses values "similar" to the precautionary principle, but does not espouse it directly because it is defined different ways in different places [15]. The IRGC general risk governance takes a "precautionary" approach in high-uncertainty situations [19].

The IRGC nanotechnology governance framework, in contrast, opposes use of the precautionary principle [20]. The document holds that the precautionary principle would lead to a moratorium on technology development, causing industry to move out of the country. For Davies [11] as well, the precautionary principle is equated with a ban, and he says that this is not helpful for a field in which continued development is expected to be beneficial.

3.5. FRAMEWORKS SUMMARY

Overall, out of the four categories discussed, the greatest attention is paid to the scientific and research aspects of nanomaterial regulation. The knowledge needed to conduct risk assessment—and the research needed to create new knowledge—is discussed to some extent in virtually all of the documents reviewed. Every document expresses concerns about nanomaterials' possible adverse effects on human health, and the more comprehensive documents reviewed define a framework for the assessment of such risks, despite significant uncertainties in necessary information.

Less attention is paid to regulatory tools. The ED-DuPont Nano Risk Framework is a good example of a framework for voluntary best practices [15], and a few sources discuss regulation tools such as labeling or tax breaks, but many agencies have not begun to write in terms of using new tools to regulate nanomaterials. USEPA and Royal Society, for example, both discuss existing regulations that are expected to be sufficient to cover nanomaterials, perhaps with slight modification [38, 48].

The social engagement/partnerships and leadership/governance categories share the trait of being often recommended in general terms. Agencies are aware that it is important to communicate risk and engage stakeholders, so they recommend doing these things; they know that it is important to be transparent and to be adaptive, so they recommend these qualities as well, although often without clear directions for how to achieve transparency or adaptiveness. More concrete examples are given for social engagement tools than for governance tools, and the lack of specific tools makes leadership/governance another knowledge gap for nanomaterial regulation.

Table 2 summarizes the areas discussed by each document.

4. Regulatory Gaps and Possible Solutions

Our review indicates that many nanomaterial management frameworks primarily focus on scientific and research aspects and, to a somewhat lesser degree, on social engagement and partnerships. Legal and regulatory aspects, as well as governance, have received comparably little discussion. The following section provides an overview of issues and approaches discussed in the peer-reviewed literature that could help in bridging these gaps. Specifically, we introduce the regulatory pyramid approach originally proposed by Ayres and Braithwaite [2] and adopted for nanomaterial regulations by Bowman and Hodge [6] and Marchant et al. [29] as a guiding framework for nanotechnology regulation. We then discuss risk assessment and the precautionary principle, as well as voluntary programs, self-regulation, and other tools.

TABLE 2. Elements of Nanomaterial Regulation Frameworks Discussed in each Document (criteria are numbered 1 to 4 under each Category; for each Document and Criterion, ■ = Document Discussed the Criterion, ▪ = Document Mentioned the Criterion, and (Blank) = Document did not Address the Criterion).

	Science and research aspects				Legal and regulatory aspects				Social engagement and partnerships				Leadership and governance			
	1	2	3	4	1	2	3	4	1	2	3	4	1	2	3	4
USEPA White Paper	■	■	■	■	■	▪	▪	▪	■	■	■	▪	▪	■	▪	
FDA		▪	■	▪		■	■			▪	▪		▪			
Woodrow Wilson Center	■	■	■	▪	■	▪	■		■	■	■		■	▪	■	▪
ED-DuPont	▪	■	■		■	▪			■	▪					■	▪
Québec Commission	▪	■	■	▪	■	■	■		■	■	■	■	■	■	▪	■
Royal Society	▪	■			■				■		■					
DEFRA			▪		■					▪						
Responsible NanoCode					■				■		▪					
EC SCENIHR	■	■	■	■	▪				■	■	■		▪	■		
EC Action Plan	▪			▪					▪		▪					
IRGC Policy Brief	▪	▪	■	■	■	▪	▪	▪	■	■	■			▪	▪	▪
IRGC White Paper 1	▪	▪	■	■	▪		▪		■	■	■			▪	■	■
IRGC White Paper 2	▪		■	■	▪		▪		■	■	■				■	■

Subcriteria for the table are as follows:

- **Science and research aspects**
  1. Development of methods for detection/characterization/data collection
  2. Assessment of environmental fate and transport/impacts
  3. Assessment of toxicology/human health impacts
  4. Assessment of health and environmental exposure
- **Legal and regulatory aspects**
  1. Voluntary regulatory and best-practices measures
  2. Information-based regulatory tools (e.g., labeling)
  3. Economics-based regulatory tools (e.g., tax or fee for safety testing)
  4. Liability-based regulatory tools (e.g., penalty for pollution)

- **Social engagement and partnerships**
  1. Promotion of education and distribution of information/use of risk communication tools
  2. Use of stakeholder engagement tools
  3. Partnerships with academia, industry, public organizations, provinces, and international regulators
  4. Emphasis on ethical conduct
- **Leadership and governance**
  1. Transparency in nanotechnology-related decisions
  2. Consideration of the benefits of nanotechnology
  3. Adaptive modification of existing or development of new legislation
  4. Consideration of precautionary principle

This section concludes with a discussion of a framework and supporting methods and tools applicable to governance of nanotechnology.

4.1. REGULATORY PYRAMID APPROACH

Our review identified multiple regulatory policy instruments (e.g., voluntary programs, labeling, tax incentives). A regulatory pyramid approach and responsive regulations [2] provide a good framework for classifying these regulatory policy instruments and associated tools (Figure 1). The underlying idea of responsive regulation is that the degree of regulatory intervention and supervision is based on a dynamic assessment of market conditions and regulated community performance, rather than a one-size-fits-all prescription. Self-regulation and best practices are characteristic of the base of the pyramid, representing the bulk of matters that can be handled informally without oversight by regulatory agencies. The regulatory approach becomes more prescriptive and punitive at the top of the pyramid. The regulatory response depends on the effectiveness of individual firms' self-regulation activities, as well as on how successfully they have responded to hazards and risks.

Bowman and Hodge [6] adjusted the regulatory pyramid approach for nano-material regulations. Here the pyramid has been replaced by a hexagon that

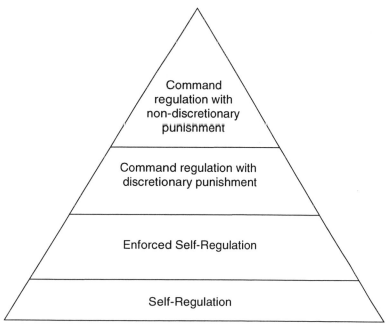

Figure 1. Regulatory Pyramid [2].

includes six regulatory *frontiers*: product safety, privacy and civil liberties, occupational health and safety (OH&S), intellectual property (IP), international law, and environmental law. In each of these areas, a range of regulatory mechanisms and tools is available to regulators, from *hard law* at the top, through licensing, codes of practice, guidelines, and other *soft law* options at the base.

Marchant et al. [29], however, maintain that the approaches of both Ayers and Braithwaite [2] and Bowman and Hodge [6] are static, while the field of nanotechnology requires dynamic and adaptive views. Thus, incremental nanotechnology regulation is proposed in their paper and depicted in their own pyramid (Figure 2). Marchant et al. [29] argue that nanotechnology regulatory activities should start with information gathering and self-regulation should and move towards hard law/legislation once more information is collected. This framework is supposed to provide an adaptive approach for addressing changes in the regulatory environment and an increasing knowledge base in the regulated community.

4.2. REGULATORY TOOLS OPERATING FROM THE APEX OF THE PYRAMID

Risk assessment and the precautionary principle have been used by regulatory agencies worldwide in various settings. This section provides an overview of the difficulties in applying these tools to nanomaterial risk management.

4.2.1. Traditional Risk Assessment Framework

Risk assessment has been practiced by USEPA and other agencies as a tool to evaluate risks associated with chemicals in the environment. Risk assessment approaches and procedures have been formulated by the US National Academy of Sciences [50] and subsequently tailored to specific applications by USEPA [46, 47] and other agencies in the US and worldwide. Risk management was initially separated from risk assessment; risk assessment was perceived as a scientific activity while risk management was dealt with in a policy framework. A risk assessment is generally constructed to have four components: hazard identification, toxicity assessment, exposure assessment, and risk characterization. Most of the documents we reviewed attempted to adjust the traditional scientific risk assessment framework to the regulation of nanomaterials.

4.2.2. Difficulties in Applying Traditional Risk Assessment Framework

Recent articles, as well as the frameworks reviewed in this study, generally use several different characteristics in their assessment of nanomaterial risk. These characteristics include chemical composition, size/shape, surface chemistry and reactivity, solubility/environmental mobility, and agglomeration

[3–5, 8, 18, 22, 30, 32, 34, 43]. In fact, there are many subcategories and other characteristics that may well prove critical to both the benefits and the risks of any given nanotechnology.

Thus, even though the risk assessment paradigm successfully used by the scientific community since the early 1980s may be generally applicable, its application to nanotechnology requires a significant information base. As described, nanomaterial exposure and toxicity assessment are complicated by the need to take several variables into account, and they require incorporating an uncertainty in basic knowledge that at present seems much larger than the uncertainty for macromaterials. Even given estimates of exposure and toxicity, risk characterizations must be developed separately for each nanomaterial, or even similar nanomaterials with different functionalization or at different environmental lifecycle stages. Because of the required effort, detailed risk characterizations may not always be possible. In some cases, knowledge of a similar compound or class of compounds may be available, but methods for incorporating information on broad toxicity and exposure classes into the traditional risk assessment regulatory framework have not been discussed in the literature.

For the most part, it is still too early to know what specific endpoints constitute evidence of harm with regard to nanoparticles. Effects of various kinds have been reported from in vivo and in vitro studies [39, 40] (and many others), and concern that use of products containing nanomaterials may lead to chronic health risks has been expressed (Peters et al. [52] and others). Fundamentally, we still do not know enough about the toxic potentials of most nanoparticles to apply traditional risk assessment techniques.

Regulatory agencies, as well as the popular and scientific media, are thus shifting their focus from the initial euphoria about the potential of the technology to concern about possible deleterious effects resulting from nanomaterial manufacture and use. Uncertainty regarding the health impacts associated with nanotechnologies and their potentially uncontrolled market growth has resulted in calls from environmental and political bodies to limit the use of nanomaterials, increase the stringency of governmental regulations, and—in extreme cases—to ban the use of nanomaterials completely. As noted in our review, the *Québec Commission de l'éthique de la science et de la technologie* believes that the precautionary principle is an essential method for ensuring that no harm is caused in situations where nanotechnology risk information is uncertain [35], and other documents make similar recommendations [38]. However, the precautionary principle is not always seen as a helpful approach [20]. As also noted in our review, Davies [11] does not believe that the technology slowdown resulting from a regulatory implementation of the precautionary principle would be helpful for nanotechnology, since development

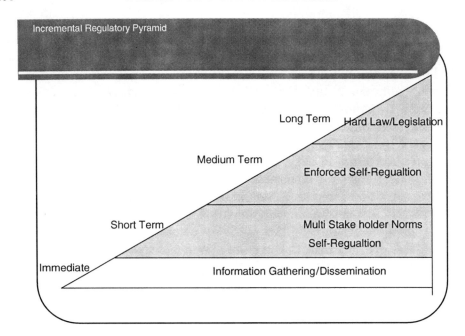

Figure 2. Incremental Regulatory Pyramid [29].

of the field should be beneficial. A responsible risk-based approach for regulation of the developing field would thus be ideal.

In fact, recent risk assessment literature and applications show that risk assessment is evolving toward integration with risk management and decision support. Risk assessment is becoming a participatory process where multiple stakeholders and their views on risks are explicitly or implicitly incorporated in the assessment. The IRGC general risk governance framework explicitly calls for inclusion of the societal context and categorization of risk-related knowledge to deal with data uncertainty [19]. In a sense, this trend indicates movement from the top of the regulatory pyramid toward its base. Such a move requires new methods and tools that are discussed in the next section.

4.3. REGULATORY TOOLS OPERATING FROM THE BASE
OF THE PYRAMID

As noted above, the base of the regulatory pyramid is self-regulation. In industry, one example is the Environmental Defense-DuPont Responsible Nano Code framework. Davies [12] believes that the Responsible Care program of the American Chemistry Council (ACC) may also be a useful example for the nanotechnology industry. Reponsible Care requires member companies to measure and publicly report performance, as well as obtain independent third-party verification that their operations are up to standards, however defined [1]. Voluntary programs have also been initiated by government, such as the DEFRA Voluntary Reporting Scheme [44], and USEPA's voluntary nanomaterial stewardship program [48].

Davies [12] notes that voluntary codes often suffer from lack of participation, as well as lack of transparency and specificity. Indeed, public opinion surveys reveal skepticism about self-regulatory programs alone [33]; failures of self-regulation could damage public acceptance of nanotechnology. Effective self-regulation with the threat of external pressure has been found to be more effective [17]. Selecting the appropriate regulatory tools for this external preassure may be crucial. Ayers and Braithwaite recommend engagement of public interest groups in this process [2], and Marchant et al. [29] expand the recommendation to include multiple stakeholder groups.

Information-based tools may play a role in applying external pressure. As Davies [12] discusses, two examples of programs that use this strategy in the U.S. are the Toxics Release Inventory (TRI) and California Proposition 65. Under the TRI, companies that release more than de minimis amounts of potentially hazardous chemicals must inform USEPA, which then publicly releases the information. In California, Proposition 65 established a state-maintained list of chemicals known to cause cancer, birth defects, or reproductive harm [9]. A product's label must declare if it contains any of the chemicals on the list (again, above de minimis amounts). Davies notes that the enforcement of these regulations is not always straightforward. Nonetheless, the tools are conceptually simple, and they inform the public of a company's or chemical's behavior, applying pressure on companies to seek safer substitutes, as appropriate.

External pressure could also be applied in the form of economic incentives. Davies [11] suggests tools such as tax breaks and tax penalties to promote adherence of companies to their industry code of conduct while penalizing those that fall behind. Another economic tool is acceleration of the review and approval process for environmentally beneficial new products. These actions would provide real incentive for companies in the nanotechnology industry to follow a code of conduct and act in an environmentally responsible manner. Davies [12] is less enthusiastic about liability tools, since these require the enforcement of tort law and are applied only after some demonstrable environmental or health harm has been committed.

Any method selected for applying external pressure to self-regulation should include information gathering tools. Information requests could help build databases of nanomaterial properties, as well as allowing the communication of risks associated with nanomaterials.

5. Risk-Informed Decision Framework for Nanotechnology Governance

The emergence of nanotechnology products has occurred much faster than the generation of corresponding environmental health and safety (EHS) data [27]. Moreover, the ability of regulatory agencies to use the EHS data also lags (Figure 3) due to the lack of data and limited resources. Given that the shelf life of new nanotechnology products is about two years or less, approaches to regulate these materials should be adjusted to the evolving nature of the field.

Our review indicates that there are many existing tools for assessing toxicities and risks; however, their application to new materials may be difficult. Traditional risk assessment boils down to comparison of exposures associated with specific hazards to regulatory benchmarks corresponding to safe exposure levels expressed in units of concentration, dose, or risk. Although agencies have tried to apply the traditional risk assessment paradigm to emerging materials, its application to nanomaterials requires dealing with a very large uncertainty in basic knowledge, while tools that are currently used for uncertainty analysis may not be easily applied to emerging threats. Integrating the heterogeneous and uncertain information in nanomaterial risk management therefore demands a systematic and understandable framework to organize the scarce technical information and expert judgment.

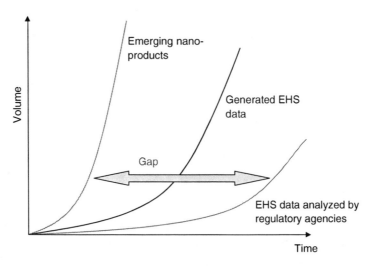

Figure 3. Schematic Representation of Emergence of Nanotechnology Products in Comparison to Generated EHS Data (based on breakout group meeting, Canadian Workshop, Edmonton 2008). This Diagram is Purely Qualitative and is Meant to Illustrate the Relative Amount of time between the Emergence of Nanoproducts, the Generation of EHS Data, and the Analysis of those EHS Data by Regulatory Agencies.

Multicriteria decision analysis (MCDA) methods provide a sound approach for decision making and management in the face of heterogeneous information, uncertainty, and risk [16, 23, 25, 41]. MCDA is recognized as legitimate and useful by organizations such as the IRGC [19], and it has been applied to multiple environmental management programs [24]. It has been recommended as one of the most promising risk governance tools [37], and an example application to nanomaterials has been reported [26, 42]. The advantages of using MCDA techniques over other less structured decision-making methods are numerous: MCDA provides a clear and transparent methodology for making decisions and also provides a formal way for combining information from disparate sources. These qualities make decisions made through MCDA more thorough and defensible than decisions made through less structured methods. The US Army Corps of Engineers is currently working on integrating risk assessment and MCDA in a joint framework (risk-informed decision framework, or RIDF) and is applying it to highly contentious restoration planning in areas affected by Hurricane Katrina [7].

Nanomaterial regulatory frameworks could be built on existing approaches with the added rigorous and transparent method for integrating technical information and expert judgment offered by MCDA. Scientific aspects of risk management are well covered by existing frameworks, and gaps in current knowledge are spelled out by many groups, including the US National Nanotechnology Initiative [38]. However, actual methods for ranking alternative management options and selecting a best option are lacking.

An MCDA approach for ranking alternative risk management tools and making efficient decisions on other issues would allow joint consideration of the benefits and risks along with associated uncertainties relevant to the decision. A generalized MCDA process follows two basic themes: (i) generating alternative options, success criteria, and value judgments and (ii) ranking the alternatives by applying value weights. The first part of the process generates and defines choices, performance levels, and preferences. The latter section methodically prunes nonfeasible alternatives by first applying screening mechanisms (e.g., harmful environmental or health effects, excessive cost) and then ranking in detail the remaining management alternatives by MCDA techniques that use the various criteria levels generated by fate and transport models, risk assessment, experimental data, or expert judgment.

Decision analysis tools can help to generate and map technical data as well as individual judgments into organized structures that can be linked with other technical tools from risk analysis, modeling, monitoring, and cost estimation. Decision analysis software can also provide useful graphical techniques and visualization methods to express the gathered information in understandable formats. When changes occur in the requirements or the decision process, decision analysis tools can respond efficiently to reprocess and iterate with the new inputs. This integration of decision tools and scientific and engineering tools allows users to have a unique and valuable role in the decision process without attempting to apply either type of tool beyond its intended scope.

The result of MCDA application is a comprehensive, structured process for selecting the optimal alternative in any given situation, drawing from stakeholder preferences and value judgments as well as scientific modeling and risk analysis. This structured process would be of great benefit to decision making in risk management, where there is currently no structured approach for making justifiable and transparent decisions with explicit tradeoffs between social and technical factors. Regulatory agencies could employ MCDA in many different situations, such as selecting the best regulatory tool to use in certain situations, prioritizing gaps in knowledge, or selecting the optimal allocation of funding.

6. Conclusions

We have reviewed current nanomaterial risk management frameworks and related documents, with a focus on identifying and assessing gaps in their coverage. We found that regulatory tools, especially from the base of the regulatory pyramid, are an important gap in the knowledge necessary for nanomaterial regulation. Current tools recommended in the literature that help fill this gap are self-regulation and enforced self-regulation; information-based tools or economics-based tools can be used to exert pressure for enforcement. These tools would help to regulate the nanotechnology industry from the bottom up, in addition to the top-down approach offered by traditional risk assessment.

Acknowledgements

The authors are grateful to Drs. Karkan, Raphold, Gemar, Maynard, Green, and Steevens, and Mr. Monica for fruitful discussions. The views and opinions expressed in this paper are those of the individual authors and not those of the US Army Corps of Engineers or US Army.

References

1. American Chemistry Council (ACC), 2007. Responsible Care. Available at: http://www.americanchemistry.com/s_responsiblecare/sec.asp?CID=1298&DID=4841.
2. Ayres, I., Braithwaite, J., 1992. Responsive Regulation: Transcending the Deregulation Debate. Oxford University Press, Oxford.

3. Biswas, P., Wu, C.-Y., 2005. Nanoparticles and the environment. Journal of the Air& Waste Management Association 55, 708–746.
4. Borm, P., Müller-Schulte, D., 2006. Nanoparticles in drug delivery and environmental exposure: same size, same risks? Nanomedicine 1 (2), 235–249.
5. Borm, P., Robbins, D., Haubold, S., Kuhlbusch, T., Fissan, H., Donaldson, K., Schins, R., Stone, V., Kreyling, W., Lademann, J., Krutmann, J., Warheit, D., Oberdorster, E., 2006. The potential risks of nanomaterials: a review carried out for ECETOC. Particle and Fibre Toxicology 3, 11.
6. Bowman, D.M., Hodge, G.A., 2007. A small matter of regulation: an international review of nanotechnology regulation. The Columbia Science and Technology Law Review VIII, 1–36. http://www.stlr.org/html/volume8/bowman.pdf.
7. Bridges, T.S., Suedel, B.C., Kim, J., Kiker, G., Schultz, M., Banks C., Payne, B.S., Harper, B., Linkov, I., 2008 (in preparation). Risk-Informed Decision Making Applied to Coastal Systems: Sustainable Management of Flood Risks and the Environment.
8. Brunner, T., Wick, P., Manser, P., Spohn, P., Grass, R., Limbach, L., Bruinink, A., Stark, W., 2006. In vitro cytotoxicity of oxide nanoparticles: comparison to asbestos, silica, and the effect of particle solubility. Environmental Science & Technology 40 (14), 4374–4381.
9. California Office of Environmental Health Hazard Assessment (CA OEHHA), 2007. OEHHA Proposition 65. http://www.oehha.ca.gov/prop65.html.
10. Canadian Workshop, Edmonton, 2008. Break-out Group Members: Drs. I. Linkov (US Army Corps of Engineers), T.A. Davis (Environment Canada), G. Goss (University of Alberta), J. Illes (University of BC), T. Medley (DuPont) and others.
11. Davies, J. Clarence, 2006. Managing the Effects of Nanotechnology. Woodrow Wilson International Center for Scholars Project on Emerging Nanotechnologies. Washington, DC 20004-3027. Available at: http://www.nanotechproject.org/process/files/2708/30_pen2_mngeffects.pdf.
12. Davies, J. Clarence, 2007. EPA and Nanotechnology: Oversight for the 21st Century. Woodrow Wilson International Center for Scholars Project on Emerging Nanotechnologies. Washington, DC 20004-3027. Available at: http://www.nanotechproject.org/file_download/files/Nano&EPA_PEN9.pdf.
13. European Commission (EC), 2005. Communication from the Commission to the Council, the European Parliament and the Economic and Social Committee: Nanosciences and Nanotechnologies: An Action Plan for Europe 2005–2009. B-1050, Brussels. Available at: http://ec.europa.eu/research/industrial_technologies/pdf/nano_action_plan_en.pdf.
14. EC Scientific Committee on Emerging and Newly-Identified Health Risks (EC SCENIHR), 2007. Opinion on the Appropriateness of the Risk Assessment Methodology in Accordance with the Technical Guidance Documents for New and Existing Substances for Assessing the Risks of Nanomaterials. B-1049 Brussels, Belgium. Available at: http://ec.europa.eu/health/ph_risk/committees/04_scenihr/docs/scenihr_o_010.pdf.
15. Environmental Defense-DuPont Nano Partnership (ED – DuPont), 2007. Nano Risk Framework. Available at: http://www.environmentaldefense.org/documents/6496_Nano%20Risk%20Framework.pdf.
16. Figueira, J., Greco, S., Ehrgott, M. (Eds.), 2005. Multiple Criteria Decision Analysis: State of the Art Surveys. Springer Science+Business Media, New York.
17. Gunningham, N., Grabosky, P., 1998. Smart Regulation: Designing Environmental Regulation. Oxford University Press, Oxford.
18. Gwinn, M., Vallyathan, V., 2006. Nanoparticles: health effects – pros and cons. Environmental Health Perspectives 114 (2), 1818–1825.
19. International Risk Governance Council (IRGC), 2005. White Paper on Risk Governance: Towards an Integrative Approach. By Ortwin Renn with Annexes by Peter Graham. CH-1219 Geneva, Switzerland. Available at: http://www.irgc.org/IMG/pdf/ IRGC_WP_No_1_Risk_Governance__reprinted_version_.pdf.
20. International Risk Governance Council (IRGC), 2006. White Paper on Nanotechnology Risk Governance. By Ortwin Renn and Mike Roco with Annexes by Mike Roco and

Emily Litten. CH-1219 Geneva, Switzerland. Available at: http://www.irgc.org/IMG/pdf/ IRGC_white_paper_2_PDF_final_version-2.pdf.

21. International Risk Governance Council (IRGC), 2007. Policy Brief: Nanotechnology Risk Governance: Recommendations for a Global, Coordinated Approach to the Governance of Potential Risks. CH-1219 Geneva, Switzerland. Available at: http://www. irgc.org/IMG/pdf/PB_nanoFINAL2_2_.pdf.

22. Kreyling, W., Semmler-Behnke, M., Möller, W., 2006. Health implications of nanoparticles. Journal of Nanomaterial Research 8, 543–562.

23. Lahdelma, R., Miettinen, K., Salminen, P., 2003. Ordinal criteria in stochastic multicriteria acceptability analysis (SMAA). European Journal of Operational Research 147 (1), 117–127.

24. Linkov, I., Satterstrom, F.K., Kiker, G., Seager, T.P., Bridges, T., Gardner, K.H., Rogers, S.H., Belluck, D.A., Meyer, A., 2006. Multicriteria decision analysis: a comprehensive decision approach for management of contaminated sediments. Risk Analysis 26 (1), 61–78.

25. Linkov, I., Satterstrom, K., Kiker, Batchelor, C., G., Bridges, T., 2006. From comparative risk assessment to multi-criteria decision analysis and adaptive management: recent developments and applications. Environment International 32, 1072–1093.

26. Linkov, I., Satterstrom, F.K., Steevens, J., Ferguson, E., Pleus, R.C., 2007. Multi-criteria decision analysis and environmental risk assessment for nanomaterials. Journal of Nanoparticle Research 9 (4), 543–554.

27. Lux Research, 2006. Taking Action on Nanotech Environmental, Health, and Safety Risks. Lux Research, New York.

28. Macoubrie, J., 2005. Informed Public Perceptions of Nanotechnology and Trust in Government. Woodrow Wilson International Center for Scholars Project on Emerging Nanotechnologies. Washington, DC 20004-3027. Available at: http://www.nanotech-project.org/process/files/2662/ informed_public_perceptions_of_nanotechnology_and_ trust_in_government.pdf.

29. Marchant, G.E., Sylvester, D.J., Abbott, K.W., 2008. A New Approach to Risk Management for Nanotechnology. Nanoethics 2, 43–60.

30. Medina, C., Santos-Martinez, M., Radomski, A., Corrigan, O., Radomski, M., 2007. Nanoparticles: pharmacological and toxicological significance. British Journal of Pharmacology 150, 552–558.

31. Moghimi, S.M., Hunter, A.C., Murray, J.C., 2005. Nanomedicine: current status and future prospects. FASEB J 19, 311–330.

32. Nel, A., Xia, T., Mädler, L., Li, N., 2006. Toxic potential of materials at the nanolevel. Science 311, 622–627.

33. National Nanotechnology Initiative (NNI), 2006. Environmental, Health, and Safety Research Needs for Engineered Nanoscale Materials. National Science and Technology Council, Committee on Technology, Subcommittee on Nanoscale Science, Engineering, and Technology. Available at: http://www.nano.gov/NNI_EHS_research_needs.pdf.

34. Oberdörster, G., Oberdörster, E., Oberdörster, J., 2005. Nanotoxicology: an emerging discipline evolving from studies of ultrafine particles. Environmental Health Perspectives 113, 823–839.

35. Québec Commission de l'éthique de la science et de la technologie (QC), 2006. Position Statement: Ethics and Nanotechnology: A Basis for Action. Québec G1V 4Z2. Available at: http://www.ethique.gouv.qc.ca/IMG/pdf/Avis-anglaisfinal-2.pdf.

36. Responsible NanoCode (RNC), 2006. Workshop report: How Can Business Respond to the Technical, Social and Commercial Uncertainties of Nanotechnology? Available at: http://www.responsiblenanocode.org/documents/Workshop-Report_07112006.pdf

37. Roco, M.C., 2008. Possibilities for global governance of converging technologies. Journal of Nanoparticle Research 10, 11–29.

38. Royal Society and Royal Academy of Engineering (RS & RAE), 2004. Nanoscience and Nanotechnologies: Opportunities and Uncertainties. Science Policy Section, The Royal Society, London SW1Y 5AG. Available at: http://www.nanotec.org.uk/finalReport.htm.

39. Seaton, A., Donaldson, K., 2005. Nanoscience, nanotoxicology, and the need to think small. Lancet 365, 923–924.

40. Shvedova, A.A., Kisin, E.R., Mercer, R., Murray, A.R., Johnson, V.J., Potapovich, A.I., et al., 2005. Unusual inflammatory and fibrogenic pulmonary responses to single-walled carbon nanotubes in mice. American Journal of Physiology. Lung Cellular and Molecular Physiology 289, L698–708.

41. Tervonen, T., Lahdelma, R., 2007. Implementing stochastic multicriteria acceptability analysis. European Journal of Operational Research 178 (2), 500–513.

42. Tervonen, T., Figueira, J., Steevens, J., Kim, J., Linkov, I., 2008 (in preparation). Risk-based Classification System of Nanomaterials. Journal of Nanoparticle Research.

43. Thomas, K., Sayre, P., 2005. Research strategies for safety evaluation of nanomaterials, part i: evaluating the human health implications of exposure to nanoscale materials. Toxicological Sciences 87 (2), 316–321.

44. UK Department for Environment, Food and Rural Affairs (UK DEFRA), 2006. UK Voluntary Reporting Scheme for Engineered Nanoscale Materials. London SW1P 3JR. Available at: http://www.defra.gov.uk/environment/nanotech/policy/pdf/vrs-nanoscale.pdf.

45. UK Department for Environment, Food and Rural Affairs (UK DEFRA), 2007. The UK Voluntary Reporting Scheme for Engineered Nanoscale Materials: Fifth Quarterly Report. London SW1P 3JR. Available at: http://www.defra.gov.uk/environment/nanotech/pdf/vrs-5.pdf.

46. US Environmental Protection Agency (USEPA), 1989. Risk Assessment Guidance for Superfund, Volume I, Human Health Evaluation Manual (Part A). Office of Emergency and Remedial Response, December, 1989. EPA/540/1-89/002.

47. US Environmental Protection Agency (USEPA), 1998. Guidelines for Ecological Risk Assessment. US EPA, Risk Assessment Forum, Washington, DC, EPA/630/R095/002F. Available at: http://cfpub.epa.gov/ncea/cfm/recordisplay.cfm?deid=12460.

48. US Environmental Protection Agency (USEPA), 2007. Nanotechnology White Paper. Prepared for the US EPA by Members of the Nanotechnology Workgroup, a Group of EPA's Science Policy Council, Washington, DC 20460. Available at: http://www.epa.gov/OSA/pdfs/nanotech/epa-nanotechnology-whitepaper-0207.pdf.

49. US Food and Drug Administration (US FDA), 2007. Nanotechnology: A Report of the U.S. Food and Drug Administration Nanotechnology Task Force, July 25, 2007. Available at: http://www.fda.gov/nanotechnology/taskforce/report2007.html.

50. US National Academy of Sciences (US NAS), 1983. Risk Assessment in the Federal Government: Managing the Process. National Academy Press, Washington, DC.

51. US National Science and Technology Council (US NSTC), 2007. Prioritization of Environmental, Health, and Safety Research Needs for Engineered Nanoscale Materials. NSTC Nanoscale Science, Engineering, and Technology (NSET) Subcommittee. Available at: http://www.nano.gov/Prioritization_EHS_Research_Needs_Engineered_Nanoscale_Materials.pdf.

52. Peters, K., Unger, R.E., Kirkpatrick, C.J., Gatti, A.M., Monari, E., 2004. Effects of nano-scaled particles on endothelial cell function in vitro: studies on viability, proliferation and inflammation: selected papers from the 18th European Conference on Biomaterials (ESB2003), Stuttgart, Germany, 2003 (guest editors: Michael Doser and Heinrich Planck). Journal of Materials Science: Materials in Medicine, 15 (4), 321–325.

ESTIMATION OF EFFECT THRESHOLDS
FOR THE DEVELOPMENT OF WATER QUALITY CRITERIA

S.M. CORMIER, P. SHAW-ALLEN

U.S. EPA Office of Research and Development
Cincinnati, OH, USA
cormier.susan@epa.gov

J.F. PAUL

U.S. EPA Office of Research and Development
Research Triangle Park, NC, USA

R.L. SPEHAR

U.S. EPA Office of Research and Development
Duluth, MN, USA

Abstract: Biological and ecological effect thresholds can be used for determining safe levels of nontraditional stressors. The U.S. EPA Framework for Developing Suspended and Bedded Sediments (SABS) Water Quality Criteria (WQC) [36] uses a risk assessment approach to estimate effect thresholds for unacceptable levels of SABS in water bodies. Sources of SABS include:

1. Erosion from agricultural, construction, forestry practices, and stream banks

2. Resuspension of deposited sediment

3. Direct discharge from municipal, industrial, and agricultural sources

Excessive levels of SABS can destroy habitat for plants and animals, reduce the quality of drinking water, impair the quality and safety of recreational waters, increase the costs associated with irrigation and navigation, and decrease aesthetics. The SABS Framework is intended as a guide to the development of water quality criteria (WQC) and restoration targets. The *SABS Framework* uses an eco-epidemiological perspective to incorporate information from field observations with data from controlled laboratory experiments. The combined information is used to develop relationships that estimate the levels of SABS that will impair aquatic life or pollute sources

I. Linkov et al. (eds.), Real-Time and Deliberative Decision Making.
© Springer Science + Business Media B.V. 2008

intended for drinking water. The *SABS Framework* uses several statistical procedures to compare the estimated effects levels derived from field and laboratory data. Protective levels and restoration goals are recommended based on scientific precedent, logical argument, and statistical resolution. The risk estimates that result from using this approach are readily applicable for use in future emergency situations.

1. Introduction

Any substance or agent has the potential to cause environmental harm. The detrimental effects of a limited number of substances are characterized in criteria documents and existing, completed risk assessments [42]. Based on these prior assessments, risk managers are able to develop possible actions for protecting and restoring environmental conditions. These actions can include controlling releases or limiting exposure to waste streams or other media. Proposed releases can also be evaluated to determine whether the actual releases are acceptable in the environment or if they need to be regulated in some way. If the substance to be released is well studied, assessors can adapt existing assessments to evaluate the new situation [13, 10, 27]. When the release is a mixture of known compounds or substances having similar properties and suspected modes of action, assessors can reapply stressor-response relationships found in existing assessments to address the new situation. Information and lessons learned from completed assessments can also contribute to the development of emergency response plans with standard operating procedures. Applying accumulated knowledge ensures an efficient, reliable reaction process that restricts the spread of a pollutant and reduces exposure or harm from the unexpected releases. This knowledge also helps the assessor and manager later, when evaluating the release, to select a remedial action that minimizes unacceptable exposures or harm from the release and from the remediation process itself.

Access to completed assessments and a mechanism for applying them to new situations are essential for emergency preparedness. For aquatic systems, this has been accomplished by agencies in the U.S. and other countries that have adopted criteria for the protection of drinking water sources, recreational waters, wildlife, and other designated uses [8, 14, 20, 42]. Regulations that require setting acceptable levels of pollutants and that require monitoring to ensure that designated uses are retained have been enormously successful in improving or maintaining water quality despite allowing permitted discharges [40]. However, many pollutants enter the waterways from overland flow or from unregulated discharges, also referred to as pollution. In the U.S., programs instituted to reduce damage from unregulated discharges of a wide range of physical, chemical, and biological agents include the U.S. Department of Agriculture's incentive programs and the U.S. EPA's nonpoint source program [32] and total maximum daily load (TMDL) program [34]. Guidance for addressing chemical agents with toxicological modes of action dates back to the early years of environmental protection but is still evolving. Guidance for determining acceptable levels of agents with physical and biological modes of action have only recently been developed and applied. One of the most recent is the *U.S. EPA Framework for the Development of Suspended and Bedded Sediments* (*SABS Framework*) [36].

The U.S. EPA specifically developed the *SABS Framework* for uncontaminated sediment; however, assessors can adapt the overall process to any stressor and thereby develop WQC or set restoration goals. The foundation for the development of WQC was originally limited to controlled laboratory toxicity tests using fish, invertebrate, and plant species [26]. More recently, the criteria values have been fine-tuned by interpreting causal relationships developed from toxicity tests in the context of body burdens and wildlife exposures [28–31, 33]. The *SABS Framework* recommends using these methods but also encourages assessors to use knowledge from causal associations developed from field studies.

This more inclusive approach retains laboratory-derived knowledge about exposure-response relationships that is independent from other influences while also evaluating more types of effects than are practicable in controlled laboratory experiments alone. Field studies can include routine seasonal biological surveys or observations of field manipulations, such as changes following restoration. Because interventions have already achieved environmental goals in other places, using stressor-response relationships observed from previous field manipulations increases confidence that criteria or restoration goals will protect and improve aquatic resources. When the agent is already in the environment, an adaptive management approach can use monitoring results to inform and improve the assessment and the resulting criteria or restoration goals.

In order to combine different types of knowledge to evaluate options for criteria values or restoration goals, the SABS Framework recommends comparing results from several analytical methods applied to different datasets and endpoints. This approach is outlined below and can be considered a general method for developing criteria to be protective and restorative for any environmental resource subject to the detrimental effects of an agent. Then an abbreviated, hypothetical example (the development of WQC for sediments deposited on moderately steep-gradient streambeds with a gravel or cobble substrate) illustrates key steps and shows how that process can be applied.

Although sediment is a natural part of aquatic habitats, sediment quantity and characteristics can affect the physical, chemical, and biological integrity of streams, lakes, rivers, estuaries, wetlands, and coastal waters [2, 3, 36, 38, 43, 44]. Suspended sediments can impair a wide range of water uses:

- Suspended sediments clog filters that are used to finish drinking water and often reduce water clarity, thereby interfering with recreational uses.

- Decreased water clarity impairs visibility and affects many animal behaviors such as prey capture and predator avoidance, recognition of reproductive cues, and other behaviors that alter reproduction and survival [17, 18].

- At very high levels, suspended sediments can cause physical abrasion and clogging of filtration and respiratory organs [1].

- Suspended particles also decrease light penetration required for photosynthesis.

Excessive levels of suspended and bedded sediment and in some circumstances insufficient levels of those sediments can cause deleterious effects [25]. When sediments are contaminated, the combination of physical effects of sediment and toxic effect of contaminants are evaluated as distinct but related causes. However, because the development of chemical criteria for contaminated sediment already have well developed methodologies and applications [37], this chapter deals with only the physical effects of excess depositions of both inorganic and organic sediment to a stream bed (deposited and bedded sediment).

Sources of deposited and bedded sediments are soils and topsoil from land in the watershed or suspended sediment removed from stream banks and from the bed of an upstream channel. Some soils, such as volcanic ash, are more susceptible to movement. Generally, smaller, lighter particles move more readily and are easily resuspended. Slope, stream gradient, channel morphology, and other natural factors affect stream flow and, therefore, the ability to move sediments. Changes in watershed land cover may increase watershed erosion by increasing overland flow and the susceptibility of soil to movement. For example, during construction, vegetation is removed and soils are compacted, reducing permeability and increasing overland flow that carries disturbed soils from uncompacted areas into waterways [25].

2. Methodology

The SABS Framework [36] is a form of ecological risk assessment described in seven steps [20 21]. These seven steps (Figure 1) can be condensed into three phases: a Planning Phase, an Analysis Phase, and a final Synthesis Phase [5].

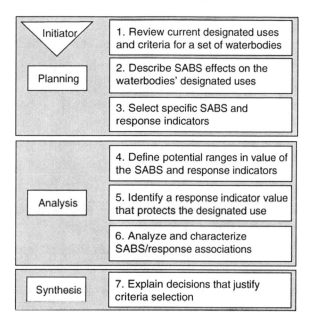

Figure 1. Phases of Assessment are Listed on the Left of the Seven Steps for Developing WQC [37].

The general process, as described here, primarily applies to the development of WQC but may also be considered a process to develop remediation goals.

Effect thresholds are selected based on scientific or legal precedent, stakeholder values, or other rationales. The effect threshold should protect the resource, retain its desired functions, and ensure safe conditions for wildlife and humans. The assessors should seek out readily available sources of information as well as datasets having the types of measurements that can be used to model stressor-response relationships. In some cases, new laboratory, field, or pilot studies may be necessary. Separate, independent studies are sought so that risk estimates can be compared and critiqued. For example, it is useful to compare results from different datasets, timeframes, or sub-samplings of datasets. The decisions of the planning phase are described in an analysis plan that guides the analysis phase. The plan should describe the objectives, datasets, and analytical approaches to be used. It should be appropriate for the environmental context of the assessment, the environmental value or use to be protected or remedied, the ecosystem type, and the measurements that represent the stressors and effects.

2.1. ANALYSIS PHASE

The objective of the Analysis Phase is to model the stressor-response relationship(s), develop an understanding of the mechanisms behind these relationships, and interpret their relevance to the environmental goals. To meet these goals, analysis results are used to answer questions like:

- What concentration of suspended sediment may occur without clogging filtration systems for a drinking water facility?
- What level of siltation can occur without adversely reducing fish spawning?
- When dredging a shipping channel, which timeframe would impose the least impact on commercially important species or their prey?

During the Analysis Phase, assessors:

1. Characterize the range and the relative acceptability of values for existing biological, environmental, and stressor conditions.
2. Quantitatively model the relationship between the stressor intensity and effects using data from laboratory studies or field observations.
3. Estimate candidate criterion values that are expected to protect against unacceptable conditions.

2.2. SYNTHESIS PHASE

In the Synthesis Phase, assessors compare the relationships developed from different datasets or study designs that result from the Analysis Phase with the effect thresholds that were identified in the Planning Phase. Decision makers can use the values of the stressor at the effect thresholds to determine acceptable levels for WQC or restoration goals.

3. Hypothetical Example

In this example, we develop WQC to regulate the amount of sediment deposited on moderately steep-gradient streambeds having a gravel or cobble substrate. The dataset used in this example is from the U.S. EPA Environmental Monitoring Assessment Program (EMAP) conducted in the Mid-Atlantic Highlands Assessment (MAHA) during the summers of 1993–1996 [39]. Data from laboratory tests were not included in this example because relevant test results were not found that could be used to estimate risks from deposited and bedded sediments.

3.1. PLANNING PHASE

In this example case, we reviewed several publications [1, 11, 43, 44] to study the effects of SABS on aquatic organisms. We used information from the reviews to develop a conceptual model that shows how SABS can affect invertebrate assemblages (Figure 2).

We considered four modes of action that lead to impaired invertebrate assemblages from increased levels of bedded sediment:

- Loss of suitable habitat
- Decreased dissolved oxygen
- Smothering
- Increased drift and predation

We developed deposited and bedded sediment criterion values for two levels of protection: aquatic life uses (ALU) and minimally acceptable aquatic life uses (MALU). We chose percent fines on the substrate as the bedded sediment metric because it is commonly used by many states. Also, good quality data were available, and acquisition protocols had been consistently applied across the entire dataset [35].

The metric of *Ephemeroptera*, *Plecoptera*, and *Trichoptera* (EPT) taxa richness was selected as the response measure because their diversity is a valued attribute and benthic aquatic invertebrates are prey for valued fish stocks [6, 12, 19, 23]. EPT taxa richness is strongly affected by sediment levels.

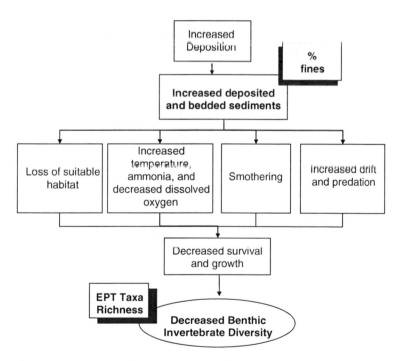

Figure 2. Conceptual Model of the Causal Relationship between Deposited and Bedded Sediments and Decreased Benthic Invertebrate Diversity.

It is accepted by regulatory agencies in most parts of the U.S., Canada, South Africa, New Zealand, Australia, Europe, and other places where it has become a commonly used metric within bioassessment indices that assess the condition of aquatic life [4, 15, 16]. Furthermore, data for EPT taxa richness were readily available for analysis and were judged to be of high quality, and the measures of EPT taxa richness could be compared with equally good quality measures of bedded sediment. Since the example is not based on actual state programs, there were no predefined biological criteria that quantitatively identified when aquatic life uses are not met. However, we did consider analyses from two independent datasets, one from the West Virginia Department of Natural Resources (DNR) that has a maximum of 15 EPT taxa at any site and another from the EMAP MAHA dataset with a maximum of 29 EPT [9]. The West Virginia DNR identifies 13 EPT taxa as meeting 100% use within its biocriteria index [9, 39]. Analyses of the EMAP MAHA dataset by Stoddard suggested characterization of condition based on ≤9 EPT taxa as poor, between 9 and 17 marginal, and ≥17 as good [7]. Because the EMAP MAHA data were used in this study and because that dataset had a greater observed maxima of EPT at sites, the values of ≤17 were applied to analyses of ALU and ≤9 EPT to MALU.

Biological effects thresholds for aquatic life uses were based on regulatory precedent, relative loss, and quantitative changepoints in stressor-response relationships (Table 1). Table 1 lists the type of evidence, the analytical method, and the risk estimation method.

TABLE 1. Example Candidate Thresholds of Biological Effect as Used in Hypothetical Example for SABS.

Basis	Evidence	Analytical method	Risk estimation method
Precedent [30]	SABS level for a proportion of streams with a given level of EPT taxa	Percentile	75% of streams ≥17 EPT taxa
Precedent [30]		Percentile	75% of streams ≥9 EPT taxa
Precedent [26]	Proportion of species affected	Species sensitivity distribution	5% of species reduced by 20%
Relative loss	Maximum expected for a SABS level	Quantile regression 90% level	5% reduction from y Intercept
Relative loss	Commonly achieved (mean) for a SABS level	Linear regression	20% reduction from y intercept
Changepoints	Statistical difference in slope (deviance reduction)	Conditional probability analysis	Change in slope from zero to >0

Three of the effect thresholds were based on current regulatory precedent; that is, threshold estimation methods that have been accepted and used by the U.S. EPA for criteria development. The percentile method is simply the SABS level measured at a stream that represents the 75th percentile of streams with an acceptable biological condition and was originally developed to derive WQC for nutrients [30]. In the hypothetical example, two effect thresholds were calculated using the percentile method: one for better quality (ALU) and one for fair quality (MALU) of biological conditions. Another method supported by precedent, species sensitivity distribution (SSD), has been used extensively for WQC for chemicals [26]. We developed a cumulative SSD for aquatic species based on field studies and calculated the level of SABS at which the 5th percentile of species are estimated to show a 20% reduction of abundance as observed in the data set for MAHA streams. This derivation used field associations and departs from the method of Stephan et al. [26], which uses laboratory toxicity tests to derive SSDs.

Biological effect thresholds that compared relative losses of species richness were calculated using linear and quantile regression methods. A 5% change was selected as a loss likely to be within a range of natural variation from forested areas (mean loss and maximum expected loss) and was applied to the ALU evaluation of the linear and quantile regression models. The effect threshold for MALU was set at 20% loss from currently attained conditions (mean and expected maximum).

Changepoints derived from conditional probability analysis (CCPA) plots were used to estimate when the probability of observing ≤17 for ALU and ≤9 for MALU began to increase. The changepoint was determined either from a change in slope of zero to a strong, positive slope (visually derived) or from a change that could be statistically detected.

3.2. ANALYSIS PHASE

3.2.1. Characterizing Biological and Exposure Conditions

Using the methods and thresholds chosen in the Planning Phase, we calculated the thresholds and analyzed the MAHA data to evaluate whether more than one criterion was necessary for different sizes of streams and stream types (Comment 3). Most values ranged from 0–36% fines for sites with >9 EPT taxa (Figure 3) and 0–10% fines for sites with >17 EPT taxa.

EPT taxa-richness values were also similar for drainage areas including those greater than $30\,km^2$ (Figure 4).

Therefore, we judged that sites could be grouped for the three drainage classes: <5, >5 <30, and $>30\,km^2$. The range of values for heavily forested areas was from 0–50% fines compared to 0–100% fines when all sites were included (Figure 5).

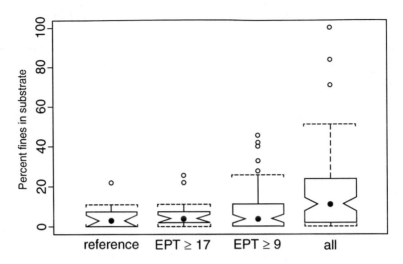

Figure 3. Notched Box and Whisker Plot. Reference Sites are Based on Land Use and Water Chemistry Parameters. The 75th Percentile for EPT >17 is 9.2% Fines and for EPT> 9 is 12.6% Fines.

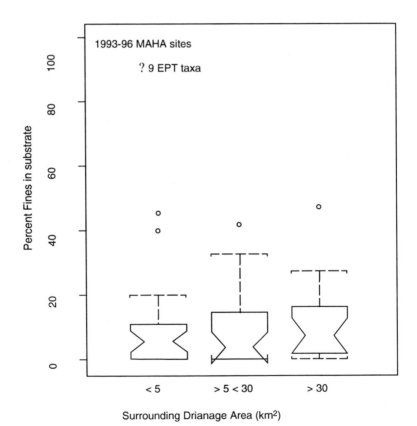

Surrounding Drianage Area (km²)

Figure 4. Notched Box and Whisker Plot of Percent Fines for Three Classes of Streams Based on Drainage Area for Lypothetical Example.

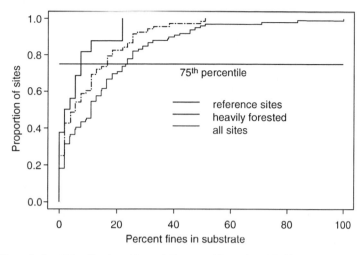

Figure 5. Cumulative Distribution Plot of Percent Fines for All Sites, Reference Sites, and Heavily Forested Sites [20, 21].

In developed but less stressed systems, we do not expect to see many sites with values in the upper range. The difference between observed amounts of percent fines in heavily forested areas and other areas suggests that both ALU and MALU criteria are necessary for a comprehensive management strategy; that is, distinct criteria for intact ecosystems and developed areas, which may need to set achievable restoration targets.

3.2.2. Develop Stressor-Response Models

As mentioned previously, we considered reviews [1, 11, 43, 44] for biological effects to invertebrates, fish, and plants from settled particles and bedded sediments. However, we could not find any suitable papers in these reviews or other published papers that quantitatively modeled for EPT taxa richness and were relevant to the MAHA data set. Therefore, we developed several stressor-response models to determine if bedded sediments were great enough to account for reductions of EPT taxa richness in streams of the mid-Atlantic and to estimate effect thresholds (Tables 1 and 2).

We estimated the proportion of streams that were affected by different levels of percent fines using the percentile method. Values were determined from cumulative distribution plots but could also have been estimated from box plots (Figure 3). The fraction of total streams was plotted against percent fines for sites with EPT taxa scores >17 and >9, and the level of percent fines at the 75th percentile of EPT taxa was determined (Figure 3). The effect threshold for ALU was 9.2% fines and for MALU was 12.6% fines.

TABLE 2. Evidence, Methods, Risk Estimation Methods for Developing Effect Levels Using Different Analytical Methods.

Evidence	Analytical method	Risk estimation method	% fines effect level	Risk estimation method	% fines effect level
		ALU		MALU	
Proportion of streams	Percentile	75th percentile	9.2	75th percentile	12.6%
Proportion of species affected	Species sensitivity distribution	5th percentile	7	Not selected	—
Maximum achievable	Quantile regression, 90% percentile	5 and 10%	5.8 and 11.5	15, 20 and 25%	17.3, 23.0, 28.8
Commonly achieved	Linear regression	5 and 10%	3.9 and 7.9	15, 20 and 25%	11.8, 15.7, and 19.7
Changepoint analysis	Conditional probability analysis	Deviance reduction	8.2	Deviance reduction	10.1

We constructed an SSD to estimate the level of percent fines that could occur while still being protective of 95% of invertebrate species observed in MAHA streams [24]. We obtained the estimate from the cumulative distribution function of effect levels of species observed in MAHA streams (Figure 6).

The effect level was the value of percent fines at which each taxon's abundance was reduced by 20%. The maximum abundance was taken from quantile regression plots that modeled the 90th percentile of the relative abundance of several species of invertebrates [24]. The effect threshold for ALU was 7% fines. There was no precedent of a threshold for MALU; therefore, no effect level was estimated.

We determined the number of EPT taxa that were commonly observed at stream sites with different levels of percent fines by plotting the number of EPT taxa observed against percent fines and modeled using least squares linear regression analysis. We modeled the expected maximum number of EPT taxa that were likely to be observed at a site with different levels of percent fines using the 90th percentile from a quantile regression. We estimated the number of EPT taxa commonly encountered for a given SABS level from the linear regression curve. The amount of sediment associated with 5, 10, 15, 20, and 25% reduction from the y-intercept was determined from the 90th quantile and linear regression curves (Figure 7).

For ALU, a 5% reduction from the number of EPT taxa commonly observed was estimated to occur at 3.9% fines. A 5% reduction from the maximum

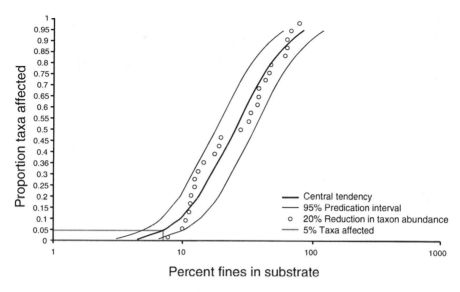

Figure 6. SSD Plots. The Abundance of 5% of the Species are Reduced by 20% at 7% Fines.

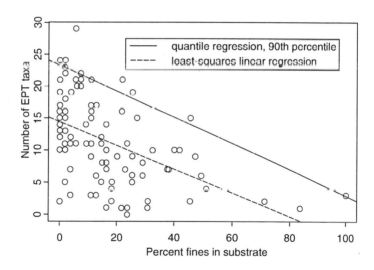

Figure 7. Scatter Wlot with 90th Percentile Quantile Regression (solid line) and Least-Squares Regression (Dashed Line).

number of EPT taxa was estimated to occur at 5.8% fines. For MALU, the 20% reduction from the number of EPT taxa commonly observed was estimated to occur at 15.7% fines. Also for MALU, a 20% reduction from the maximum number of EPT taxa was estimated to occur at 23% fines.

We used CCPA to estimate the probability of observing <17 and <9 EPT taxa richness for observed levels of percent fines. For ALU, the conditional probabilities for observing <17 EPT had a slope of zero from 0–7% fines (Figure 8).

From deviance reduction analysis, the changepoint occurred at 8.2% fines. For MALU, the slope of the probabilities of observing <9 sharply increased from 0% to about 17% fines; a statistically distinct difference was determined at 10.1% fines (Figure 9). Note that the point at the far left of Figures 8 and 9 represents the probability for observing <17 or <9 EPA for the entire range of percent fines (0–50% fines) and not the probability of observing <17 or <9 EPT at zero percent fines.

3.3. SYNTHESIS PHASE

3.3.1. *Compare Risk Estimates*

The recommendation for criterion values for the hypothetical case includes:

- **Aquatic life use (ALU)—criterion of no more than 7% fines.** This criterion is similar to existing precedents. Based on the proportion of species affected, 75% of sites with >17 EPT had an effect threshold at 9.2% fines. According to the results of the SSD analysis, 95% of EPT taxa would be protected most of the time when levels remained below 7% fines. There was an estimated

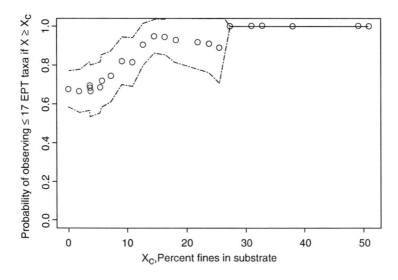

Figure 8. CCPA Plot of Probabilities of Observing <17 EPT Taxa for Different Levels of Percent Fines. Confidence Intervals Indicated by Hashed Lines.

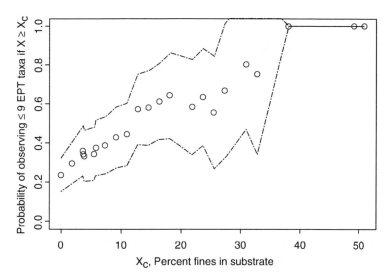

Figure 9. CCPA Plot of Probabilities of Observing <9 EPT Taxa for Different Levels of Percent Fines. Confidence Intervals Indicated by Hashed Lines.

loss of 5% from the maximum attainable number of EPT taxa at 5.8% fines and a 5% reduction for EPT values commonly observed at 3.9% fines. Furthermore, there was an increased probability of observing <17 EPT taxa at sites above 8.2% fines. Values for all methods were from 3.9–9.2% fines. A value of 7% fines was judged to be protective of the resource and conform to the most protective precedent, which was from the SSD method.

- **Minimally acceptable aquatic life use (MALU)—criterion of no more than 15% fines.** The minimal marginal conditions (<9 EPT taxa) were observed in 75% of the streams below 12.6% fines. There was an estimated loss of 20% from the maximum attainable numbers of EPT taxa at 23% fines, and a 20% reduction was estimated to be commonly observed at 15.7% fines. Furthermore, there was an increasing probability of observing <9 EPT taxa at sites from 0–17% fines and a statistically significant change-point at 10.1% fines. Values ranged from 10.1–23.0% fines. The mean effect threshold of all methods was 15% fines.

Table 3 summarizes the estimated effect level for percent fines based on the several methods evaluated for aquatic life use and marginal aquatic life use.

No criteria were developed for regulatory use in this case because this is a hypothetical example. Although real data were used in the examples, the resulting "criteria" should not be construed as a rigorous recommendation. Moreover, the criteria values were derived for bedded sediments and using only benthic invertebrates (EPT taxa richness), which is not likely to

TABLE 3. Summary of Effect Levels of Percent Fines Based on Five Analytical Methods.

Method	ALU	MALU
Percentile	7	12
SSD	7	
Quantile regression	5.8–11.5	17.3–28.8
Linear regression	3.9–7.9	11.8–19.7
Change point-conditional probability	8.2	10.1
Hypothetical candidate criteria values	7% fines	15% fines

be protective of overall designated use, even for Mid-Atlantic high-gradient streams. Additional assessment endpoints (e.g., coldwater fish production) would need to be considered along with EPT taxa richness and potentially other response measures to be confident that the criteria would be protective of all desired designated uses. Also downstream effects from transported sediment were not evaluated in a full risk assessment before selecting criteria.

4. Discussion

The SABS Framework provides a scientifically defensible approach for identifying effect thresholds that is useful for nontraditional modes of action and risks. Because the approach compares results from several analytical methods, there may be greater confidence in the decision, and the expectations of potential outcomes from actions may be more realistic. Also, for nontraditional stressors, statutory and legal precedents have not been time-tested and knowledge from several corroborating methods strengthens an assessor's credibility and the resulting decision.

The percentile method has precedent for nutrients, a nontraditional stressor [30]. The precedents for SSDs are strongly supported by legal and statutory precedent; however, the precedent is based on controlled laboratory toxicity tests while the analyses described here were based on a novel application using field observations [24]. As such, the precedent of the 5th percentile is reasonable and informative but not a precedent that has been fully reviewed by either the scientific community or the courts. Likewise, estimates based on relative loss were not grounded in legal precedent but do provide reasonably objective technical information for evaluating the impact of selecting different levels of percent fines as criteria. The values based on deviations from maxima and median values were comparable with other estimates. The statistically based changepoint analysis is objective and repeatable, but there is no known legal or statutory precedent for its use.

Both rapid and deliberative decision making can be informed by predeveloped risk estimates of known or commonly occurring stressors. When there is an emergency threatening life, property, or irreplaceable natural resources, assessors can expeditiously use available criteria, and the stressor-response models on which they are based, to estimate immediate effects and continued risks as management actions attempt to control deleterious effects. When time is not crucial, a slower, more deliberative gathering of information to support decision making is possible and preferred. This process can accommodate time to find and assure the quality of datasets, seek published stressor-response models, and even implement new data collection and analysis. This is the approach illustrated in this chapter. However, we recognize that the selected criteria could also include thresholds for total loss of the resource, which could be valuable for emergency situations. Also, the stressor-response models could be quickly reanalyzed for other purposes that might not be recognized until the situation arises. Therefore, it is good scientific practice to make stressor-response models and datasets open to others rather than to simply publish final values.

Most existing risk estimates assess exposures to single chemicals [26]. However, wildlife can be harmed by nontraditional stressors for which most toxicity test methods are not suitable. The SABS Framework was developed for determining effect thresholds for an agent with a mode of action that causes physical abrasion, reduction in water transparency, burial, and alteration of substrates that make them unsuitable habitats for aquatic life. Laboratory toxicity tests are not capable of evaluating these modes of action. Therefore, the SABS Framework combines techniques using toxicity tests developed by the U.S. EPA's WQC program along with an expanded repertoire of analytical tools and approaches. By using different datasets, different endpoints, and different analytical methods, systematic biases, which might have been overlooked, can be qualitatively evaluated in the synthesis phase. Overall, this approach ensures that credible scientific input will inform decision making that is more likely to protect the environment and the functions it provides to protect all life.

Acknowledgements

The authors are indebted to the many federal and state scientists who helped to develop methods for developing stressor-response associations and guidance for developing WQC. The chapter was greatly improved by editorial suggestions from Christopher Broyles and Michael Griffith. The research described in this paper was funded by the U.S. EPA (the Agency).

This paper has not been subjected to Agency review; therefore, it does not necessarily reflect the views of the Agency. Mention of trade names or commercial products does not constitute endorsement or recommendation for use.

References

1. Berry, W., N. Rubinstein, B. Melzian, and B. Hill. 2003. The Biological Effects of Suspended and Bedded Sediment (SABS) in Aquatic Systems: A Review. Internal Report of the U.S. EPA Office of Research and Development, Narragansett, RI. Available at: http://www.epa.gov/waterscience/criteria/sediment/appendix1.pdf.
2. Caux, P.Y., D.R.J. Moore, and D. MacDonald. 1997a. Ambient Water Quality Guidelines (Criteria) for Turbidity, Suspended and Benthic Sediments: Technical Appendix. Prepared for BC Ministry of Environment, Land and Parks (now called Ministry of Water, Land and Air Protection). April 1997. Available at: http://www.env.gov.bc.ca/wat/.
3. Caux, P.Y., D.R.J. Moore, and D. MacDonald. 1997b. Sampling Strategy for Turbidity, Suspended and Benthic Sediments: Technical Appendix Addendum. Prepared for BC Ministry of Environment, Lands and Parks (now called Ministry of Water, Land and Air Protection). April 1997. Available at: http://www.env.gov.bc.ca/wat/.
4. Cormier, S. M. and J. J. Messer. 2004. Opportunities and challenges in surface water quality monitoring. In Environmental Monitoring, G. Bruce Wiersma, ed., pp. 217–238, Boca Raton, FL: Lewis.
5. Cormier, S. M. and G. W. Suter II 2008. A framework for fully integrating environmental assessment. Environmental Management, 4(4).
6. Davis, W. S. 1995. Biological Assessment and Criteria: Building on the Past. In Biological Assessment and Criteria, W. S. Davis and T. P. Simon, eds., pp. 7–14, Boca Raton, FL: Lewis.
7. Davis, W. and J. Scott. 2000. Mid-Atlantic Highlands Streams Assessment: Technical Support Document. EPA/903/B-00/004. Mid-Atlantic Integrated Assessment Program, Region 3, U.S. Environmental Protection Agency, Ft. Meade, MD.
8. Environment Canada. 2004. Canadian Water Quality Guidelines. Available at: http://www.ec.gc.ca/CEQG-RCQE/English/Ceqg/Water/default.cfm.
9. Gerritsen, J, J. Burton, and M. T. Barbour. 2000. A stream condition index for West Virginia wadeable streams. Prepared for U.S. EPA Office of Water, U.S. EPA Region 3, and West Virginia Department of Environmental Quality.
10. Jardine, C., S. Hrudey, J. Shortreed, L. Craig, D. Krewski, C. Furgal, and S. McColl. 2003. Risk management frameworks for human health and environmental risks. Journal of Toxicology and Environmental Health B, 6:569–641.
11. Jha, M. and W. Swietlik. 2003. Ecological and Toxicological Effects of Suspended and Bedded Sediments on Aquatic Habitats - A Concise Review for Developing Water Quality Criteria for Suspended and Bedded Sediments (SABS). U.S. EPA, Office of Water draft report, August, 2003.
12. Klemm, D. J., K. A. Blocksom, W. T. Thoeny, F. A. Fulk, A. T. Herlihy, P. R. Kaufmann, and S. M. Cormier. 2002. Methods development and use of macroinvertebrates as indicators of ecological conditions for streams in the Mid-Atlantic Highlands region. Environmental Monitoring and Assessment 78(2):169–212.
13. Linkov, I., F. K. Satterstrom, G. Kiker, T. P. Seager, T. Bridges, K. H. Gardner, S. H. Rogers, D. A. Belluck, and A. Meyer. 2006. Multicriteria decision analysis: a comprehensive decision approach for management of contaminated sediments. Risk Analysis 26:61–78.

14. Marchant, R., F. Wells, and P. Newall. 2000. Assessment of an ecoregion approach for classifying macroinvertebrate assemblages from streams in Victoria, Australia.. Journal of the North American Benthological Society 19:497–500.

15. Maxted, J., B. Evans, and M. R. Scarsbrook. 2005. Development of macroinvertebrate protocols for soft-bottomed streams in New Zealand. Journal of Marine and Freshwater Research 37:793–807.

16. Metcalfe-Smith, J. 1994. Biological water-quality assessment of rivers: Use of macroinvertebrate communities. In The Rivers Handbook, Hydrological and Ecological Principles, P. Calow and G. Petts, eds., pp. 144–170, Cambridge, MA: Blackwell Science.

17. Newcombe, C. P. 2003. Impact assessment model for clear water fishes exposed to excessively cloudy water. Journal of the American Water Resources Association 39:529–544.

18. Newcombe, C.P. and J. O. T. Jensen. 1996. Channel suspended sediment and fisheries: a synthesis for quantitative assessment of risk and impact. North American Journal of Fisheries Management 16:693–727.

19. Ohio Environmental Protection Agency. 1987. Biological Criteria for the Protection of Aquatic Life: Volume II: Users Manual for Biological Assessment of Ohio Surface Waters. Division of Water Quality Planning and Assessment, Ecological Assessment Section, Columbus, OH, WQMA-SWS-6.

20. Organisation for Economic Co-operation and Development. 2007. Homepage. Guidance on Hazards to the Aquatic Environment: Proposal for revision of Annex 9 (A9.1-A9.3 and Appendix VI) accessed April 2008. Available at: http://www.oecd.org/data oecd/44/24/39638556.doc.

21. Paul, J. F., S. M. Cormier, W. Berry, P. Kaufmann, R. Spehar, D. Norton, R. Cantilli, R. Stevens, W. Swietlik, and B. Jessup. 2008. Developing water quality criteria for suspended and bedded sediments. Water Practices 2:2–17.

22. Paul, J. F., S. M. Cormier, W. Berry, et al. 2007. Developing water quality criteria for suspended and bedded sediments - illustrative example application. Water Environment Federation TMDL 2007 Conference, Bellevue, Washington, Water Environment Federation.

23. Plafkin, J. L., M. T. Barbour, and K. D. Porter. 1989. Rapid Bioassessment Protocols for Use in Rivers and Streams: Benthic Macroinvertebrates and Fish. Office of Water Regulations and Standards, Washington, DC, EPA-440-4-89-001.

24. Shaw-Allen, P., M. Griffith, S. Niemela, J. Chirhart, and S. Cormier. 2006. Using biological survey data to develop sensitivity distributions captures exposures and effects in complex environments. Society For Environmental Toxicology and Chemistry, Montreal, Canada, November, 5–9, 2006.

25. Spehar, R., S. M. Cormier, D. L Taylor. 2007. Candidate Causes. Sediments. In Causal Analysis, Diagnosis Decision Information System, Available at: www.epa.gov/caddis

26. Stephan, C. E., D. I. Mount, D. J. Hansen, J. H. Gentile, G. A. Chapman, and W. A. Brungs. 1985. Guidelines for deriving numerical national water quality criteria for the protection of aquatic organisms and their uses. PB 85-227049. National Technical Information Services, Springfield, VA.

27. Suter, G. 2007. Ecological Risk Assessment. CRC Press. Taylor and Francis Group, Boca Raton, FL. EPA 1980. Water Quality Criteria Documents; Availability. Guidelines for deriving numerical national water quality criteria for the protection of aquatic organisms and their uses. Appendix B. Fed. Reg. 45, No. 231.

28. U.S. EPA. 1994. Interim guidance on determination and use of water-effect ratio for metals. EPA-823-B-94-001. Office of Water/Office of Science and Technology. Washington, DC.

29. U.S. EPA. 2000a. Ambient aquatic life water quality criteria for dissolved oxygen (salt water) Cape Cod to Cape Hatteras. EPA-822-R-00-012. Office of Water, Office of Science and Technology, Washington, DC and Office of Research and Development, National Environmental Effects Research Laboratory, Atlantic Ecology Division, Narragansett, RI.

30. U.S EPA. 2000b. Nutrient Criteria Technical Guidance Manual for Rivers and Streams (Nutrient Guidance) EPA–822–B–00–002, 256 pages. Available at: http://www.epa.gov/waterscience/criteria/nutrient/guidance/rivers/index.html

31. U.S. EPA. 2003a. Ambient water quality criteria for dissolved oxygen, water clarity, and chlorophyll a for the Chesapeake Bay and its Tidal Tributaries. Region III, Chesapeake Bay Program, Annapolis MD, Region III, Water Protection Division, Philadelphia PA Office of Water, Office of Science and Technology, Washington, DC.
32. U.S. EPA. 2003b. Non-point Source Program and Grants Guidelines for States and Territories. Fed. Reg. 68, No. 205:60653–60674.
33. U.S. EPA. 2004a. Notice of Draft Aquatic Life Criteria for Selenium and Request for Scientific Information, Data, and Views, W-FRL-7849-4. Fed. Reg.: December 17, 2004, 69(242):75541–75546.
34. U.S. EPA. 2004b.Total Maximum Daily Loads: National Section 303(d) List Fact Sheet. U.S. EPA Office of Water. Available at: http://oaspub.epa.gov/waters/national_rept.control#TOP_IMP.
35. U.S. EPA. 2005. Use of Biological Information to Better Define Designated Aquatic Life Uses in State and Tribal Water Quality Standards: Tiered Aquatic Life Uses. U.S. EPA, Washington, DC, EPA-822-R-05-001.
36. U.S. EPA. 2006. U.S. Environmental Protection Agency. 2006. Framework for Developing Suspended and Bedded Sediments (SABS) Water Quality Criteria, U.S. EPA, Washington, DC, EPA-822-R-06-001, p. 150, May.
37. U.S. EPA. 2006b. Contaminated Sediment in Water. Available at: http://epa.gov/water-science/cs/.
38. U.S. EPA. 2007a. Causal Analysis, Diagnosis Decision Information System. Available at: www.epa.gov/caddis.
39. U.S. EPA 2007b. Environmental Monitoring and Assessment Program (EMAP). Available at: www.epa.gov/emap/html/data.html.
40. U.S. EPA. 2007c. Biocriteria. Available at: http://www.epa.gov/waterscience/biocriteria/.
41. U.S. EPA. 2008a. Contaminated Sediment in Water. Available at: http://epa.gov/water-science/cs/.
42. U.S. EPA. 2008b. Water Science. Available at: http://www.epa.gov/waterscience/.
43. Waters, T. F. 1995. Sediment in streams- sources, biological effects and control. American Fisheries Society Monograph 7. American Fisheries Society, Bethesda, MD.
44. Wood, P. J., and P. D. Armitage. 1997. Biological effects of fine sediment in the lotic environment. Environmental Management 21(2):203–217.

COMPREHENSIVE RISK ASSESSMENT

Applying the Cultural Property Risk Analysis Model to the Canadian Museum of Nature

R. WALLER

Chief, Conservation
Canadian Museum of Nature
Ottawa, Canada
rwaller@mus-nature.ca

Abstract: Comprehensive environmental assessments of risks to the collections of the Canadian Museum of Nature were completed in 1993, 1998, and 2003. The assessments are based on comprehensive identification of specific risks within a framework of sources of hazards, called agents of deterioration, and expected frequency of risk events, ranging from continuous to less than one event per century. Between these assessments, numerous projects were undertaken to mitigate risks to collections. These activities have resulted in a significant net reduction in total risk to collections but not all changes in assessed risks relate to changes in actual risk.
Comparison of results among the three risk assessments indicates that differences result from:

- Changes in perception of risks

- Changes resulting from improved understanding of, or ability to quantify, risks

- Changes to magnitudes of specific risks as a result of risk treatments

In addition to enabling priority setting for further collection care and conservation research activities, repeated risk assessment has greatly increased staff, management and governance awareness of collection care issues and of changes in risks to collections over time. The results allow estimation of the benefits of proposed risk treatments and of the expected benefit of further risk characterization.

1. Introduction

The Canadian Museum of Nature (CMN) completed comprehensive assessments of risks to the collections in 1993, 1998, and 2003, using a method now termed the Cultural Property Risk Analysis Method (CPRAM) [7–9]. During

I. Linkov et al. (eds.), Real-Time and Deliberative Decision Making.   179

Figure 1. The CMN's Natural Heritage Building (NHB), 1997. This purpose-built collection holding facility is a tangible result of collection risk analysis. (Photo and Copyright: Martin Lipman.)

the five years between the first two assessments the CMN designed, constructed, and moved into a purpose-built collection-holding institution (Figure 1). In addition, numerous collection management and conservation projects were undertaken to mitigate risks to collections. These activities resulted in both total risk reduction and improved understanding of remaining risks.

Maintaining collections requires that potential risks to collections be considered comprehensively [9]. The requirement for a comprehensive assessment leads to this being an environmental risk assessment in the sense of considering the whole environment affecting collections. Once identified, risks need to be evaluated rationally. Comprehensiveness, clarity in purpose and scope, and rationality (minimally semi-quantitative and preferably quantitative) are characteristics of any good risk assessment method [2, 3]. In this paper, special attention is given to identification of risks and to lessons learned from repeated risk assessments.

2. Cultural Property Risk Analysis Model

The basic steps involved in the CMN's CPRAM are:

- Define project scope, including—for example—collection contents and values considered.
- Divide overall institution collection holdings into units for assessment.

- Identify specific risks to assess.
- Quantify risks.
- Analyze and present results.
- Plan collection care projects.
- Refine estimates of uncertain risks through research.

3. Scope and Assessment Unit Divisions

Collections were defined as being all, and only, formally accessioned objects. Excluded from the scope was material in temporary custody for research, consignment, etc., and material for consumptive use (Category 5 material within the value classification system of Price and Fitzgerald [5]). The period of time over which risks were projected was one century. This is an arbitrary choice, but one that is appropriate in a museum collection context for several reasons. Most simply, large museums have existed for one or two centuries. One century is about three curatorial career spans and is an easily conceptualized timeframe for collection care professionals. Finally, planning to deliver collections with minimal expected losses (risk) to a time 100 years in the future is equivalent to assuming a discount rate of about 1%, which is appropriate for protecting a property that is highly valued for the public good.

Overall collection holdings were divided into 19 collection units according to a range of criteria, including administration, nature of specimen material, primary storage hardware, and storage environment.

4. Risk Identification

Risks were comprehensively identified within a framework of ten sources of risk, called "agents of deterioration" in the museum sector, and three types of risk. The agents of deterioration [6] are:

- Physical forces
- Fire
- Water
- Criminals
- Pests
- Contaminants

- Light and ultraviolet radiation and electromagnetic fields
- Incorrect temperature
- Incorrect relative humidity
- Dissociation

This set of agents has been shown through many years of application to be comprehensive in incorporating all sources of risk (e.g., hazards, threats) to museum collections. Other groupings of sources of risk could be used. An essential characteristic of an acceptable "sources of risk" framework is that it be comprehensive. Desirable features of a framework include minimal ambiguity and minimal requirement for arbitrary decisions about where a specific risk belongs. Due to multiple dependencies of expected losses, some arbitrary assignments of specific risks to these categorical sources of risk are inevitable. For example, the embrittlement of cellulose and cellulose ester films is strongly dependent on temperature, relative humidity, and contaminants. The eventual crumbling of these embrittled materials will be the result of a physical force-related risk. Eventually a museum risk assessment may be able to keep track of mutual interdependencies and prorate the risk appropriately among categorical sources of risk. At present, however, an institution will choose one of these categorical sources of risk within which it will identify and evaluate the risk.

Because most agents of deterioration can manifest over a wide range of frequency and severity, three types of risk are distinguished. These range from Type 1, rare and catastrophic events, to Type 3, constant but persistent processes (Figure 2). Recognizing different types of risks facilitates both identifying risks comprehensively and finding sources of authoritative information on hazards and risks.

Frequency ⟍ Intensity	Continual	Sporadic	Rare
Catastrophic		·······Type 1·······	
Severe		·······Type 2·······	
Gradual/Mild	·······Type 3·······		

Figure 2. Three types of risk range in frequency and severity.

Risk identification is as much, or more, art than science. Within CPRAM a combination of source of hazard and type of risk, such as physical forces—Type 2—is termed a generic risk. Within each generic risk, a number of specific risks are defined to reflect more particular sources of risk or vulnerabilities of cultural properties. This hierarchical approach of describing sets of specific risks within each generic risk enables comprehensive risk identification while minimizing double-counting of risks. Using comprehensiveness as the dominant goal in initial stages of risk identification, brainstorming with diverse groups of stakeholders followed by inventive thinking about how framework-structured checklists can be completed has proven most useful for museum collection risk assessment as it has for nature preserves [4].

Table 1 illustrates, with examples of risks to CMN collections, how a source of risk is combined with a type of risk to arrive at a "generic risk" and then how a clear scenario is described to establish a "specific risk."

The goal in risk identification is to identify enough of the most significant specific risks within each generic risk to be confident that most (perhaps 90% or more) of the total magnitude of the generic risk is captured. In the case of Type 3 risks—and to a lesser extent Type 2 risks—this can be achieved with some confidence. This concept is shown in Figure 3, where the non-shaded portions of each rectangle represent suspected portions of each generic risk that have not been identified.

Combining the suspected proportion of unidentified risks with the estimated magnitude of risk gives a rough estimate of the magnitude of risk being overlooked, and hence the importance of investing in more thorough

TABLE 1. Selected examples of specific risks within the three generic risks resulting from the "physical forces" source of risk.

Generic risk	Specific risk
Physical forces—Type 1	Earthquake causing building collapse resulting in breakage or crushing of collection objects
	Earthquake causing toppling of storage units or objects resulting in breakage or crushing of collection objects
	Snow loading causing roof collapse resulting in breakage or crushing of collection objects
Physical forces—Type 2	Accidental physical damage to collection objects during use
Physical forces—Type 3	Poor support causing distortion of collection objects
	Overcrowded storage causing abrasion, breakage, etc. to collection objects

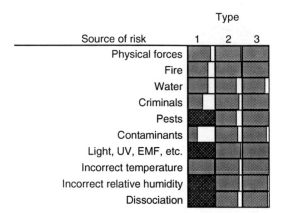

Figure 3. Gray area of each rectangle depicts suspected proportion of actual magnitude of generic risks represented by identified and assessed specific risks. Darkest shaded rectangles reflect implausible combinations of source and type of risk.

risk identification. For many Type 1 risks, uncertainty of risk identification remains a major challenge. In practice, one continues to identify and roughly estimate specific risks within a generic risk until it is clear that additional risks have magnitudes much less than 10% of the highest specific risk within that generic risk.

Following comprehensive risk identification, several cycles of qualitative or semi-quantitative screening are conducted. In the first cycle, risks judged as irrelevant or implausible are noted and set aside. For example, snow loading causing roof collapse, a significant risk for flat-roofed buildings in Canada, would be excluded from further consideration for an assessment of a museum in Lisbon. Risks that are considered potentially significant are then quantified.

5. Quantify Risks

The Magnitude of Risk (MR) was defined as the expected loss in value of the collection over the next 100 years, considering other factors such as collection growth, use, societal value changes, and so on, to be constant over that time. The use of ratio scales with clearly defined upper and lower endpoints allowed the mathematical operations of addition and multiplication to be properly applied [1]. In addition, ratio scales enable a precautionary approach through conservative estimation of probable upper bounds for each risk variable. Although any number of ratio variables can be multiplied together, four variables were always employed in the determination of MR as shown in Eq. (1).

$$\textbf{MR} \text{ (Magnitude of Risk)} = \textbf{FS} \times \textbf{LV} \times \textbf{P} \times \textbf{E} \qquad (1)$$

Where

FS = Fraction Susceptible

LV = Loss in Value

P   = Probability

E   = Extent

Each of these variables is determined for every plausible combination of specific risk and collection unit. First, the Fraction (of the collection) Susceptible (FS) to the specific risk is determined. Next, considering objects characteristic of the FS, Loss in Value (LV) that could result from a worst-case occurrence of the risk is estimated. The product of FS × LV can be considered the maximum "theoretical" part of the collection value subject to loss from that specific risk. The Probability for Type 1 risks is the chance of at least one event of a specified severity occurring over the next century. It is determined with help from and in collaboration with appropriate national or international agencies and organizations. The Extent reflects the amount of the FS that will be affected, the degree to which the LV will be realized, or both. It is estimated by projecting the effect of one century of exposure to the current setting, collection care, and use circumstances. Simple multiplication of the four variables, which are all fractions between 0 and 1 inclusive, gives the Magnitude of Risk, which itself is a fraction between 0 and 1 inclusive. The Magnitude of Risk is the expected loss in utility value of the collection over the next century, assuming the current collection care situation to continue.

For most collection units, the magnitudes of risks range over many orders of magnitude, even though only those risks considered relevant and plausible enough to identify and estimate were evaluated. Overall results of the 1998 CMN collection risk assessment (Figure 4) demonstrate the complex relations of risks to collection units.

The total risk to CMN's collection holdings due to each generic risk (combination of agent of deterioration and type of risk) was estimated by summing for each generic risk across all collection units after normalizing to express risks as risk to the total (number of objects in the) CMN collections. Figure 5 shows a comparison of generic risks to total CMN holdings as assessed in 1993 and again in 1998. All but two generic risks were reduced, primarily by the building project but also by other collection management projects. The two generic risks that increased slightly were Water-2, anticipated sporadic leaks through the roof causing water damage to exposed objects, and RH-2, incorrect fluid preservative levels and concentrations. The Water-2 risk increased as a result of all collection

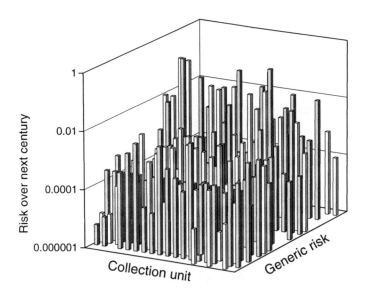

Figure 4. Magnitudes of 22 generic risks affecting 19 collection units (CMN 1998 NHB risk assessment). Note the overall complexity of the risk management situation and that magnitudes of risks range over more than five orders of magnitude.

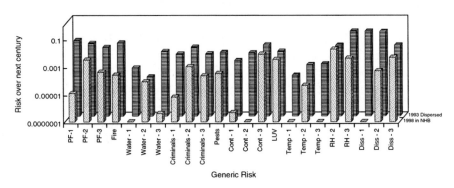

Figure 5. Comparison of generic risk, as fraction of entire Canadian Museum of Nature collection holdings, between 1993, back bars, and 1998, front bars. PF-1, Physical forces—Type 1; e.g., earthquake causing breakage. PF-2, Physical forces—Type 2; e.g., accidental breakage. PF-3, Physical forces—Type 3; e.g., poor support causing distortion. Fire—Consumption by fire; Water—Type 1; e.g., inundation by river flood. Water—Type 2; e.g., roof leaking. Water—Type 3; e.g., rising damp. Criminals—Type 1; e.g., major theft. Criminals—Type 2; e.g., isolated vandalism. Criminals—Type 3; e.g., pilfering. Pests, e.g.; insects and rodents. Cont-1, Contaminants—Type 1; e.g., smoke from a nearby disaster. Cont-2, Contaminants—Type 2; e.g., dust from construction activity. Cont-3, Contaminants—Type 3; e.g., Permanent gaseous pollutants. LUV, light and radiation. Temp-1, Temperature—Type 1; e.g., melting of an ice core collection. Temp-2, Temperature—Type 2; e.g., incorrect temperature causing softening or melting. Temp-3, Temperature—Type 3; e.g., higher than ideal. RH-2, Relative Humidity—Type 2; e.g., drastic change leading to fracture. RH-3, Relative Humidity—Type 3; e.g., too high, accelerating paper degradation. Diss-1, Dissociation—Type 1; e.g., collection abandonment. Diss-2, Dissociation—Type 2; e.g., misfiling an object. Diss-3, Dissociation—Type 3; e.g., failure to ensure transfer of legal title for gifts.

now being under a flat roof where previously some were held in multistory buildings. The RH-2 risk increase was a result of reduced levels of routine maintenance activities while staff attention was diverted to building planning and move preparation activities.

6. Risk Treatments and Their Results

In 1993, CMN collections occupied 12 leased warehouse spaces. Some collections were held in inferior storage hardware. In 1996–1997, during the time between the first two assessments, the CMN designed, had built, and occupied a purpose-built collection housing building (Figure 1). At the same time, storage hardware was upgraded to modern museum standards. In addition, following the move, and before the 1998 risk assessment, a collection emergency preparedness plan was developed and disseminated. Training in emergency response procedures and methods was conducted. A number of smaller, targeted risk remediation projects were also undertaken.

Of particular interest to consideration of emergency preparedness is evaluation of the changes in Type 1 risk and the relative contribution of Type 1 risks to total risk. These comparisons are shown in Figure 6.

Over the period 1993–1998, Type 1 risks were the most reduced of the three types of risk. There are several reasons for this. First, much protection against the effects of Type 1 risks is afforded at the levels of location, site characteristics, and building construction and systems. These were considerably improved by the building project. Second, many systems that provide life safety protection also contribute to property protection from Type 1 risks. Buildings designed and built to the most modern building codes will afford better property protection against Type 1 risks. Finally, consolidating staff in

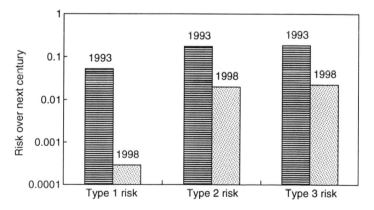

Figure 6. Comparison of risk to the CMN's collections by type of risk for 1993 and 1998.

one location, completing an emergency preparedness plan, and training staff in emergency response further mitigated against the effects of Type 1 risks.

It is also evident in Figure 6 that the totals of both Type 2 and Type 3 risks are approximately two orders of magnitude (100 times) greater than the total of Type 1 risks. Consequently, current collection care priorities are now focused on reducing Type 2 and 3 risks.

7. Actual and Understood Changes in Assessed Risks

Although differences in perceptions, understanding, and methods of assessing risks produced some of the differences, most of the changes between the 1998 and 2003 risk assessments reflect real reductions in levels of risk. Without the influence of a major capital project, differences in risk assessments conducted in 1998 and 2003 were much reduced and were of comparable magnitude for changes in understanding and for changes due to risk treatments, rates of collection use, or other objective, quantifiable measures (Figure 7). It is evident in Figure 7 that in terms of gross change in assessed risk, the total changes due to understanding only are of comparable magnitude to, but 50% higher than, total actual changes in risk. When considered as net differences, the sum of actual changes in risk is negative and nearly five times greater than the sum of changes due to understanding only. Most changes in actual

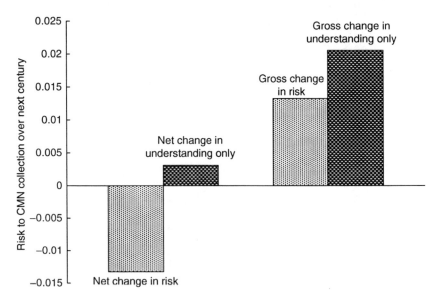

Figure 7. Comparison of net and gross (sum of absolute values) changes in risk for both actual changes in risk due to risk treatments and perceived changes due to changes in understanding only.

risk were reductions as a result of projects intended to reduce risks. In contrast, changes in understanding are more likely to either increase or decrease assessed risk. For change in understanding, the fact that net change in risk is about seven times smaller than gross change indicates near balance between changes in knowledge causing increases or decreases in risk.

8. Conclusion and Lessons of Interest to General Environmental Risk Assessment

Application of the CPRAM to the CMN has led to more rational allocations of resources for collection preservation. Senior management could be presented with reports offering opportunities for risk reduction instead of just petitions for more resources. This resulted in improved resource commitments to collection preservation. By anticipating those risks that will become priorities for treatment in coming years, it is possible to plan research activities to provide knowledge of key issues as and when required.

The risk assessment system was defined such that cultural property collections are considered static rather than dynamic systems. This great simplification permitted a comprehensive "snapshot" view of risks to be developed. It is understood that risk assessments must be conducted in a regularly repeated fashion to account for changes in the current collection environment as well as changes in understanding of risks. A second critically enabling simplification was to consider only the proportion of total value at risk. Doing this allowed sidestepping of the very difficult and controversial issue of valuation of cultural property.

Although the CPRAM was developed specifically for application to cultural property, certain lessons learned during its development are thought to be of interest to the broader risk analysis community. These lessons include:

- Risks to cultural heritage, despite sparse relevant knowledge, can be identified within a comprehensive framework of sources and types of risks.

- Exploring varied perspectives of risk information allows risk treatment and risk research priorities to be identified.

- Risks to cultural heritage can and should be considered as part of any comprehensive environmental assessment.

Acknowledgements

The author is grateful to the many staff of CMN who have contributed to the development of this method and to providing information and data needed to complete assessments.

References

1. Barzilai, J., 2005. Measurement and preference function modeling. *International Transactions in Operational Research* 12:173–183.
2. Boroush, M., 1998. "Understanding risk analysis." American Chemical Society (ACS) Resources for the Future (RFF). Available at: http://www.rff.org/misc_docs/risk_book.pdf
3. Buchanan, M., Porter, N., Goodwin, D., MacDiarmid, S., and Knight, K. (Eds.), 1999. Guidelines for managing risk in the Australian and New Zealand public sector. HB 143:1999. Standards Association of Australia, Strathfield, Australia.
4. Carey, J., Beilin, R., Boxshall, Burgman, M., and Flander, L., 2007. Risk-based approaches to deal with uncertainty in a data-poor system: stakeholder involvement in hazard identification for marine national parks and marine sanctuaries in Victoria. Australia. *Risk Analysis* 27(1):271—281.
5. Price, J. C., and Fitzgerald, G. R., 1996. Categories of specimens: a collection management tool. *Collection Forum* 12(1): 8–13.
6. Michalski, S., 1990. An overall framework for preventive conservation and remedial conservation. In K. Grimstad (Ed.), *9th Triennial Meeting: Preprints*. ICOM Committee for Conservation, Los Angeles, CA, 589–591.
7. Waller, R. R., 1994. Conservation risk assessment: a strategy for managing resources for preventive conservation. In A. Roy and P. Smith (Eds.), *Preventive Conservation Practice, Theory and Research*, preprints of the contributions to the Ottawa Congress, 12–16 September 1994. The International Institute for Conservation of Historic and Artistic Works, London. Available at: http://www.museum-sos.org/docs/WallerOttawa1994.pdf
8. Waller, R. R., 1995. Risk management applied to preventive conservation. In C. L. Rose, C. A. Hawks and H. H. Genoways (Eds.), *Storage of Natural History Collections: A Preventive Conservation Approach,* 21–28. Society for the Preservation of Natural History Collections, Iowa City. Available at: http://www.museum-sos.org/docs/WallerSPNHC1995.pdf
9. Waller, R. R., 2003. *Cultural Property Risk Analysis Model: Development and Application to Preventive Conservation at the Canadian Museum of Nature.* Göteborg Studies in Conservation 13. *Göteborg Acta Universitatis Gothoburgensis*, Göteborg.
10. Waller, R., and Michalski, S., 2004. Effective preservation: from reaction to prediction. *Conservation, the GCI Newsletter* 19(1):4–9. Available at: http://www.getty.edu/conservation/publications/newsletters/19_1/feature.html

DPSIR AND RISK ASSESSMENT OF DUMPED CHEMICAL WARFARE AGENTS IN THE BALTIC SEA

H. SANDERSON

National Environmental Research Institute
Aarhus University, Department of Policy Analysis
Frederiksborgvej 399, Post Box 358
DK-4000 Roskilde, Denmark
Tel: +45-4630-1822
Fax: +45-4630-1114
hasa@dmu.dk

M. THOMSEN

National Environmental Research Institute
Aarhus University, Department of Policy Analysis
Frederiksborgvej 399, Post Box 358
DK-4000 Roskilde, Denmark

P. FAUSER

National Environmental Research Institute
Aarhus University, Department of Policy Analysis
Frederiksborgvej 399, Post Box 358
DK-4000 Roskilde, Denmark

Abstract: This paper presents screening-level assessment as a conservative model-based scope assessment of potential risks to help guide environmental risk assessment of chemical warfare agents (CWAs) in fish communities. The paper also presents the DPSIR approach to complex environmental issues. The two approaches are applied to a case study involving CWA munitions dumped in the Baltic Sea after the Second World War.

1. Introduction

As a result of the disarmament of Germany following the Second World War, approximately 65,000 t of stockpiled chemical warfare agents (CWAs) munitions were ordered to be disposed of and a significant portion of these

I. Linkov et al. (eds.), Real-Time and Deliberative Decision Making.
© Springer Science + Business Media B.V. 2008

were dumped at sea during 1947–1948 [3, 6]. The Baltic Sea alone received more than 50% of Germany's CWA arsenal, with the largest amount, approximately 32,000 t, dumped east of the Danish island of Bornholm, thus presenting from a tonnage perspective the potential worst-case exposure scenario. The weapons have now been resting on the seabed and in the sediment of the Baltic Sea for almost six decades, and the extent of corrosion of the shells and thus release of the toxic chemicals has raised environmental concerns [5]. Some shells have leaked their contents, whereas others might still be intact [5, 6, 10], but the probability of environmental releases and exposure increases with time as the containers corrode. However, there are no comprehensive publicly available records regarding the environmental concentrations of the parent CWA compounds dumped east of Bornholm, which would be needed to evaluate the potential environmental risk associated with CWAs dumped in the Baltic Sea [2]. Moreover, until recently there were also large gaps in the available compiled screening-level environmental toxicity data and general environmental hazard information concerning CWAs [10].

The aims of this paper are:

- To present screening-level assessment as a conservative model-based scope assessment of potential risks to help guide a potential subsequent site-specific environmental risk assessment of CWAs in the fish community. The assessment is detailed in Sanderson et al. [10]. In the present paper only the main findings are reported.

- To present the DPSIR approach for grasping complex environmental issues. This approach considers Driving forces (D), Pressures (P), State (S), Impact (I), and Response (R). The DPSIR approach facilitates the expression of potential risks, the creation of indicators to monitor potentially developing threats to environmental safety, and the necessity of societal response. The munitions dumped in the Baltic Sea following the Second World War are a case study for the DPSIR approach as applied in the EU sixth framework programme within the project "Modelling of Ecological Risks Related to Sea-Dumped Chemical Weapons" (MERCW).

2. Case Study 1: Fish Community Risk Assessment

The German CWA munitions were dumped mostly in a primary, designated site, located in a circular area with a radius of 3 nautical miles in the Baltic Sea east of the Danish island of Bornholm (located at 55E 21'0 N and 15E 37'02 E), covering an area of 97 km^2. The water depth at this location ranges from 70 to 105 m. The seabed is covered by a 5–6-m-thick layer of mud. However, not all CWA was dumped at the designated site; hence a secondary, and more realistic dump area has been identified, covering 791 km^2 (located at 55°10'00–55°23 N and 15°24'–15°55' E) (Figure 1).

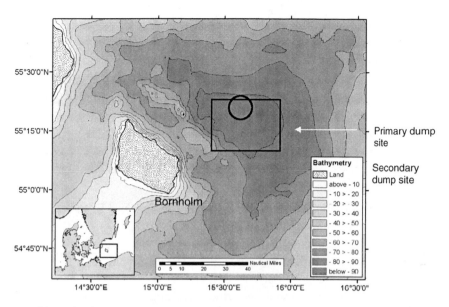

Figure 1. Map of Primary (Small) and Secondary (Large) Dump Sites in the Bornholm Basin.

TABLE 1. Confirmed Dumped Chemical Warfare Agents East of Bornholm [5].

Compound	CAS number	Dumped (t)
Chloroacetophenone (CAP)[a]	532-27-4	515
Sulphur mustard gas (Yperite)[b]	505-60-2	7,027
Adamsite[c]	578-94-9	1,428
Clark I[d]	712-48-1	711.5
Triphenylarsine[d]	603-32-7	101.5
Phenyldichloroarsine[d]	696-28-6	1,017
Trichloroarsine[d]	7784-34-1	101.5
Zyklon B[e]	74-90-8	74
Monochlorobenzene[f]	108-90-7	1,405

[a]Riot control agent; [b]blistering agent; [c]organoarsenic blistering agent; [d]arsine oil constituents – organoarsenic blistering agent; [e]blood agent; [f]additive.

The confirmed dumped chemical warfare agents and their amounts are reported in Table 1.

2.1. METHODS

We considered two potential exposure scenarios of CWA in the fish community:

A. A totally mixed box comprising the water volume enclosed by the primary and secondary dumping areas, respectively. The release and mixing of the total mass in the bulk water is assumed instantaneous. Accumulation in the sediment is included and the influence of degradation (hydrolysis), in the water is evaluated.

B. A continuous release over 60 years of the total mass from the bomb-shells at the seabed to the water phase. The water current is 5 cm/s, which induces a turbulent mixing of the bulk water and an advective transport of agents. Sedimentation, diffusion to sediment, degradation (hydrolysis) and accumulation in sediment is also included in the analysis. Simple first order dissipation models were applied to calculate the predicted environ-mental concentration (PEC).

For the assessment of fish toxicity we used the ECOSAR model provided by the U.S. Environmental Protection Agency (USEPA), and for the inter-species extrapolation of measured or predicted fish toxicities we applied the Interspecies Correlation Estimation (ICE) program, also provided by USEPA. We used species sensitivity distributions (SSDs) for determination of the fish community's no-observed-effect concentration (NOEC). SSDs are statistical distributions that describe the variation among a set of species in toxicity of a particular chemical or mixture. Aquatic quality criteria can be based on SSDs using the 5th percentile as a cut-off concentration (HC5) [1], where 95% of the community is protected, translating to a community NOEC.

2.2. RESULTS

The most realistic assessment result is Scenario B: 70 m; secondary dump site; chronic toxicity; at 0–20 cm above the sediment, with a total mixture toxic unit (TU)[1] [9] of 0.62. Triphenylarsine is the CWA with the highest realistic risk profile at 0.2 TU, followed by Adamsite at 0.17, Clark I at 0.086 and Yperite at 0.083 TU. The horizontal and vertical extent of poten-tial risk is illustrated in Figure 2 where the summed TU concentrations (risk areas) in the seawater are depicted for the primary and secondary dump sites, respectively. The risk volume increases in the direction of current as long as there is a release, and the greatest risk occurs at the outer boundary of the dump site.

In Figure 2 the risk volumes for the mixture of dumped agents along the direction of sea current are shown for the primary dump site, with continu-ous release in the first 10 km, and for the secondary dump site, with continuous

---

TU > 1 indicates a risk. The predicted environmental concentration exceeds the predicted no observed effect concentration.

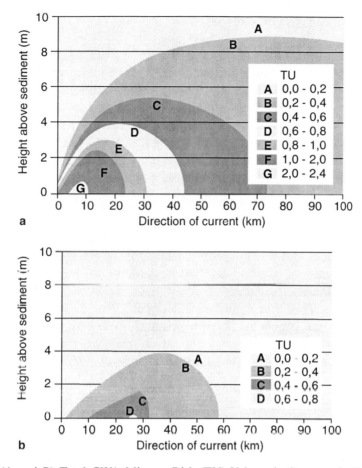

Figure 2. (A and B) Total CWA Mixture Risk (TU) Volume in Seawater in Primary and Secondary Dump Sites.

release in the first 28 km. Based on the primary dump site conditions, the risk volume (TU > 1) is <2.5 m above sediment and <23 km down current. Risks are not predicted for the secondary dump site; however, with an increase in the margin of exposure to a minimum of five (TU < 0.2) the risk boundaries would be <4 m above sediment and <58 km down current.

2.3. CONCLUSION

It can be concluded that environmental risks towards the fish community in the Bornholm basin from dumped CWAs cannot be ruled out based on this screening-level assessment, hence further empirical risk assessment is warranted.

3. Case Study 2: Science-Policy Interface

Indicators play integral roles in the translation of scientific results to measurable and tangible policy targets and aims and vice versa (bottom-up), operationalization of political targets (top-down) to measurable endpoints verifying progress and feasibility of management strategies. Indicators may be used for communicating complex scientific assessments of ecosystem health in a simplified and aggregated manner and as rapid aggregated assessment parameters. Two frameworks for monitoring and assessment are addressed within the MERCW project: DPSIR and the Ecosystem Approach [4, 7].

3.1. DPSIR

The DPSIR approach is used for extracting existing knowledge regarding Drivers, Pressures and resulting State, Impacts and Responses in relation to the dump site east of Bornholm. The available historical and local knowledge was used in the cruise design, surveys to be conducted within the project period of 2005–2008, and in support of uncertainty evaluation of the final ecological risk assessment of the Bornholm dumping site due to release of CWAs. An overall monitoring and evaluation framework in accordance with the DPSIR cycle is presented in Figure 3.

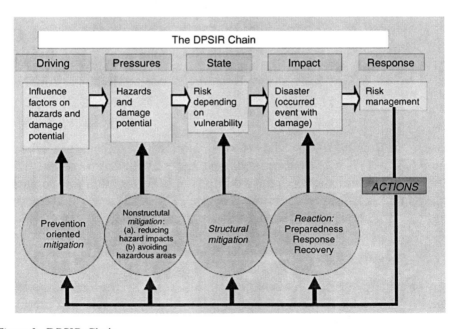

Figure 3. DPSIR Chain.

TABLE 2. Elements of the DPSIR Chain and Related Key Uncertainties.

DPSIR element	Key uncertainties
D (Driving force/-s)	• Construction of a new gas pipeline through the dump site (uncertainty of quantifying effect of driving force) • New unknown driving forces; e.g., trawling and other perturbations of the seafloor
P (Pressure/-s)	• Exact position of dumped objects (uncertainty of location) • State of dumped objects (uncertainty of leaching rates) • Accurate records of dumped amounts
S (State/-s)	• Concentration of CWAs (uncertainty of exposure, fate and transport parameters towards humans and the environment) • Site characteristics (uncertainty of site specific parameters)
I (Impact/-s)	• Impacts on ecosystem (uncertainty of most relevant ecosystem effect endpoints) • Impacts on humans (uncertainty of relevant effect endpoints—long term exposure and effects)
R (Response/-s)	• Defining actions (uncertainty and ambiguity of communicating and implementing scientific risk analysis in policy actions) • Development of CWA hazard indicators (uncertainty of linking critical DPSI parameters with indicators, and aggregation of associated uncertainties)

Drivers (D) are the underlying factors leading to the potential risk problem; Pressures (P) are human activities/interferences directly affecting the environment; i.e., in this case dumped munitions; State (S) is defined as the conditions of the environment in terms of the level, quality and/or quantity of physical, biological, chemical phenomena in time and space; Impacts (I) are defined as the effects of changes in the quality of the environment on ecosystem and human health; and Response (R) is in this case efforts of the society (different actors) to mitigate the problem.

Importantly, DPSIR analysis should also include identification of key uncertainties, which also may accumulate in increasing aggregative indicators. For the MERCW project, we identified the following in Table 2.

These uncertainties should be considered and communicated, and if possible quantified, in relation to the MERCW project.

ECOSYSTEM APPROACH

The PSI elements of the DPSIR cycle (Figure 3) were used for developing a science-policy interface; hence, monitoring variables included in the MERCW cruise activities were used to develop the Ecosystem Approach [7].

TABLE 3. Available Knowledge from Existing Monitoring Surveys used for Identification of Possible Hotspot Areas.

Potential descriptive indicators	Elements of the DPSIR chain
Sediment characteristics indicative of corrosion release, transport	P
Presence of buried objects based on magnetometry	P
Presence of CWA-resistant microbe characterization and distribution	S-I
Concentration of CWAs and total arsenic	S
Heavy metals indicative of presence of munitions	S
Observed *in situ* biota effects; e.g., in benthos, bioaccumulation, fish consumer exposure	S-I
Local layman expert knowledge and various authorities' context-specific reports	P-S-I

Based on multivariate pattern recognition techniques of existing knowledge and the results of the model-based screening-level fish community risk assessment, selected indicator-based assessment parameters have been agreed upon to be included in the monitoring program of MERCW (Table 3). Indicators are designed to be specific to the source of CWA exposure and at the same time related to the health of the Bornholm deep marine ecosystem. In other words, an ecosystem-based monitoring and evaluation framework is being developed specifically for CWAs in the Baltic Sea.

The objective of the ecosystem approach is to identify the most important ecosystem properties and components, and the subsequent development of ecosystem-based management objectives, using a top-down (a few very integrative often politically driven indicators) or bottom-up (multiple more specific indicators, often science-driven) approach.

In relation to using the indicators as arguments in the Response (R) phase, several political actions have been taken to improve the general state of the Baltic Sea. The Helsinki Commission (HELCOM) is dealing with strategies for improvement of the state of the Baltic Sea [8]. The Baltic Sea Action Plan was prepared by HELCOM, including four key issues requiring action:

- Eutrophication
- Hazardous substances
- Maritime activities
- Biodiversity

The MERCW project deals with an ecosystem-based approach for assessing any potential negative impacts on the ecosystem health caused by the CWAs dumped after the Second World War. The results uncovered under MERCW— e.g., the screening- level assessment is and will be communicated to HELCOM

[10] and the DPSIR approach was used for defining ongoing MERCW activities as well as upcoming cruises for the purpose of monitoring parameters that may be used as performance indicators—e.g., within the HELCOM Baltic Sea assessment system—and thus support the overall goal of the Baltic Sea life being undisturbed by hazardous substances. Specifically for hazardous substances there are three goals included in the Baltic Sea Strategy [4]:

- No risk of effects on the marine ecosystem
- No risk to consumers of fish; i.e., no bioaccumulation
- Achievement of background level concentrations

These three goals of the marine strategy and the conclusions from the screening-level assessment of the Bornholm deep marine ecosystem calls for a monitoring strategy able to detect any changes and developments in environmental characteristics related to a change in the Bornholm deep marine ecosystem health. The effects of munitions-related chemical exposure on the Baltic Sea ecosystem and human health need to be investigated as part of MERCW.

Monitoring of exposure and biological effects on species representing community disturbances will contribute to the indicator-based assessment system. At this point the selected indicators are defined descriptive indicators of changes in the state of the Baltic Sea environmental and ecosystem health. Once target values are defined, the indicators become performance indicators, or distance-to-target indicators.

4. Discussion and Conclusions

Environmental risks to Baltic Sea fish communities and the ecosystem in general cannot be dismissed based on a conservative screening-level assessment. Further empirical assessments are warranted.

The DPSIR and Ecosystem approaches have been very valuable as mechanisms to steer the MERCW project both in the empirical phase as well as in the reporting and communication phase with authorities. At this moment, approximately 20 indicators have been identified and included in existing monitoring programs in a top-down approach. These indicators may be included in the indicator-based description of the state of the Baltic Sea ecosystem. However, a bottom-up approach as developed in this project seems more appropriate for developing indicator-based proxies for ecosystem health. The HELCOM-selected indicators are to be provided with target values reflecting good ecological status, analogous to the EU Habitat Directive and the new Water Framework Directive. At this point, no target values have been included for munitions-related compounds [12]. However, the ecosystem approach calls for identifying sources that threaten the health of the

Baltic Sea. This will result in planning and implementation of management actions to protect and promote the health of the Baltic Sea ecosystem.

In 2008 the measured MERCW variables will be evaluated, and if a risk problem still cannot be excluded, performance indicators capturing the most important ecosystem properties and components related to dumped munitions will be proposed for distance-to-target monitoring in accordance with management objectives. The results will be communicated to the HELCOM secretariat and via the MERCW web page [11]. Depending on the results of the monitoring survey, validated indicators given target values may be included in the final ecosystem-based assessment system which needs to be operational by 2010 [4].

Acknowledgements

EU Commission Sixth Framework Programme Priority project: *MERCW*, Modelling of Ecological Risks related to Sea-dumped Chemical Weapons (Contract No. 013408).

References

1. Dyer, S. D., Versteeg, D. J., Belanger, S. E., Chaney, J. G., and Mayer, F. L., 2006. Interspecies correlation estimation predict protective environmental concentrations. *Environ. Sci. Technol.* 40:3102–3111.
2. Garnaga, G., and Stankevicius, A., 2005. Arsenic and other environmental parameters at the chemical munitions dumpsite in the Lithuanian economic zone of the Baltic Sea. *Environ. Res. Eng. Manage.* 33:24–31.
3. Glasby, G. P., 1997. Disposal of chemical weapons in the Baltic Sea. *Sci. Total Environ.* 206:267–273.
4. HELCOM. 2006. Ecological objectives for an ecosystem approach. HELCOM Stakeholder Conference on the Baltic Sea Action Plan. Available at: http://www.helcom.fi/stc/files/ BSAP/FINAL%20Ecological%20Objectives.pdf
5. HELCOM. 1993. Complex analysis of the hazard related to the captured German chemical weapon dumped in the Baltic Sea. HELCOM CHEMU 2/2/1/Rev. 1 27. Sept. 1993, Vilnius Lithuania.
6. HELCOM. 1994. Report on chemical munitions dumped in the Baltic Sea. Report to the 15th meeting of the Helsinki Commission. Danish Environmental Protection Agency. Available at: http://www.rurociagi.com/spis_art/2005_4/raport.pdf
7. UNESCO. 2006. *A Handbook for Measuring the Progress and Outcomes of Integrated Coastal and Ocean Management,* IOC Manuals and Guides, 46; ICAM Dossier, 2. UNESCO, Paris (English).
8. HELCOM. Undated. Baltic Sea Action Plan. Available at: http://www.bsap.pl/eu/bsap_ eu.html
9. Mitchell, J. A. K., Burgess, J. E., and Stuetz, R. M., 2002. Developments in ecotoxicity testing. *Rev. Environ. Bio./Technol.* 1:169–198.
10. Sanderson, H., Fauser, P., Thomsen, M., and Sørensen, P. B., 2008. Screening level fish community risk assessment of chemical warfare agents in the Baltic Sea. *J. Hazard. Mater.* 154:846–857.

11. MERCW page. Available at: http://www.fimr.fi/en/tutkimus/muututkimus/mercw.html
12. HELCOM. 2007. chapter 3 p. 21. Available at: http://meeting.helcom.fi/c/document_library/get_file?folderId=72398&name=DLFE-28884.pdf

SOCIAL AND ECOLOGICAL CHALLENGES WITHIN THE REALM OF ENVIRONMENTAL SECURITY

L. KAPUSTKA, R. MCCORMICK, K. FROESE

Golder Associates, Ltd.
1000, 840-6th Avenue SW
Calgary, Alberta T2P 3T1 Canada
Larry_Kapustka@golder.com

Abstract: Here we provide a brief discussion of the key principles of ecology and risk required to successfully manage natural resources and maintain environmental security. These ecological principles represent the context within which all economies function; the set of rules that—even if violated in the short-term—over the long haul will define the boundaries of our actions. The risk assessment framework offers a way of organizing information, which then allows us to pose critical questions and find the answers needed to effectively manage our pursuit of triple-bottom-line sustainability.

1. Introduction

As governments and societies evolve their understanding of prospective acts of terror, environmental security has come to the fore as a prominent topic. This public discourse has led to a broader appreciation of the deep connection between societal well being and ecological systems. On one level, we know sustained economic prosperity links directly to "surplus" ecological resources; that is, goods and services acquired by society from ecological systems. Surplus resources are those that are regenerated within some reasonable period through conversion of solar energy into food, fiber, or some other humanly accessible commodity, such as timber, or service, such as the assimilation of human-generated nutrients by wetlands. Thoughtful extraction of minerals and fossil fuels also can be managed to provide sustainable social structures and minimize impacts to ecological systems.

On another level, there exists only a poor recognition by societies of how ecological systems function, even those functions tightly woven into the global economy. Confrontations over energy supplies and clean water dominate contemporary news. Managers responsible for supply of these resources are

I. Linkov et al. (eds.), Real-Time and Deliberative Decision Making.
© Springer Science + Business Media B.V. 2008

acutely aware of the vulnerability of energy supply routes, surface waters, and aquifers to intentional acts of aggression, mischief, or accidents.

Here we provide a brief discussion of the key principles of ecology and risk required to successfully manage natural resources and maintain environmental security. These ecological principles represent the context within which all economies function; the set of rules that—even if violated in the short-term—over the long haul will define the boundaries of our actions. The risk assessment framework offers a way of organizing information, which then allows us to pose critical questions and find the answers needed to effectively manage our pursuit of triple-bottom-line sustainability.

2. Ecological Imperatives

Ecology, like other sciences, is value neutral [8]. Perceptions of utility, aesthetics, and worth emerge from our sense of place, cultural legacy, and contemporary traditions. Influenced by life experiences and new knowledge, perceptions change. This dynamism of perception leads to differences in values among different peoples, uncompleted reflections of our varied cultural, ethnic, class, gender, and age-related experiences. Implicit in this understanding is the sobering realization that ecological systems will function at some level with or without humans. The recognized societal value of a mangrove swamp, coastal salt marsh, king salmon, or any other ecological entity is just that: a societal value. If a mangrove swamp is eliminated, some other vegetative cover type, with a different suite of plant and animal associates, will occupy that space. The altered landscape will have different properties, different rates of productivity, and different quantities of surplus materials that we might exploit, but there will be a functional ecological system.

Economies, whether explicitly acknowledged by society or not, are based on the flow of ecological goods and services. The relationships between economic and cultural prosperity and ecological systems become clear when catastrophic events such as droughts, floods, unseasonal cold snaps, excessive heat, and other events disrupt food production. More subtle disturbances also occur, often with great economic consequences, such as the spread of disease or the introduction of exotic species [9].

Insights that emerged following discoveries of May [14] and others (as cited by Gleick [6]) provide the foundation for understanding ecological systems as chaotic entities that demand new ways of thinking about predictions of ecological conditions. We now know that ecological systems are self-organizing, complex, multidimensional, nonlinear, and dynamic entities. Equilibrium is never attained; one part of a system may appear to be in stasis, but other parts of the system are not. Historical events determine current and future structures and past conditions cannot be repeated. Collectively, these conditions render the forecasting of future system states tenuous at best [13].

Several aspects of ecological systems confound our ability to make predictions. A key factor relates to the rate at which ecological processes play out. The ebb and flow of populations and species assemblages across a landscape are tempered by multiple internal and external factors including climate, weather, predation, disease, and competition for limited resources. Collectively, these dynamic responses can give a sense of direction to the resource, such as progression to a long-lived forest type from the time of the last disturbance. However, short-term trajectories (in an ecological sense) may give false indications of long-term trends (predictions desired by an economic society). Coincidental "fortuitous environmental changes" that align nicely with a particular policy hypothesis also can be misleading.

Predictions also are made difficult due to actions of multiple stressors. In any environmental setting, multiple parameters influence organisms, populations, and ecosystems. For example, there can be several metal and organic substances, biotic interactions, and physical conditions present at any given time. Across a landscape, exposure to these parameters varies from locale to locale and over time.

Arguably, no organism resides at the optimum position for all of its niche parameters. In other words "stress" is a constant. However, physiological mechanisms provide organisms with the means of finessing the effects of specific stressors through the adjustment or realignment of baseline optimal conditions. For example, as the weather changes from spring through fall, plants effectively shift their response to temperature, gradually adjusting to warmer conditions in spring and then reversing this trend in the fall. In northern temperate climates, temperatures readily tolerated by plants in April or May can be fatal when they occur in July or August. Anticipation [18] and acclimation are important survival mechanisms for organisms. In addition, the cumulative effects of stressors confound predictions of their effects.

Complex stressors are those that cause different effects under different circumstances [4]. Examples of this include differential responses to essential nutrients across the range of concentrations from deficiency through sufficiency and finally to toxicity. With essential nutrients, there are differential responses to a given nutrient depending on the co-occurrence of paired nutrients (e.g., copper and molybdenum). Similarly, response to a stressor depends on the degree to which the exposed organisms are acclimated or adapted to the particular stressor. Most interesting are the situations in which the sequence of exposure to different stressors results in different ecosystem-level responses [5].

The implications of responses to a set of complex stressors in ecological risk assessment can be quite profound. Though some of the better studied relationships (such as elemental pairs copper:molybdenum or zinc:cadmium, as well as pH:ammonium) are often considered, responses to complex stressors, if acknowledged at all, are seldom incorporated into risk assessments. When monitoring an ecological system as a means of evaluating the predicted consequences of a release, complex stressor interactions could be highly significant.

Slight variations in initial conditions of a population, community, or ecological system and the magnitude of a stressor can have profound consequences. In other words, responses are not proportional to the magnitude of stress across the full range of possibilities. Most of us are aware of common examples that illustrate this point—a 5°C temperature change (say from 25°C to 20°C) on a given day would not be terribly disruptive to us, but a shift from +1°C to −4°C would have much larger importance to organisms unaccustomed to freezing conditions, such as citrus trees. Similarly, we can observe major changes with over-harvest of timber, excess harvest of fish, diversion of water from estuaries, and many other scenarios. The concept of a tipping point [2] applies to societies and to ecology; societies after all are subcomponents of ecological systems. We should anticipate that terrorist activities, as well as the presumably benign actions of others, can have profound consequences to ecological systems, especially when the actions occur near a tipping point.

A common concern of environmental management and the focus of ecological risk assessments is the establishment of a reference baseline condition that can be used to evaluate pre- and post-conditions for specified endpoints. The rationale for establishing a reference baseline carries some intriguing philosophical baggage [10, 11, 20]. Implicit in the pursuit of the baseline is an assumption that a stable ecological condition would exist, but for the actions of humans. With or without humans acting on the landscape, climate-driven ecological succession has been occurring; in the Northern Hemisphere the most recent episode is being shaped by changes since the last glacial epoch waned some 10–15,000 years ago.

Humans have a philosophical penchant for embracing constancy [17] even when compelling data to the contrary exist. The search for a reference baseline reflects this penchant, but need not be crippling. Though the search for this elusive ecological baseline is difficult, we can describe a snapshot view, a fixed point in time, in which we characterize static conditions [12]. After selecting some desired prior landscape condition, ecologists, it would seem, have an obligation to clearly describe those conditions that are possible and those that are unattainable. Even so, there remain many challenges in monitoring the changing status of those conditions, and many other confounding factors need to be acknowledged [7].

The linkage of economy to ecology relies on the rates of change of critical resources and ecosystem functions [15, 19]. As a change in availability of a resource occurs, there must be corresponding changes in the economy (such as price changes or restraints on demand). If the ecological changes are rapid, there are likely to be disruptions within the economic system before society develops a means of coping with the new realities. In this sense, one aspect of environmental security is the alignment of economics with the anticipated flow of ecological goods and services—including preparedness for the range of scenarios that entail the need for rapid response (for example after an act of terrorism or a natural disaster) and more deliberate responses (such as those associated with climate change or invasive species).

3. Risk Assessment

Risk assessment is one of the most powerful tools available to manage environmental security. An assessment begins with problem formulation, a phase of work that explicitly acknowledges the management goals and decisions that are to be made [21]. Diverse stakeholder input is central to developing a good risk assessment and this should begin with the earliest stages of work.

Tools are available to guide the elicitation of stakeholder values as well as to engage stakeholders in the decision process [1]. The degree of success in reaching consensus is directly related to the timing of communications and clarity of each stakeholder's role. If the goal genuinely is to obtain meaningful input from stakeholders, it is critical to engage in dialogue before decisions have been made; *telling* is much less successful than *asking*. In the arena of environmental security, emotions are already high, so efforts to conduct dialogues calmly become most important.

The steps of problem formulation normally articulate the boundary conditions of a risk assessment through unambiguous statements regarding the values to be protected, also known as assessment endpoints. The assessment endpoints guide the selection of measurements (measurement endpoints) and models that are used to characterize the risks. Typically, pictorial and narrative descriptions are organized into one or more conceptual models that depict the functional relationships between pressures or stressors and the values to be protected. Iterative passes through the conceptual model are needed to refine it and hone in on the assessment and measurement endpoints. Determination of the information needed to complete a valid risk assessment provides the foundation for defining data quality objectives, producing a sampling and analysis plan, and proceeding through the analysis and characterization stages of the assessment. A crucial aspect of problem formulation is the understanding of the ecological setting or context that the analysis is to address; here the ecological imperatives described previously are paramount.

The risk assessment approach has particular application for environmental security. Risk, in the final analysis, is about evaluating scenarios. The outcome of a risk assessment is an estimate of the probability of different scenarios occurring. Input to the analysis of scenarios can be an admixture of quantitative and qualitative observations and direct measurements or modeled projections. The output can be organized to feed these data into multicriteria decision analysis programs. In the end, by varying different input parameters of the scenarios, the sensitivity of the various inputs can be evaluated.

In the interplay of policy and regulatory actions, varying degrees of tension inevitably arise due to differences in stakeholders' tolerance or acceptance of environmental risk. These tensions often are created as a direct consequence of the processes followed in reaching decisions, but there is much more. Explorations from nearly two decades ago into risk perception have provided powerful insights into the way people handle multiple forms of information as they make decisions (Table 1). In general, we can conclude that scientific or technical descriptions of a risk event or activity form only a small part of the body of information that people process as they consider accepting or rejecting the risk. Those science-based or technological features are largely limited to understanding the mechanisms and characterization of uncertainty. From the regulatory side, the most critical feature influencing public acceptance of decisions is trust in the responsible institution. Many of the remaining features relate in one form or another to communications and the degree of control that the public feels it can exercise, either directly or indirectly. Historically, public notice and public hearings and comment periods have been the primary means for public input into the environmental management regulatory process. As the regulatory process evolves to meet current challenges, there are opportunities to achieve the goals of public input in ways that are more satisfying to all stakeholders and simultaneously streamline the process so that efforts can be focused on issues in proportion to the importance of the issues. One approach that is effective in gaining trust and transferring an appropriate level of control to public groups is the Consensus Based Environmental Decision (CBED) procedure standardized recently by the American Society for Testing Materials International [1].

Within the regulatory arena, many well intentioned policies and laws are operational. Unfortunately, most, if not all, policies and laws were crafted without consideration of contemporary ecological insights. As such, policies often have unintended consequences. In some instances, rigidity of prescriptive measures stifles innovation or at least provides the foil for inaction.

As we move into the next generation of environmental management, we ought to embrace the insights of contemporary ecology and related socioeconomic advances. In her delightful commentary on the Hierarchy of Influences, Meadows [16] describes the 12 touch points or nodes where people can attempt

TABLE 1. Factors Important in Risk Perception and Evaluation [3].

Factor	Conditions associated with increased public concern	Conditions associated with decreased public concern
Catastrophic potential	Fatalities and injuries grouped in time and space	Fatalities and injuries scattered and random
Familiarity	Unfamiliar	Familiar
Understanding	Mechanisms or process not understood	Mechanisms or process understood
Uncertainty	Risks scientifically unknown or uncertain	Risks known to science
Controllability (personal)	Uncontrollable	Controllable
Voluntariness of exposure	Involuntary	Voluntary
Effects on children	Children specifically at risk	Children not specifically at risk
Effects manifestation	Delayed effects	Immediate effects
Effects on future generations	Risk to future generations	No risk to future generations
Victim identity	Identifiable victims	Statistical victims
Dread	Effects dreaded	Effects not dreaded
Trust in institutions	Lack of trust in responsible institutions	Trust in responsible institutions
Media attention	Much media attention	Little media attention
Accident history	Major and sometimes minor accidents	No major or minor accidents
Equity (also related to environmental justice)	Inequitable distribution of risks and benefits	Equitable distribution of risks and benefits
Benefits	Unclear benefits	Clear benefits
Reversibility	Effects irreversible	Effects reversible
Origin	Caused by human actions or failures	Caused by acts of nature or God

to manage systems (Figure 1). She points out that, ironically, people tend to focus first on the nodes of lesser importance (i.e., 12, then 11, and so on), to the overall drivers of the system, probably because they are easier.

The current paradigm of energy security is "protect the source at all costs," and the only source considered is fossil sun. Energy security becomes national security, and national security has morphed to include environmental security. The current atmosphere in which we discuss environmental security focuses much on numbers (attacks, cells, immigrants), stocks (monetary, energy), on down to the rules of the (current) system. The perceived risks of direct attacks on our supporting ecological systems bind tightly to those numbers and rules. Meadows' key point, and the reason that the power to transcend paradigms comes first in the hierarchy, refers to the unspoken certainty that the current paradigm of environmental security thinking is the only paradigm. Letting go of that certainty and opening up to the idea of

12. Constants, parameters, numbers

11. The sizes of buffers and other stabilizing stocks, relative to their flows

10. The structure of material stocks and flows and nodes of intersection

9. The lengths of delays, relative to the rate of system changes

8. The strength of negative feedback loops, relative to the impacts they are trying to correct against

7. The gain around driving positive feedback loops

6. The structure of information flows

5. The rules of the system

4. The power to add, change, evolve, or self-organize system structure

3. The goals of the system

2. The mindset or paradigm out of which the system arises

1. The power to transcend paradigms

Figure 1. Hierarchy of Influences [16].

multiple, interlinked ways of dealing with our interactions with local ecologies and global ideologies reveals where the power of paradigm resides.

Surfacing our assumptions about the ruling paradigm allows for the depiction of an even greater set of possible system transformations. The strategies used by systems ecologists in framing complex socioecological interactions mesh cleanly with the formalisms of integrated risk management. Managing for environmental security presents many challenges. To be successful, it is essential that the ecological context of different scenarios be understood and that a diverse range of affected stakeholders is engaged. An integrated risk assessment approach can provide the framework for exploring various scenarios.

References

1. ASTM-I (American Society for Testing and Materials-International). 2006. E2348-06 Standard Guide for Framework for Consensus Based Environmental Decision Process. *Annual Book of Standards*, ASTM-I, West Conshohocken, PA.

2. Gladwell M. 2000. *The Tipping Point: How Little Things Can Make a Big Difference.* Little, Brown, New York.

3. Covello VT, Sandman PM, Slovic P. 1988. Risk communication, risk statistics, and risk comparisons: a manual for plant managers. Chemical Manufacturers Association, Washington, DC. Accessed 16 February 2007. Available at: http://www.psandman.com/articles/cma-appc.htm

4. Dorward-King EJ, Suter II GW, Kapustka LA, Mount DR, Reed-Judkins DK, Cormier SM, Dyer SD, Luxon MG, Parrish R, Burton Jr GA. 2001. Distinguishing among factors that influence ecosystems. pp 1–26 in Baird DJ, Burton Jr GA (eds). *Ecological Variability: Separating Natural from Anthropogenic Causes of Ecosystem Impairment.* SETAC Press, Pensacola, FL.

5. de Ruiter PC, Moore JC, Griffiths B. 2001. Food webs: interactions and redundancy in ecosystems. pp 123–139 in Baird DJ, Burton Jr GA (eds). *Ecological Variability: Separating Natural from Anthropogenic Causes of Ecosystem Impairment.* SETAC Press, Pensacola, FL.

6. Gleick J. 1987. *Chaos: Making a New Science.* Penguin Books, New York.

7. Kapustka L. 2008. Limitations of the current practices used to perform ecological risk assessment. *Integr. Environ. Assess. Manage.* 4:290–298.

8. Kapustka LA, Landis WG. 1998. Ecology: the science versus the myth. *Hum. Ecol. Risk Assess.* 4:829–838.

9. Kapustka L, Linder G. 2007. Invasive species: a real but largely ignored threat to environmental security. In Linkov, I. et al. (eds). *Environmental Security at Ports and Harbors.* Springer, Dordrecht.

10. Lackey RT. 2001. Values, policy, and ecosystem health. *BioScience* 51:437–443.

11. Lackey RT. 2007. Science, scientist, and policy advocacy. *Conserv. Biol.* 21:12–17.

12. Landis WG, McLaughlin JF. 2000. Design criteria and derivation of indicators for ecological position, direction and risk. *Environ. Toxicol. Chem.* 19:1059–1065.

13. Matthews RA, Landis WG, Matthews GB. 1996. The community conditioning hypothesis and its application to environmental toxicology. *Environ. Toxicol. Chem.* 15:597–603.

14. May RM. 1976. Simple mathematical models with very complicated dynamics. *Nature* 261:459–467.

15. McCormick RJ, Zellmer AJ, Allen TFH. 2004. Type, scale, and adaptive narrative: keeping models of salmon, toxicology and risk alive to the world. In Kapustka LA, Galbraith H, Luxon M, Biddinger GR (eds). *Landscape Ecology and Wildlife Habitat Evaluation: Critical Information for Ecological Risk Assessment, Land-Use Management Activities, and Biodiversity Enhancement Practices,* ASTM STP 1458. ASTM International, West Conshohocken, PA.

16. Meadows DH. 1999. *Leverage Points: Places to Intervene in a System.* The Sustainability Institute, Hartland Four Corners, VT.

17. Norton B. 1996. Change, constancy, and creativity: the new ecology and some old problems. *First Annual Cummings Colloquium on Environmental Law: Beyond the Balance of Nature: Environmental Law Faces the New Ecology. Duke Environ. Law Policy Forum* 7:49–70.

18. Rosen R. 1980. Anticipatory systems in retrospect and prospect. *Behav. Sci.* 11–23.

19. Stahl R, Kapustka L, Bruins RJF, Munns WR. (eds). 2007. *Valuation of Ecological Resources: Integration of Ecology and Socioeconomics in Environmental Decision Making* Taylor & Francis, Boca Raton, FL.

20. Suter GW. 1993. A critique of ecosystem health concepts and indices. *Environ. Toxicol. Chem.* 12:1533–1539.

21. USEPA (United States Environmental Protection Agency). 1998. *Guidelines for Ecological Risk Assessment.* EPA/630/R-95/002F. Risk Assessment Forum, Washington, DC.

ADAPTIVE MANAGEMENT AND THE COMPREHENSIVE EVERGLADES RESTORATION PLAN

D. GUINTO and R. REED

HydroPlan
Portland, Oregon, USA
dguinto@q.com

Abstract: The Comprehensive Everglades Restoration Plan (CERP) was authorized by the Water Resources Development Act (WRDA) of 2000 as a framework for the restoration, preservation, and protection of the South Florida ecosystem while providing for the other water-related needs of the region. CERP explicitly acknowledged shortfalls in achieving planning objectives that could not be addressed due to project constraints, risks and uncertainties, technological limitations and inadequate evaluation methodologies at that time. Given these constraints and the limited level of detail accomplished in the feasibility study, CERP deferred specific details for achieving planning objectives and long-term project implementation. Consequently, successful CERP implementation relies on effective adaptive management strategies. This article provides a brief overview of CERP, discusses the current adaptive management strategy and presents a case study, which highlights challenges and issues.

1. Background

Florida faces major water management challenges driven in large part by a state population that is projected to increase from nearly 16 million in the year 2000 to 26.5 million by 2030 [1]. The seasonal conditions in South Florida result in either too much or too little rainfall. This variability coupled with limited storage capacity causes water shortages, environmental degradation, and an average of 1.7 billion gallons of water a day lost to tide [2]. The Everglades is now considered to be the most threatened ecosystem in the nation [3].

The Central and Southern Florida (C&SF) Project Comprehensive Review Study, also known as the "Restudy," was authorized by Congress in 1992 to reexamine the C&SF Project and to determine the feasibility of modifying the project to restore the South Florida ecosystem while providing for other water-related needs of the region. The authorizing legislation required

I. Linkov et al. (eds.), Real-Time and Deliberative Decision Making.
© Springer Science + Business Media B.V. 2008

the study to investigate making structural or operational modifications to the C&SF Project for improving the quality of the environment; protecting water quality in the south Florida ecosystem; improving protection of the aquifer; improving the integrity, capability, and conservation of urban and agricultural water supplies; and improving other water-related purposes [3]. This study resulted in the authorization of the Comprehensive Everglades Restoration Plan (CERP). Currently estimated at $14.8 billion, CERP is the largest restoration initiative ever undertaken. CERP is composed of 68 major components that involve creation of approximately 217,000 acres of reservoirs and wetland-based water treatment areas, two wastewater reuse plants, seepage management, underground storage for approximately 1.6 billion gallons of water per day, and removal of more than 240 miles of levees and canals in natural areas [2]. Figure 1 illustrates the features of CERP.

These components vastly increase storage and water supply for the natural system, as well as for urban and agricultural needs, while maintaining current Central and Southern Florida Project purposes. CERP proposes to increase the water budget of the area from 1.7 billion gallons per day to 2.4 billion gallons per day. Specifically, the plan will improve the functioning of more than 2.4 million acres of the south Florida ecosystem; improve Lake Okeechobee water levels for littoral zone health; eliminate almost all damaging freshwater releases to the Caloosahatchee and St. Lucie estuaries; improve urban and agricultural water supply; improve water deliveries to Florida Bay, Biscayne Bay, and other estuaries; improve regional water quality conditions; and maintain existing levels of flood protection [1]. CERP remains a conceptual plan, however, and efforts to implement and execute this ambitious project are characterized by risk, uncertainty and debate. The scope and magnitude of CERP present obvious challenges in planning, policy making, and implementation. It is enormously difficult to characterize and assess progress toward ecosystem restoration at the large geographic and temporal scale of CERP.

2. Implementing CERP

WRDA 2000 authorized CERP as a framework, yet recognized the unparalleled technical uncertainties and political challenges. Given the level of detail provided in the authorized document, it was anticipated that CERP would be modified periodically to achieve its goals and purposes more effectively and precisely. Consequently, WRDA 2000 required the development of programmatic regulations to ensure that CERP goals and purposes are achieved and provided funding for an adaptive assessment and monitoring program. The programmatic regulations (33 CFR, Part 385) were promulgated in 2003 and establish a framework and process for

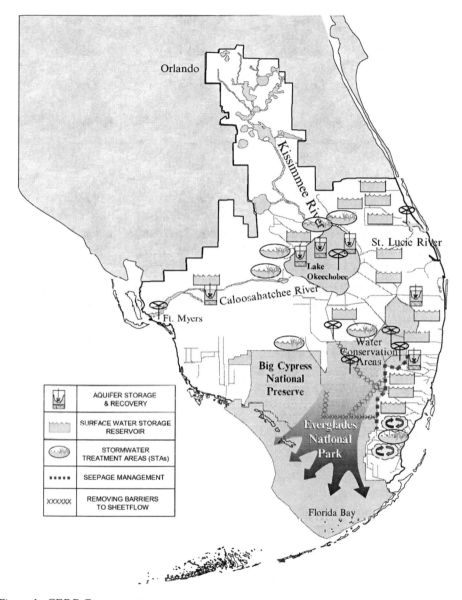

Figure 1. CERP Components.

integration of new information throughout CERP's 30-year implementation. Plan modifications and refinements recommended based on new and/or improved information were to be achieved through individual project implementation reports (PIRs), systemwide monitoring, and assessment strategies.

2.1. PROJECT IMPLEMENTATION REPORTS

The CERP program is composed of 68 major components that are grouped into more than 40 projects. Each project is developed by an interagency, multidisciplinary team responsible each project's PIR. The programmatic regulations require that the PIRs:

formulate and evaluate alternative plans to optimize the project's contributions towards achieving the goals and purposes of the Plan, and to develop justified and cost effective ways to achieve the benefits of the Plan.

Interim guidance has been developed to assist project delivery teams (PDTs) in the plan formulation activities during the development of a PIR. The guidance provides a means of formulating projects while maintaining a system perspective. The guidance identifies the goal of CERP formulation and evaluation as to "reasonably maximize the project's contribution toward the system-wide benefits of CERP compared to cost."

Further, the interim guidance directs PDTs to formulate alternative projects to better define, refine, and/or optimize projects and/or to investigate more cost-effective ways to achieve the same or greater benefits at a lesser cost compared to that predicted for CERP identified by the Restudy. While this guidance generally captures the intent of the Programmatic Regulations, it does not define a process that would encourage or even allow PDTs to investigate alternative projects outside their project boundaries to achieve CERP benefits at a lower cost.

2.2. SYSTEMWIDE AND PROJECT-LEVEL ANALYSIS

The benefit and impact analysis conducted for each CERP project is accomplished at both the local and systemwide scale. For example, a reservoir project could have adverse impacts to wetlands within the footprint of the project while the storage function of the reservoir (in combination with other CERP features) could have significant ecologic benefits by restoring sheetflow across vast areas of the Everglades and downstream estuaries. The impacted wetlands are generally considered a local effect, while the ecological benefits to the Everglades and downstream estuaries are considered systemwide effects. Regional models are used to assess impacts to sheetflow and estuaries, while subregional models are used to assess impacts to the footprint and in the vicinity of the project. Figure 2 displays the terms used for system and project level analysis.

2.3. CERP ADAPTIVE MANAGEMENT STRATEGY

CERP was designed to facilitate project modifications based on lessons learned from system responses, both expected and unexpected, and from future restoration targets as those become more refined. CERP includes

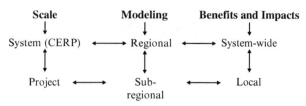

Figure 2. CERP Components.

an adaptive management strategy to ensure that new information about the natural system, learned from continuing research and from measuring responses to implementation of plan components. CERP can be used to reduce gaps and increase the level of success without significant increases in implementation costs. Specifically, adaptive assessment uses a well focused, regional monitoring program to measure how well each CERP component accomplishes its objectives. This, in turn, sets up opportunities for refinement of succeeding components. Such adaptive assessment and regional monitoring are essential features of CERP. Various documents have been developed by RECOVER[1] to frame the adaptive assessment program for CERP. For more information on CERP's adaptive management strategies, see the Comprehensive Everglades Restoration Plan Adaptive Management Strategy [4].

Adaptive management is a science- and performance-based approach to ecosystem management in situations where predicted outcomes have a high level of uncertainty [4]. Adaptive management has been an integral component of CERP. The Restudy identified specific shortfalls which were to be addressed during plan refinement in order to fully achieve CERP planning objectives [2].

The RECOVER team is responsible for the development and implementation of the CERP Adaptive Management Program. This program comprises four elements: CERP planning, performance assessment, update process, and management and science integration. Figure 3 displays these four elements and their relationships.

---

[1] The Restoration Coordination and Verification (RECOVER) team ensures the application of scientific and technical information in ways that are most effective in supporting the objectives of CERP.

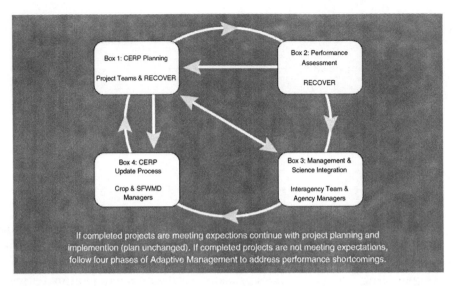

Figure 3. CERP Adaptive Management Framework [4].

3.   Case Study: Application of Adaptive Management in CERP

The Restudy identified the array of components necessary to achieve the Everglades' restoration and other water resources objectives based on information available at that time. However, there are shortfalls in the plan that could not be addressed due to project constraints, risks and uncertainties, and the limits of knowledge, technology and evaluation methodologies at that time. Currently, the established objectives fall short of targeted levels by a significant amount: 40–90%.

To help address these shortfalls, it is essential that the original guiding principles that framed the vision be viewed in the context of CERP's potential role in achieving objectives and maintaining flexibility for more effective and efficient implementation. The guiding principles state:
Project Delivery Teams and RECOVER will actively coordinate in the formulation and evaluation of project designs, in order to identify the plans that can improve on the predicted performance of the version of the Plan approved in 1999. The success of CERP will depend on a thorough understanding of the relationships between the contribution of each project and the overall goals of the Plan. [5]
Further,

CERP implementation will include the application of a system-wide science strategy and adaptive management program, designed to maximize the effective use of existing knowledge and incorporate new scientific and technical information, as a basis for continually improving the design, operation and performance of the Plan. [5]

From these guiding principles, success in reducing CERP's shortfalls based on achieving the established objectives is contingent on three important factors:

1. Successful identification of system solutions to problems through system planning

2. Effective design and implementation of appropriate restoration projects and

3. Comprehensive monitoring or tracking of improvements and shortfalls toward the desired goals and targets, which can be used to make adjustments to the plan and to reach agreement on objectives and priorities

While WRDA 2000 approved CERP, it is expected that the plan will be modified periodically to achieve its goals and purposes more effectively and precisely. These modifications and refinements were to be achieved initially through the PIRs and later systemwide monitoring and assessment strategies. Therefore, addressing CERP shortfalls during planning of CERP projects is key to immediately improving plan performance. Individual project teams must look for creative opportunities to address the critical shortfalls. Ultimately it is the PDTs, working with RECOVER, that will identify cost-effective means of achieving the restoration objectives. For example, one of the shortfalls identified during the Restudy was achievement of the restoration target for the St. Lucie Estuary. Consequently, the IRL-South PDT reevaluated alternative plans and identified a project that significantly improved CERP performance within the St. Lucie Estuary. In addition, the project addressed the spatial extent shortfall by restoring wetland areas within the drainage basin of the estuary. This innovative approach reduced the amount of reservoir and stormwater treatment areas needed for the project while significantly contributing to the spatial extent objective.

However, the mechanisms in place to deal with recommendations for plan improvements outside project boundaries do not appear to be functioning. For example, 100,000 acre-feet of additional water was found in the course of preparing the Indian River Lagoon-South (IRL-S) PIR [6] but was not utilized in either the PIR or subsequently, to date, through an adaptive management strategy by RECOVER. While the intent of the IRL-S PDT was to support the goal of systematically improving CERP based on new information, it appears that the current implementation process falls short in supporting plan improvements.

4. Conclusions

Restoration of what remains of the Everglades ecosystems represents one of the most ambitious ecosystem restoration initiatives ever conceived [7]. Despite significant progress in program management, scientific understanding, and project evaluations, no CERP projects have been completed to date. Budgetary constraints coupled with scientific and technical uncertainties have caused significant delays in project implementation. Moreover, federal funding, inflation, and unanticipated coordination costs contribute to increased scrutiny, additional reporting requirements and skepticism among the extensive consortium of partners and stakeholders. NRC [7] has completed a review of the progress in restoring the Everglades and has determined that the monitoring and assessment plan documents describe a well designed, statistically defensible monitoring program and an ambitious assessment strategy. However, implementation of the monitoring plan is occurring more slowly than planned. A coordinated approach is necessary to improve modeling tools and focus modeling efforts toward direct support of the CERP adaptive management process. Astute monitoring coupled with effective and timely response and refinement is key to the successful implementation of the plan.

Consistent with recommendations from the National Research Council [7], an Incremental Adaptive Restoration Strategy to formulate projects within CERP and address some of the issues encountered with CERP implementation has been developed. The current draft is available for online review [8].

References

1. Office of Economic and Development Research, the Florida Legislature, Demographic Estimating Conference Database, updated August 2007. Available at: http://edr.state.fl.us/population.htm
2. U.S. Army Corps of Engineers and South Florida Water Management District, 1999. Central and Southern Florida Project Comprehensive Review Study Final Integrated Feasibility Report and Programmatic Environmental Impact Statement. Jacksonville District, U.S. Army Corps of Engineers, Jacksonville, FL, and South Florida Water Management District, West Palm Beach, FL.
3. Light, S., and Dineen, J., 1994. Water Control in the Everglades: A Historical Perspective, in Ogden and Davis, eds., *Everglades: The Ecosystem and Its Restoration,* St. Lucie Press, Delray Beach, FL.
4. Comprehensive Everglades Restoration Plan: Adaptive Management Strategy. Restoration Coordination and Verification Team (RECOVER), 2006. U.S. Army Corps of Engineers, Jacksonville District, Jacksonville, FL and South Florida Water Management District, West Palm Beach, FL.
5. CERP Vision Statement. Available at: http://www.evergladesplan.org/pm/program_docs/cerp_vision_statement.cfm

6. Central and Southern Florida Project, Indian River Lagoon—South, Final Integrated Project Implementation Report and Environmental Impact Statement, 2004. U.S. Army Corps of Engineers and South Florida Water Management District. Jacksonville District, U.S Army Corps of Engineers, Jacksonville, FL and South Florida Water Management District, West Palm Beach, FL.
7. Progress Toward Restoring the Everglades: The First Biennial Review, 2006. U.S. Army Corps of Engineers and South Florida Water Management District, 2007. National Academy Press, Washington, DC.
8. Adaptive Management. Available at: http://www.evergladesplan.org/pm/program_docs/adaptive_mgmt.aspx.

PART 3

MULTICRITERIA DECISION ANALYSIS: METHODOLOGY AND TOOLS

UNCERTAINTY MODELING WITH IMPRECISE STATISTICAL REASONING AND THE PRECAUTIONARY PRINCIPLE IN DECISION MAKING

I. KOZINE

Department of Management Engineering
Technical University of Denmark
Roskilde, Denmark
igor.kozine@risoe.dk

Abstract: A number of unconventional formal approaches to decision making have been developed to provide mathematical foundations for rational choices under both aleatory and epistemic uncertainty. They challenge a central assumption of the Bayesian theory, that uncertainty should always be gauged by a single (additive) measure, and values should always be gauged by a precise utility function [3].

Decision-making theorists have presented approaches for arriving at rational decisions in spite of imprecision and indeterminacy [4–8, 10]. This paper introduces the theory of upper and lower previsions, provides examples, discusses how to account for unreliable statistical judgements, and reviews the relationships between the Precautionary Principle, indecision, and imprecise statistical reasoning.

1. Introduction

One of the gravest errors in any type of risk management process is the presentation of risk estimates which convey a false impression of accuracy and confidence – disregarding the uncertainties inherent in basic understanding, data acquisition, and statistical analysis. [1]

Decision making concerning human activities with potentially harmful consequences and high uncertainties is based on both scientific findings of the risk assessment and societal norms such as the Precautionary Principle (PP). However, risk assessments along with uncertainty measures complemented by the need to comply with the PP do not compel adoption of a particular course of action. This is usually left to the discretion of decision makers. As the stakes rise, the lack of scientific consistency among all systems analysis constituents preceding the option selection may result in failing to select an acceptable option. Systems analysis constituents include hazard/threat

I. Linkov et al. (eds.), Real-Time and Deliberative Decision Making.
© Springer Science + Business Media B.V. 2008

identification, risk assessment, uncertainty assessment, account of societal norms, and decision making. Studying each of them as a separate component is necessary, but this is no longer sufficient. An integrated approach, binding them in a formally consistent framework, is a coveted target for risk analysts.

Conceptual and computational structure of analyses of complex systems involves a division of uncertainty into aleatory uncertainty, which arises because the system under study can potentially behave in many different ways; and epistemic uncertainty, which arises from a lack of knowledge about quantities that have fixed but poorly known values. Aleatory uncertainty is also called stochastic and irreducible, while epistemic is called reducible. Such separation plays a particularly important role in risk analyses, where aleatory uncertainty arises from many possible adverse outcomes or consequences and epistemic uncertainty arises from a lack of knowledge with respect to quantities required in the characterization of the frequency, evolution, or consequences of individual potential adverse effects [2].

A number of unconventional formal approaches to decision making have been developed to provide mathematical foundations for rational choices under both aleatory and epistemic uncertainty. They allow for the limited cognitive abilities of human beings and could be regarded as formal variants of the PP. They also give a perspective of how the integrated framework could be built.

Though different in detail, they have a very important point in common. They challenge a central assumption of the Bayesian theory, that uncertainty should always be gauged by a single (additive) probability measure, and values should always be gauged by a precise utility function [3]. This assumption has been referred to as the Bayesian dogma of precision.[1] The opponents of the dogma of precision claim that imprecision, indeterminacy, and indecision are compatible with rational choice [4].

One unconventional theory of rational choice is discussed by Gårdenfors and Sahlin [5]. The point of departure from the conventional theory of rational choice—Bayesian decision theory—is that the amount and quality of information the decision maker has concerning the possible states and outcomes of the decision situation in many cases constitute an important factor when making decisions. To describe this aspect of the decision situation, the authors say that the information available concerning the possible states and outcomes of a decision situation has different degrees of epistemic reliability. The second step is to recognize that the reliability of

---

[1] Perhaps the most noticeable calls to revise the Bayesian theory for making rational choices were pronounced by Herbert A. Simon (Nobel Prize winner). See, for example, Simon [9].

a probability assignment for states affects the risk of the decision. The less reliable the probability assignment, the more risky the decision, other things being equal.

Another theory of decision making is discussed by Levi [6]. His point of departure is that we often do not know or cannot decide what we most prefer; yet we still have to choose. In such cases, called decision making under unresolved conflict, the requirement that preferences should be logically coherent does not necessarily imply that choices should satisfy properties of consistency such as avoiding sure loss. A rational agent, Levi claims, ought to restrict his choice to the set of admissible options; within this set, any choice is allowed. The theory suggests a formal way of constructing the set of admissible choices.

There are some other developed rules of rational choice that accept as a starting point a lack of knowledge for exactly defining utilities and probability assignments for the set of outcomes. They presuppose that numerical input for decision making is interval-valued and suggest different approaches for choosing one option among those permissible. The width of the interval manifests the lack of knowledge concerning utilities and probability assignments [4, 7, 8].

Decision making is the final phase in systems analysis and to all appearances there are mathematically furnished rules to make rational choices under lack of knowledge. The question to ask now is: Are there formal frameworks for uncertainty modeling that are built on the clear distinction between aleatory and epistemic uncertainty?

Reasoning that can accommodate the both types of uncertainties is called imprecise statistical reasoning and is motivated by the idea that the dogma of precision is mistaken and imprecise probabilities are needed in statistical reasoning and decision. The pivotal concept of this reasoning is imprecise probability, which is a generic term for a range of mathematical models that measure chance or uncertainty without sharp numerical probabilities. These models include belief functions, Choquet capacities, comparative probability orderings, convex sets of probability measures, fuzzy measures, interval-valued probabilities, possibility measures, plausibility measures, and upper and lower expectations or previsions [4].

In pursuit of uncertainty representation, aggregation, and propagation through models of reliability and risk, we employ the theory of upper and lower expectations (previsions) as described by Walley [4] and Kuznetsov [10] and build interval statistical models based on it. Generally speaking, to measure aleatory uncertainty, we need some kind of probability; to measure epistemic uncertainty, we need intervals.

This paper introduces the theory of upper and lower previsions in a 'soft' way, avoiding heavy formalism. A variety of statistical evidence admitted in the framework is exemplified. A way to account for unreliable statistical judgements is also briefly described. A short passage on the relationships between the PP, indecision, and imprecise statistical reasoning concludes the paper.

2. Discrete Case

Let us look first at what kind of discrete problem can be solved in the framework of the theory of upper and lower expectations.

Assume there are three possible outcomes s_1, s_2, and s_3 in a subject matter of interest. This is an exhaustive set of events meaning that $P(s_1) + P(s_2) + P(s_3) = 1$, where $P(\cdot)$ stands for a probability. Information on the probabilities of the occurrences of these events is given as three pieces of evidence: (1) $P(s_1) \in [0.1, 0.3]$, (2) s_2 is at least two times as probable as s_3, and (3) s_2 and s_3 is at least as probable as s_1. What probabilities $P(s_2)$ and $P(s_3)$ can one derive based on the provided information?[2]

One can hardly expect that the source imprecise information can result in precise answers in the form of precise probabilities $P(s_2)$ and $P(s_3)$. What is the mechanism for arriving at an answer?

As we have three possible outcomes, the simplex representation can demonstrate well the basic ideas of the approach. In Figure 1, the vertexes 1, 2, and 3 correspond to the three states s_1, s_2, and s_3. The probability simplex is an equilateral triangle with height one unit, in which the probabilities assigned to the three elements are identified with perpendicular distances

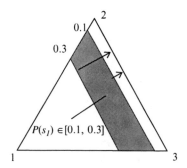

Figure 1. Presentation of the Statistical Evidence $P(s_1) \in [0.1, 0.3]$ on the Simplex.

---

[2] This Example has been Demonstrated in Greater Detail [11.]

from the three sides of the triangle. Adding up these three distances gives 1. Thus, each point inside of the simplex can be thought of as a precise probability distribution. The simplex representation is especially useful for depicting pieces of statistical evidence and studying their effects on the probabilities of outcomes.

The first piece of evidence, $P(s_1) \infty [0.1, 0.3]$, is depicted in Figure 1; Figure 2 depicts all the source information with the simplex representation.

The source evidence can be rewritten in the form of inequalities (1) $0.1 \leq P(s_1) \leq 0.3$, (2) $P(s_2) \geq 2P(s_3)$, and (3) $P(s_2) + P(s_3) \geq P(s_1)$. These inequalities and condition $P(s_1) + P(s_2) + P(s_3) = 1$ define a constrained area which is shown in black in Figure 2. The calculation of upper and lower bounds for the probabilities of interest becomes a geometric task. The calculated values of the probabilities are $\bar{P}(s_2) = 0.466$, $\bar{P}(s_2) = 0.9$, $\bar{P}(s_3) = 0$, $\bar{P}(s_3) = 0.3$, while $\bar{P}(s_1) = 0.1$ and $\bar{P}(s_1) = 0.3$ remain unchanged.

It can be noticed from Figure 2 that the evidence $P(s_2) + P(s_3) \geq P(s_1)$ does not contribute to the precision and can be discarded without influencing the result. That is, the black area, defining the lower and upper probabilities, does not change if this evidence is removed from the set of evidence. This simply supports the common-sense fact that not all information has a positive contribution to the precision of the result.[3]

The coherent imprecise probabilities are considered a particular case of the theory of imprecise coherent previsions and are based on three fundamental principles: avoiding sure loss, coherence, and natural extension. A probability model *avoids sure loss* if it cannot lead to behavior that

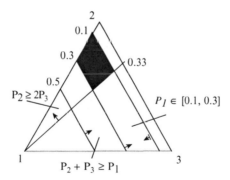

Figure 2. Presentation of All Available Statistical Evidence on the Simplex.

---

[3] Precision is considered the value of difference between the upper and lower bound of the probability of interest.

is certain to be harmful. This is a basic principle of rationality. *Coherence* is a stronger principle, which characterizes a type of self-consistency. Coherent models can be constructed from any set of probability assessments that avoid sure loss through a mathematical procedure of *natural extension* which effectively calculates the behavioral implications of the assessments [4].

The principle of avoiding sure loss for the lower and upper probabilities is equivalent to holding the following inequalities:

$$0 \leq \underline{P}(A_i) \leq \overline{P}(A_i) \leq 1 \forall i = 1, \ldots, n, \tag{1}$$

$$\underline{P}(\Omega) = \sum_{i=1}^{n} \underline{P}(A_i) \leq 1 \text{ and}$$

$$\overline{P}(\Omega) = \sum_{i=1}^{n} \overline{P}(A_i) \geq 1$$

where A_i are pairwise-disjoint subsets for any $i,j = 1, \ldots, n$ whose union is Ω, the possibility space.

The construction of coherent imprecise statistics and probabilities of events different from A_i is performed through the natural extension. The natural extension for this particular case is the solution of two linear programming problems

$$\underline{M}g = \sum_{i=1}^{n} g(A_i)P(A_i) \rightarrow \min \tag{2}$$

$$\overline{M}g = \sum_{i=1}^{n} g(A_i)P(A_i) \rightarrow \max \tag{3}$$

subject to

$$\left.\begin{array}{l} \sum_{i=1}^{n} P(A_i) = 1 \\ \underline{P}(A_i) \leq P(A_i) \leq \overline{P}(A_i), \quad i = 1, \ldots, n \end{array}\right\} \tag{4}$$

Function g can be, for example, $g = x$ and then $\underline{M}g$ and $\overline{M}$ are lower and upper mean values of x.

If g is a characteristic function of an event B, i.e., $g = I_B(A_i) = 1$ if $A_i \in B$ and $I_B(A_i) = 0$ if $A_i \notin B$, then the natural extension is

$$\underline{P}(B) = \sum_{i=1}^{n} I_B(A_i)P(A_i) \rightarrow \min \tag{5}$$

$$\overline{P}(B) = \sum_{i=1}^{n} I_B(A_i)P(A_i) \rightarrow \max \tag{6}$$

subject to (4).

The lower and upper mean values $\underline{M}g$ and $\bar{M}g$ or $\bar{p}(B)$ and $\underline{P}(B)$ obtained as the solutions of linear programming problems (2) and (3) subject to (4) and (5) and (6) subject to (4) are referred to as coherent. In [4] and [10] other definitions of the natural extension can be found.

The sense of the natural extension in precise mathematical terms is to estimate the interval $[\underline{M}g, \bar{M}g]$ of possible values of Mg for all probability distributions for which $\underline{P}(A_i) \leq P(A_i) \leq \bar{P}(A_i)$, $i = 1,\dots,n$. That is, we assume that any probability distribution consistent with the initial judgements $\underline{P}(A_i) \leq P(A_i) \leq \bar{P}(A_i)$ for $i = 1,\dots,n$ is possible and base our inferences on this assumption without preferring a particular distribution.

An example is provided below.

3. Interpretation of Upper and Lower Probabilities

For many people, the first time they heard of the Pentagon's plan to accept bets on terrorist activities was when the bizarre-sounding idea was abandoned. ...The Defence Advanced Research Projects Agency (DARPA) would have traded futures contracts that paid out if particular events, including terrorist attacks, took place. It was widely attacked as both ghoulish and nonsensical. [26]

Expressions (5) and (6) give us a formal definition (mathematical representation) of the upper and lower bounds for probabilities as maxima and minima of the objective functions subject to a set of constraints. In turn, the set of constraints also includes upper and lower probabilities. Where do they come from? How can one acquire them?

To answer these questions we need to distinguish first the issue of interpretation from that of mathematical representation. There are many kinds of mathematical models for uncertainty, such as additive probability measures, upper and lower probabilities, and comparative probability ordering. Any of these models can be given various interpretations. Similarly, any single interpretation of probability can be given various mathematical representations. De Finetti's work is a valuable example of how interpretation can profoundly affect the mathematical theory. His emphasis on finite (rather than countable) additivity and on exchangeability is a consequence of his operational interpretation [4].

Let the possibility space Ω be the set of possible states of the world that are of interest. The elements of Ω are assumed to be mutually exclusive and exhaustive. A gamble is a bounded real-valued function defined on the domain Ω. A gamble X should be interpreted as a random or uncertain reward; if the true state of the world turns out to be ω, then the reward is $X(\omega)$ units of an appropriate asset. The reward may be negative, in which case it represents a loss of $X(\Omega)$ units. The value of the reward X is uncertain, because it is uncertain which element of Ω is the true state.

Essentially, gambles are risky investments in which the utility values of the possible outcomes are known precisely [12]. The subject's uncertainty about a domain can be measured through his attitudes to gambles X defined on that domain, and particularly by determining whether he will buy or sell a gamble X for a specified price x. In principle, we could measure the subject's uncertainty concerning Ω to any desired accuracy by offering him sufficiently many gambles and observing which of them are accepted. Equivalently, we could measure the subject's lower and upper previsions for a particular Gamble X, which are defined to be the supremum acceptable buying price and infimum acceptable selling price for X. The transaction in which a Gamble X is bought at price x has reward function $X - x$, a new gamble. A subject's supremum acceptable buying price for X is the largest real number c such that he is committed to accept the gamble $X - x$ for all $x < c$. Similarly, the transaction in which a gamble X is sold for price x has reward function $x - X$, and a subject's infimum acceptable selling price for X is the smallest real number d such that he is committed to accept the gamble $x - X$ for all $x > d$. This leads to the theory of upper and lower previsions in [10]. The marginal buying and selling prices (lower and upper previsions) for a gamble may differ because the subject is indecisive or because he has little information about the gamble. As the amount of relevant information increases, the difference between the marginal buying and selling prices typically decreases. In the special case where every gamble X has a 'fair price,' meaning that the supremum acceptable buying price agrees with the infimum acceptable selling price, one obtains the theory of linear previsions [13].

Subsets of Ω, which are called events, can be identified with their indicator functions, which are gambles as well. When A is a subset of Ω, buying and selling prices (lower and upper previsions) for the indicator function A can be regarded as betting rates on and against A (lower and upper probabilities).

4. Judgements Admitted in Imprecise Statistical Reasoning: Continuous Case

The thesis that, "all available statistical evidence in risk and reliability analyses is to be utilized" is repeated in numerous guidelines in risk and reliability analysis. Everybody agrees but nobody knows how to make this true. As the remedy, Bayesian updating is usually brought up. Unfortunately, many people seem to believe that this is the only way of producing coherent statistical inferences. That is not so, for two reasons [14].

First, coherent statistical inferences need not be based on any assessment of prior probabilities. Second, even when inference proceeds by updating prior probabilities, imprecise prior probabilities can be presented by several mathematical models other than a set of prior probability distributions. In many problems it is difficult to identify a suitable prior distribution or set of prior distributions to perform Bayesian sensitivity analysis. Coherent imprecise previsions constitute an alternative method that in some problems is more convenient and traceable.

In this section I will give some examples of the judgments that can be easily utilized by the method and that are relevant for a continuous set of possible outcomes. (More on admitted judgments can be found elsewhere [15, 16].) Examples will usually involve the notion of time to failure (a continuous variable), this being a favorite target for reliability analysts. I will try to avoid giving too much mathematical formalism, but some of it cannot be avoided. To utilize a judgment it has to be represented in a mathematical form that is then used as a constraint for a properly constructed objective function.

Direct judgements on the lower and upper probabilities of events or in general—lower and upper previsions are a straightforward way to elicit the imprecise probability characteristics of interest. Constraint $\underline{a} \leq \int_{R_+} f(x)\rho(x)dx \leq \overline{a}$ is the model of a direct judgement. If, for instance, $f_i(X) = X$, then $\underline{a}_i$, $\overline{a}_i$ are the lower and upper expected values of $\underline{X}$, correspondingly. If X is time to failure, then $\underline{a}_i$, $\overline{a}_i$ are the lower and upper bounds for the mean time to failure. If $f_i(X) = I_{[t, \infty]}(X)$, where $I_{[t, \infty]}(X)$ is an indicator function such that $I_{[t, \infty]}(X) = 1$ if $X \in [t, \infty]$ and $I_{[t, \infty]}(X) = 0$ otherwise, then $\underline{a}_i$, $\overline{a}_i$ are the lower and upper bounds for the probability of failure occurrence within $[t, \infty]$.

On a general note, direct judgements can be elicited and utilized for any probability characteristic that can be represented as an expectation to a properly chosen gamble.

Being able to utilize *comparative judgements* is a good feature of the theory of imprecise previsions. They could be, for example, "the failure of component A within the time interval $[0,10]$ is at least as probable as the failure of component B within $[0,20]$," or "the mean time to failure of component B is less than the mean time to failure of component A." The first judgement is modeled as follows:

$$\int_0^\infty \int_0^\infty (I_{[0,10]}(x_A) - I_{[0,20]}(x_B))\rho(x_A, x_B)dx_A dx_B \geq 0,$$

and the second:

$$\int_0^\infty \int_0^\infty (x_A - x_B)\rho(x_A, x_B)dx_A dx_B \geq 0.$$

Another kind of judgement is a *structural judgement*. Informally, a structural judgement is a hypothetical judgement that if you were willing to accept Gamble X, then you would be willing also to accept Gamble Y[4]. Structural judgements may involve the notions (properties) of independence and permutability, and both types can be modeled.

If the objective function for computing the lower bound of the expected value of a random function g appears in a form like this

$$\bar{M}(g) = \sup_P \int_{R_+^n} g(x)\rho(x)dx, \; \underline{M}(g) = \inf_P \int_{R_+^n} g(x)\rho(x)dx,$$

where $x = (x_1, ..., x_n)$, then this models the complete ignorance with regard to independence. The infimum is sought over the set P of all possible joint probability density functions $\rho(x)$. No structural judgement is introduced here. If there is a ground on which to judge independence among x_i, then $\rho(x) = \rho(x_1)... \rho(x_n)$. It is clear that in this case set P is reduced and consists only of densities which can be represented as a product. As set P becomes smaller, then the imprecision, $\Delta = \bar{M}(g) - \underline{M}(g)$, is reduced.

In fact, the scope of judgements that can be utilized by the method is very wide (for more examples see [4], page 169). This, therefore, makes the thesis "all available statistical evidence in risk and reliability analyses is to be utilized" persuasive. This is because a tool really exists that can utilize a wide spectrum of evidence.

5. Unreliable Judgements (Hierarchical Models)

Good is prepared to define second order probability distributions..., and third order probability distributions over these, etc., until he gets tired. [17]

The quality of information that a decision maker has concerning the possible states and outcomes of a decision situation is in many cases an important factor when making decisions. Experts providing judgements have different levels of expertise and their sources of information may not be equally reliable. So it is natural to assign different degrees of plausibility or probability to opinions by different experts. To allow for this, a kind of hierarchical model can be used. In general, hierarchical models arise when there is a "correct" or "ideal" (first-order) uncertainty model about a phenomenon of interest, but the modeler is uncertain about what it is. The modeler's uncertainty is then called second-order uncertainty [12]. The hierarchical model is, in many applications, a useful assessment strategy for constructing a first-order prior distribution [14].

The most common hierarchical model is the Bayesian one, where both the first and the second-order model are (precise) probability measures [18–22]. Other models allow imprecision in the second-order model, but still assume that the first-order model is precise. Examples are the robust Bayesian models [18], models involving second-order possibility distributions [14, 23, 24], and the Gardenfors and Sahlin epistemic reliability model [5]. In [12] de Cooman introduced and studied a particular type of imprecise behavioral second-order model in terms of so-called lower desirability functions.

We have studied hierarchical uncertainty models of a general form: imprecise first- and second-order uncertainty models. Both models of uncertainty, first-order and second-order, are coherent interval statistical models.

Suppose that we have a set of unreliable interval-valued expert judgements on a parameter of interest b. To be more specific, we have n intervals $B_i = \left[b_1^i, b_2^i \right]$ provided by n experts, where b_1^i and b_2^i are the lower and upper bound of the interval B_i, respectively. The intervals provided are thought of as covering the true value of b, and are the models of uncertainty of the first order. The levels of confidence in the judgements depend on available information about experts' performance and their competences and may be subject to their own self-assessment. Suppose that each of n experts or each of their judgements is characterized by a subjective probability γ_i or, in general, by an interval-valued probability $[\underline{\gamma}_i, \overline{\gamma}_i]$, i = 1, ..., n. Now a hierarchical model can be written as follows:

$$\Pr\left\{ b_1^i \leq b \leq b_2^i \right\} \in \left[\underline{\gamma}_i, \overline{\gamma}_i \right], i = 1, \ldots, n$$

The hierarchical model is introduced to become a useful assessment strategy for constructing first-order uncertainty intervals. Its implementation is illustrated by the problem of combining expert opinions.

As given above, the information concerning a parameter b is given by a collection of n intervals B_i. Combined lower, $\underline{b}$, and upper, $\overline{b}$, bounds for b are the goals.

The result will definitely depend on the degree of credibility to each of the provided judgements. Say, the analyst is absolutely (100%) and equally confident about all the judgements. In terms of the formalism introduced above this means that $\Pr\left\{ b_1^i \leq b \leq b_2^i \right\} = 1 \forall i = 1, \ldots, n$, that is, $\underline{\gamma}_i = \overline{\gamma}_i = 1 \forall i = 1, \ldots, n$. As proven in [15], this case yields a simple rule of combination called the conjunction rule [4]:

$$\underline{b} = \max_{i=1,\ldots,n} b_1^i \text{ and } \overline{b} = \min_{i=1,\ldots,n} b_2^i$$

This rule is valid only for nonconflicting judgements ("consistent collection of intervals") and if the analyst is prepared to accept the modeling of the linguistic expression "equally credible" as $\underline{\gamma}_i = \overline{\gamma}_i = 1 \ \forall \ i = 1,\ldots, n$. Consistency as well as the absence of conflict mean that $\cap_i, {}_iB_i \neq \varnothing$.

Another rule of combination is valid if all intervals in the collection are nested ("consonant"), that is, if

$$\left[b_1^1, b_2^1\right] \subseteq \left[b_1^2, b_2^2\right] \subseteq \ldots \subseteq \left[b_1^n, b_2^n\right] \quad \text{and}$$

the credibility to the judgements is expressed in the different form $\underline{\gamma}_i = \gamma_i, \overline{\gamma}_i = 1, i = 1,\ldots n$ and $\gamma_1 \leq \gamma_2 \leq \ldots \leq \gamma_n$. A closer look at this information gives a hint that this kind of source data setup is nothing other than a possibility distribution. This case of hierarchical models was described in [12] and [14].

The combination rule for this case follows:

$$\underline{b} = \sum_{i=1}^{n} (\gamma_i - \gamma_{i-1}) b_1^i$$
$$\overline{b} = \sum_{i=1}^{n} (\gamma_i - \gamma_{i-1}) b_2^i$$

In this rule, it is assumed that $\gamma_0 = 0$ and $\gamma_n = 1$.

A model for "equally credible" judgements could be differently constructed with the hierarchical model introduced. The modeler may choose to model equal credibility in the following way:

$[\underline{\gamma}_i, \overline{\gamma}_i] = [\gamma_i, 1]$ and $\gamma_1 = \gamma_2 = \ldots = \gamma_n = \gamma$ then the last rule of combination degenerates to

$$\underline{b} = \gamma b_1^1 + (1-\gamma) b_1^n$$
$$\overline{b} = \gamma b_2^1 + (1-\gamma) b_2^n$$

If γ tends to 1, then the results of this rule coincide with the results of the conjunction rule.

The conjunction rule can also be applied to consonant intervals as this rule is valid for a consistent collection of intervals, and it is clear that nested intervals are nonconflicting pieces of evidence. But it should be kept in mind that the conjunction rule presupposes that the analyst is 100% confident about all the judgments; i.e., $\underline{\gamma}_i = \overline{\gamma}_i = 1$.

If the collection of intervals is conflicting (there is at least one pair of nonoverlapping intervals), then one way of reconciling the conflict is to accept complete ignorance concerning the level of credibility in the judgments. That is, the analyst can assume $\underline{\gamma}_i = 0, \overline{\gamma}_i = 1, \forall i = 1,\ldots n$. Using this assumption we arrive at the unanimity rule

$$\underline{b} = \min_{i=1,\dots,n} b_1^i \quad \text{and} \quad \overline{b} = \max_{i=1,\dots,n} b_2^i$$

These are simple combination rules that have been derived based on the hierarchical model, and the way they have been derived was fully predefined by the theoretical framework of coherent imprecise probabilities. This fact is worth stressing, since, in contrast, in the framework of purely Bayesian approach and point-valued probabilities only some ad-hoc combination rules are possible. An example is the linear opinion pool which is one of many others devised to combine evidence.

6. Precautionary Principle and Indecision

Determinacy and decisiveness in decision making are favored by the public and decision makers, while fuzziness and indecision in providing crisp answers are reckoned as signs of incompetence and meekness which are usually disliked. In this regard, Bayesian decision theory appears the right one as providing a clear-cut answer to what action is to be preferred.

In contrast, the approach to decision making based on imprecise (interval-valued) probabilistic criteria will reach results that, generally, do not yield an 'optimal' action that is preferred to all others. In effect, this means that there is a third alternative answer under decision making. It is indecision in saying neither 'yes' nor 'no.' The failure to determine a uniquely optimal action simply reflects the absence of information about the set of possible actions.

What would be a strategy which could be used to make a decision in case there is more than one reasonable action? One of them is to search for more information concerning the set of possible actions to make the probabilities and utilities more precise. The other is to postpone a decision until a later time, when more information may be available. For more strategies see [4], p. 239–240.

A small but growing number of authors have called for, and observed the development towards, a paradigm shift in environmental decision making. As uncertainty becomes an accepted fact by scientists on the one side and the public and politicians on the other:

this requires a change of attitude on both sides: The politicians have to accept that fuzzy answers may be the best expression of expertise. The scientists have to learn that identification of the fuzzy borderline between knowledge and ignorance may be the sign of real competence. [25]

Imprecise statistical reasoning provides models to quantify scientific incertitude that is a result of a lack of relevant information or sizable uncertainty.

When there is little information on which to base our conclusions, we cannot expect reasoning (no matter how clever or thorough) to reveal a most probable hypothesis or a uniquely reasonable course of action. There are limits to the power of reasons [4]. An educated mind should provide answers consistent with the relevant knowledge and uncertainty.

One of the important novelties of imprecise statistical reasoning approach is that we now have a formal framework in which we can articulate uncertainty and indecision.

References

1. McColl, S., Hicks, J., Craig, L. and Shortreed, J. (2000) Environmental Health Risk Management. A Primer for Canadians. NERAM.
2. Guest editorial. (2004) Alternative representation of epistemic uncertainty. *Reliability Engineering and System Safety* 85(1–3):1–369 (July–September).
3. Savage, L.J. (1954) *The Foundations of Statistics*. Wiley, New York.
4. Walley, P. (1991) *Statistical Reasoning with Imprecise Probabilities*. Chapman & Hall, New York.
5. Gårdenfors, P. and Sahlin, N.-E. (1982) Unreliable probabilities, risk taking, and decision making. *Synthese* 53:361–386.
6. Levi, I. (1986) *Hard Choices: Decision Making Under Unresolved Conflict*. Cambridge University Press, Cambridge.
7. Utkin, L. and Augustin, T. (2003) Decision Making with Imprecise Second-Order Probabilities. *In ISIPTA'03—Proceedings of the Third International Symposium on Imprecise Probabilities and their Applications*, Eds. J.-M. Bernard, T. Seidenfeld and M. Zaffalon. Carleton Scientific, Lugano, Switzerland.
8. Augustin, T. (2001) On Decision Making under Ambiguous Prior and Sampling Information. *In ISIPTA'01 – Proceedings of the Second International Symposium on Imprecise Probabilities and Their Applications*, Eds. G. de Cooman, T.L. Fine and T. Seidenfeld. Cornell University, Ithaca, NY.
9. Simon, H.A. (1983) *Reason in Human Affairs*. Stanford University Press, Stanford, CA.
10. Kuznetsov, V. (1991) *Interval statistical models* [In Russian]. Radio and Sviaz. Moscow.
11. Kozine, I. and Utkin, L.V. (2002a) Interval-valued finite Markov chains. *Reliable Computing* 8:97–113.
12. de Cooman, G. (1999) Lower desirability functions: a convenient imprecise hierarchical uncertainty model, *ISIPTA '99: Proceedings of the First International Symposium on Imprecise Probabilities and Their Applications,* eds. G. de Cooman, F. G. Cozman, S. Moral and P. Walley, pp. 111–120, Imprecise Probabilities Project, Gent.
13. De Finetti, B. (1974) *Theory of Probability. A Critical Introductory Treatment*. Wiley, Chichester.
14. Walley, P. (1997) Statistical inferences based on a second-order possibility distribution. *International Journal of General Systems* 9:337–383.
15. Kozine, I.O. and Utkin, L.V. (2002b) Processing unreliable judgements with an imprecise hierarchical model. *Risk Decision Policy* 7:1–15.
16. Kozine, I. and Utkin, L.V. (2003) Variety of judgements admitted in imprecise statistical reasoning. *Risk Decision Policy* 8:111–120.
17. Levi, I. (1973) Inductive logic and the improvement of knowledge. *Technical Report*, Columbia University.
18. Berger, J.O. (1985) *Statistical Decision Theory and Bayesian Analysis*. Springer, New York.

19. Cooke, R.M. (1991) *Experts in Uncertainty. Opinion and Subjective Probability in Science.* Oxford University Press, New York.
20. Good, I.J. (1979) Some History of the Hierarchical Bayesian Methodology. Eds. Bernardo, J.M. et al. *Bayesian Statistics. Proceedings of the First International Meeting Held in Valencia* (Spain), University Press.
21. von Winterfeldt, D. and Edwards, W. (1986) *Decision Analysis and Behavioral Research.* Cambridge University Press, Cambridge.
22. Zellner, A. (1971) *An Introduction to Bayesian Inference in Econometrics.* Wiley, New York.
23. de Cooman, G. (1998) Possibilistic Previsions. *Proceedings of IPMU'98 (Seventh Conference on Information Processing and Management of Uncertainty in Knowledge-Based Systems,* July 6–10, 1998, Paris, France), Editions E.D.K.
24. Gilbert, L., de Cooman, G. and Kerre, E.E. (2000) Practical implementation of possibilistic probability mass functions, *Proceedings of Fifth Workshop on Uncertainty Processing* (WUPES 21–24 June, Jindvrichouv Hradec, Czech Republic), pp. 90–101.
25. Harremoes, P. (2003) The need to account for uncertainty in public decision making related to technological change. *Integrated Assessment* 4(1):18–25.
26. Harford, T. (2003) All bets are off at the Pentagon. *Financial Times.* September 2.

SMAA-III

A Simulation-Based Approach for Sensitivity Analysis of ELECTRE III

T. TERVONEN AND J.R. FIGUEIRA

Centre for Management Studies, Instituto Superior Técnico
(CEG-IST) Technical University of Lisbon
2780-990 Porto Salvo, Portugal
tommi.tervonen@ist.utl.pt, figueira@ist.utl.pt

R. LAHDELMA

University of Turku, Department of Information Technology
Joukahaisenkatu 3-5, FIN-20014 Turku, Finland
risto.lahdelma@cs.utu.fi

P. SALMINEN

School of Business and Economics, University of Jyväskylä,
P.O. Box 35
FIN-40014 Jyväskylä, Finland
p.salmine@econ.jyu.fi

Abstract: ELECTRE III is a well established multiple-criteria decision-making method with a solid track record of real-world applications. It requires precise values to be specified for the parameters and criteria measurements, which in some cases might not be available. In this paper we present a method, SMAA-III, which allows ELECTRE III to be applied with imprecise parameter values. By allowing imprecise values, the method also allows an easily applicable robustness analysis. In SMAA-III, simulation is used and descriptive measures are computed to characterize stability of the results.

1. Introduction

ELECTRE III is a well established multiple-criteria decision-making (MCDM) method for ranking a discrete set of alternatives. It belongs to the ELECTRE family of methods, which are based on constructing and

I. Linkov et al. (eds.), Real-Time and Deliberative Decision Making.
© Springer Science+Business Media B.V. 2008

exploiting an outranking relation [3]. ELECTRE III has a long history of successful real-world applications in different areas. The inputs for ELECTRE III consist of criteria evaluations on a set of alternatives and preference information expressed as weights and thresholds.

ELECTRE III is a pseudocriteria-based model, and as such it uses a threshold to model indifference between pairs of alternatives. Although this threshold might be an easy concept for a typical decision maker (DM) to understand, simulation studies have shown that it causes the model be quite unstable with respect to changes in the indifference threshold value [8]. Because of this instability, robustness should always be analyzed by considering different values for the threshold.

Real-world decision-making problems in general include various types of uncertainties inherent in problem structuring and analysis [1]. Eliciting the DMs' preferences in terms of relative criteria importance coefficients or weights is usually difficult. Such weights should always be considered imprecise, because humans usually do not think about preferences as exact numerical values, but as more vague concepts [14]. In some cases, weight information may be entirely missing, which corresponds to extremely imprecise weights.

This work presents a tool for dealing with imperfect knowledge within the ELECTRE III method. It can be used either when information is poor or when a robustness analysis needs to be done. The way robustness analysis is conducted comprises intensity of exploration in the parameter space. This is achieved by applying simulation in such a way that the parameter space is explored with a high concentration of discrete values. In addition to this, the exploration is coherent with the model. This means that, for example, when exploring the weight space, the meaning of weight is taken into account. In ELECTRE III weights represent the number of "votes" criteria have.

Capability to derive robust conclusions when applying MCDM methods is nowadays of utmost importance. The main sources of imperfect knowledge that are present in complex and multifaceted decision-making situations require careful observation of the results, and make them dependent on an exploration of the neighborhood of the parameters used mainly to represent preferences or technical aspects of the problem. If an alternative almost always occupies the first position when changing simultaneously all the parameters in a certain neighborhood, it means that it can be a good choice for future implementation; these are the kind of robust conclusions we are interested in.

The method presented in this paper is based on Stochastic Multicriteria Acceptability Analysis (SMAA) [7], a family of decision support methods for aiding DMs in discrete decision-making problems. For a survey of SMAA methods, refer to Tervonen and Figueira [15]. The proposed method, SMAA-III, explores weight, criteria measurement, and threshold spaces, in order to describe which values result in certain ranks for the alternatives. It allows ELECTRE III to be used with different kinds of imprecise or partially missing information. This brings numerous advantages. Firstly, SMAA-III allows performing an initial analysis without preference information in order to eliminate "inferior" alternatives. Secondly, it allows DMs to express their preferences imprecisely, which can lower the DMs' cognitive effort compared to specifying precise weights. Thirdly, imprecise criteria measurements can be represented with arbitrary joint probability distributions, modeling imprecision in a coherent way not possible with ELECTRE III. Fourthly, it allows representation of the preferences of a group of DMs. Fifthly, the method can be used for analyzing the robustness of the results by representing the imprecision of the elicited weights as constraints or as suitable probability distributions.

In SMAA-III, robustness is analyzed with respect to weights, criteria measurements, and thresholds. Traditionally, robustness with ELECTRE methods is analyzed by considering discrete points in the weight space (see, e.g., [12]). But in the case of ELECTRE III this is not enough: weights between these points that might give contradictory results are missed. There are also simulation techniques for robustness analysis outside the SMAA methodology [2], but to the best of our knowledge, they have never before been applied to ELECTRE III.

This paper is organized as follows: ELECTRE III is briefly introduced in Section 2. SMAA-III is presented in Section 3. We skim rapidly through some computational aspects in Section 4 before proceeding to conclusions in Section 5.

2. ELECTRE III

ELECTRE III is designed for solving a discrete ranking problem. It consists of m alternatives $a_1, ..., a_i, ..., a_m$, which are evaluated in terms of n criteria $g_1, ..., g_j, ..., g_n$. We denote by J the set of criterion indices. $g_j(a_i)$ is the evaluation of criterion g_j for alternative a_i. Without loss of generality, we assume that all criteria are to be maximized.

Similarly to the other ELECTRE family methods, ELECTRE III is based on two phases. In the first phase, an outranking relation between pairs of alternatives is formed. The second phase consists of exploiting this relation, producing a final partial pre-order and a median pre-order.

S denotes the *outranking* relation, that is, aSb denotes that "alternative a is at least as good as alternative b."

ELECTRE III applies pseudocriteria in constructing the outranking relation. A pseudocriterion is defined with two thresholds for modeling preference: an indifference threshold $q_j(g_j(\cdot))$ for defining the difference in criterion g_j that the DM deems insignificant, and a preference threshold $p_j(g_j(\cdot))$ for the smallest difference that is considered absolutely preferred. Between these two is a zone of "hesitation" between indifference and strict preference. ELECTRE III also defines a third threshold: the veto threshold $v_j(g_j(\cdot))$. It is the smallest (negative) difference that completely nullifies (raises a "veto" against) the outranking relation. In addition to the thresholds, preferences are quantified through a weight vector $w = (w_1,\ldots, w_j,\ldots, w_n)$. Without loss of generality, we assume that $\sum_{j \in J} w_j = 1$.

Exploitation of the outranking relation produces a partial pre-order, in which every pair of alternatives is connected with indifference (I), incomparability (R), or preference ($\succ$) relation.

2.1. CONSTRUCTING THE OUTRANKING RELATION

The outranking relation between every pair of alternatives is constructed based on a comprehensive concordance index and partial discordance indices. The concordance index is computed by considering individually for each criterion g_j the support it provides for the assertion aSb, "a outranks b with respect to criterion g_j". The partial concordance index is computed as follows, for all $j \in J$:

$$
c_j(a,b) = \begin{cases}
1, & \text{if } g_j(b) - g_j(a) \leq q_j(g_j(a)), \\[2ex]
\dfrac{g_j(a) + p_j(g_j(a)) - g_j(b)}{p_j(g_j(a)) - q_j(g_j(a))}, & \text{if } q_j(g_j(a)) < g_j(b) - g_j(a) \leq p_j(g_j(a)), \\[2ex]
0, & \text{if } g_j(b) - g_j(a) > p_j(g_j(a)).
\end{cases}
$$

After computing the partial concordance indices, the comprehensive concordance index is computed as follows:

$$
c(a,b) = \sum_{j \in J} w_j c_j(a,b).
$$

The discordance of criterion g_j describes the veto effect this criterion imposes against the assertion aSb. The partial discordance indices are computed separately for each criterion $j \in J$:

$$d_j(a,b) = \begin{cases} 1, & \text{if } g_j(b) - g_j(a) \geq v_j(g_j(a)), \\ \dfrac{g_j(b) - g_j(a) - p_j(g_j(a))}{v_j(g_j(a)) - p_j(g_j(a))}, & \text{if } p_j(g_j(a)) \leq g_j(b) - g_j(a) < v_j(g_j(a)), \\ 0, & \text{if } g_j(b) - g_j(a) < p_j(g_j(a)). \end{cases}$$

By applying the previously mentioned indices, the degree of credibility of the outranking assertion aSb is defined as:

$$\rho(a,b) = \begin{cases} c(a,b) \displaystyle\prod_{j \in V} \dfrac{1 - d_j(a,b)}{1 - c(a,b)}, & \text{if } V \neq \varnothing \\ c(a,b), & \text{otherwise,} \end{cases}$$

with

$$V = \{ j \in J : d_j(a,b) > c(a,b) \}.$$

Notice that when $d_j(a,b) = 1$ for any $j \in J$, this implies that $\rho(a,b) = 0$.

2.2. THE EXPLOITATION PROCEDURE

The exploitation of the outranking relation consists of two phases. In the first phase, two complete pre-orders, Z_1 (descending) and Z_2 (ascending) are constructed with the so-called distillation procedures. In the second phase, a final partial pre-order or a complete median pre-order is computed based on these two pre-orders.

The distillation procedures work by iteratively cutting the fuzzy outranking relations with descending λ-cutting levels. With a given cutting level λ_*, alternative a outranks alternative b ($aS_{\lambda_*}b$) if the following holds:

$$aS^{\lambda_*}b \Leftrightarrow \begin{cases} \rho(a,b) > \lambda_*, \text{ and} \\ \rho(a,b) > \rho(b,a) + s(\rho(a,b)), \end{cases}$$

where $s(\cdot)$ is the distillation threshold, usually defined as [1]:

$$s(x) = 0.3 - 0.15x.$$

The pre-orders are constructed in an iterative manner. In each step the alternatives with the highest or lowest qualification scores are distilled, depending on whether the distillation is descending or ascending. The qualification score is computed as a difference between the number of alternatives that the selected alternative outranks and the number of alternatives that outrank it for a given cutting level. The procedure is presented in Algorithm 1.

In the original ELECTRE III, a median pre-order is computed based on the two complete pre-orders, Z_1 and Z_2, and the final partial pre-order. The final partial pre-order is computed as the intersection of the two complete pre-orders in such a way that the following relations hold:

$$a \succ b \Leftrightarrow (a \succ^{Z_1} b \wedge a \succ^{Z_2} b) \vee (aI^{Z_1}b \wedge a \succ^{Z_2} b) \vee (a \succ^{Z_1} b \wedge aI^{Z_2}b),$$

$$aIb \Leftrightarrow (aI^{Z_1}b \wedge aI^{Z_2}b),$$

$$aRb \Leftrightarrow (a \succ^{Z_1} b \wedge b \succ^{Z_2} a) \vee (b \succ^{Z_1} a \wedge a \succ^{Z_2} b).$$

Algorithm 1: Distillations

1. Determine the maximum value of the credibility indices in the set under consideration. Assign this to λ.

2. Determine $\lambda_4 = \max_{d(a,b) < \lambda - z(\lambda)} \{d(a,b)\}$, where (a,b) belong to the set under consideration.

3. If $\lambda_4 = 0$, end this distillation.

4. Determine for each alternative its *qualification* score; that is, the difference between the number of alternatives it outranks and the number of alternatives that outrank it. Outranking is determined according to λ_*.

5. The set of alternatives having the largest (or smallest, if the distillation is ascending) qualification is the current distillate.

6. If the number of alternatives in the current distillate is larger than 1, repeat the process from step 2 inside the distillate.

7. Form a new set under consideration by removing the distilled alternatives from the current one. If this set is not empty, repeat the process on the new set from step 1.

8. The final pre-orders are ranked so that the alternatives in the first distillate are given rank 1, in the second rank 2, etc.

After this, the median pre-order can be computed by removing the incompa-rabilities and calculating the differences of ranks of an alternative in the two complete pre-orders.

2.3. ROBUSTNESS ANALYSIS FOR WEIGHTS

There are numerous weight elicitation techniques proposed for ELECTRE methods; the following are among the most recent and popular:

1. DIVAPIME by Mousseau [10] produces intervals for weights.

2. Hokkanen and Salminen [5] used two different weight elicitation proc edures and found that the normalized sets of weights had minor differ-ences.

3. SRF by Figueira and Roy [4] allows weight elicitation in a user-friendly manner by using a technique based on a pack of "playing cards" to determine the relative importance of criteria coefficients. It can produce interval weights and was also designed to support multiple DMs.

4. The approach proposed by Rogers and Bruen [11] uses pairwise compari-sons to elicit the weights.

The first three techniques, which produce intervals or two-weight sets that may be used to define intervals, can be used directly in robustness analysis. With the fourth weight elicitation technique, intervals (such as ± 10%) could be defined around the original weights.

Traditionally the robustness analysis for ELECTRE methods has been an ad hoc investigation into the effect of changing values [1]. This type of investi-gation typically considers only discrete points (for example, extreme points) of the feasible weight space (e.g., weight intervals). The procedure of building the pre-orders is based on exploiting the fuzzy outranking relation, which is non-linear and discontinuous by nature. Therefore, instead of just a few discrete points, it is important to analyze the entire continuum of the weight space.

3. SMAA-III

In order to overcome the limitations of ELECTRE III, SMAA-III applies simulation and studies the effect of changing parameter values and criterion evaluations on the results. The imprecision is quantified through joint density functions in the corresponding spaces.

The weights are represented by a weight distribution with joint density function $f_W(w)$ in the feasible weight space W. The weights are non-negative and normalized. The weight space is an $n - 1$ dimensional simplex:

$$W = \left\{ w \in R^n : w \geq 0 \text{ and } \sum_{j \in J} w_j = 1 \right\}.$$

Completely missing preference information is represented by a uniform (constant) weight distribution in W; that is:

$$f_W(w) = 1 / \text{vol}(W).$$

If some kind of preference information is available, different weight distributions can be applied [7]. In practice, the preferences can usually be elicited as interval constraints for weights. In this case, a uniform distribution in the space bounded by the constraints is used. Figure 1 illustrates the restricted feasible weight space of a three-criteria problem with lower and upper bounds for w_1. In this paper the focus is on weight information provided as intervals, because:

1. If there are multiple DMs whose preferences need to be taken into account, the weight intervals in general can be determined to contain the preferences of all DMs [7].

2. Weight intervals allow simple robustness analysis even when only deterministic weights are available, by specifying, for example, a $\pm 10\%$ interval for each weight.

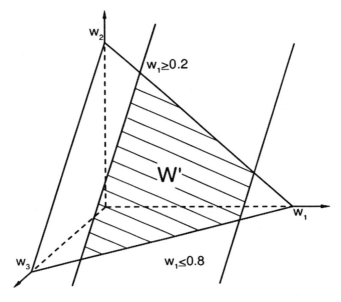

Figure 1. Feasible Weight Space of a Three-criteria Problem with Lower and Upper Bounds for w_1.

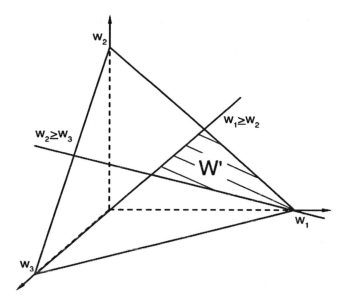

Figure 2. Feasible Weight Space of a Three-criteria Problem with Ranking of the Criteria.

It should be observed that other forms of easily elicitable preference information can be used as well, such as ranking of the criteria. A ranking can be obtained by asking the DMs to identify their most important and second most important criterion etc. Figure 2 illustrates the feasible weight space for a three-criteria problem with the ranking $w_1 \geq w_2 \geq w_3$.

Imprecise thresholds are represented by stochastic functions $\alpha_j(\cdot)$, $\beta_j(\cdot)$, and $\gamma_j(\cdot)$, corresponding to the deterministic thresholds $p_j(\cdot)$, $q_j(\cdot)$, and $v_j(\cdot)$, respectively. To simplify the notation, we define a 3-tuple of thresholds $\tau = (\alpha, \beta, \gamma)$. It has a joint density function f_τ in the space of possible values defining the functions. It should be noted that all feasible combinations of thresholds must satisfy $q_j(a_i) < p_j(a_i) < v_j(a_i)$.

Traditionally the thresholds in ELECTRE models have been used to model preferences of the DMs (e.g., differences deemed significant) as well as data imprecision. But it has been shown that the indifference threshold does not correspond to a linear imprecision interval [8]. Therefore, in SMAA-III thresholds are used only to model preferences (together with weights). Imprecision in the criteria measurements is modeled with stochastic variables.

These stochastic variables are denoted with ξ_{ij} corresponding to the deterministic evaluations $g_j(a_i)$. They have a density function $f_X(\xi)$ defined in the space $X \subseteq R^{m \times n}$. In principle, arbitrary distributions can be used, but in practice a uniform distribution in a certain interval or a Gaussian distribution is used.

Incomparabilities between alternatives can be present in the final results of ELECTRE III. This is one of the main features of ELECTRE methods in comparison with the methods applying classical multi-attribute utility theory [6]. In the late seventies, it was considered a very important theoretical advance. But, in reality when dealing with practical situations, incomparabilities in the final result are inconvenient. This aspect was soon observed [13] and partial preorders were replaced by complete pre-orders or median pre-orders. We apply median pre-orders in computing rank acceptability indices. The only information lost in using the median pre-order as the primary measure of the ranking is the incomparability. As our method is also aimed to help analysts accustomed to ELECTRE III, we will later present another index to measure incomparability.

$$a \succ b \Leftrightarrow \begin{cases} (a \succ^{Z_1} b \wedge a \succ^{Z_2} b) \vee (aI^{Z_1}b \wedge a \succ^{Z_b} b) \vee (a \succ^{Z_1} b \wedge aI^{Z_2}b) \\ (a \succ^{Z_1} b \wedge b \succ^{Z_2} a) \wedge (|r^{Z_1}(a) - r^{Z_2}(a)| < |r^{Z_1}(b) - r^{Z_2}(b)|), \\ (b \succ^{Z_1} a \wedge a \succ^{Z_2} b) \wedge (|r^{Z_1}(a) - r^{Z_2}(a)| < |r^{Z_1}(b) - r^{Z_2}(b)|) \end{cases} \quad (1)$$

$$aIb \Leftrightarrow \neg(b \succ a) \wedge \neg(a \succ b),$$

where $r(\cdot)$ is the ranking of an alternative in the superscripted pre-order. Monte Carlo simulation is used in SMAA-III to compute three types of descriptive measures: rank acceptability indices, pairwise winning indices, and incomparability indices. In order to compute these indices, let us define a *ranking function* that evaluates the rank r of the alternative a_i with the corresponding parameter values:

$$\text{rank}(i, w, \xi, \tau).$$

The evaluation of this function corresponds to executing ELECTRE III and returning the rank of the corresponding alternative in the resulting median pre-order. We will next introduce the three indices. Interpretation of their values is presented in Section 4 through various re-analyses.

3.1. RANK ACCEPTABILITY INDEX

The rank acceptability index, b_i^r, measures the share of feasible weights that grant alternative a_i rank r in the median pre-order by simultaneously taking into account imprecision in all parameters and criterion evaluations. It represents the share of all feasible parameter combinations that make the alternative acceptable for a particular rank, and it is most conveniently expressed as a percentage.

The rank acceptability index b_i^r is computed numerically as a multidimensional integral over the spaces of feasible parameter values as:

$$b_i^r = \int_{W:\text{rank}(i,w,\xi,\tau)=r} f_W(w) \int_X f_X(\xi) \int_T f_T(\tau) dT dw d\xi.$$

The most acceptable ("best") alternatives are those with high acceptability for the best ranks. Evidently, the rank acceptability indices are within the range [0,1], where 0 indicates that the alternative will never obtain a given rank and 1 indicates that it will always obtain the given rank with any feasible choice of parameters.

Using the rank acceptability indices as measures of robustness is quite straightforward: when the index is near 1, the conclusion is robust. Nevertheless, caution should be used when interpreting the results in cases where these indices are computed without weight information to characterize the problem. If an alternative obtains a low score for first-rank acceptability, it does not necessarily mean that it is "inferior". The DMs' actual preferences may well lie within the corresponding (small) set of favorable first-rank weights.

3.2. PAIRWISE WINNING INDEX

The pairwise winning index o_{ik} [9] describes the share of weights that place alternative a_i on a better rank than alternative a_k. An alternative a_i that has $o_{ik} = 1$ for some k always obtains a better rank than alternative a_k, and can thus be said to *dominate* it.

The pairwise winning index o_{ik} is computed numerically as a multidimensional integral over the space of weights that gives a lower rank for one alternative than for another:

$$o_{ik} = \int_{w \in W : \text{rank}(i,w,\xi,\tau) < \text{rank}(k,w,\xi,\tau)} f_W(w) \int_X f_X(\xi) \int_T f_T(\tau) dT dw d\xi.$$

The pairwise winning indices are especially useful when trying to distinguish between the ranking differences of two alternatives. Because the number of ranks in the median pre-order varies among different simulation runs, two alternatives might obtain similar rank acceptabilities although one is in fact inferior. In these cases looking at the pairwise winning indices between this pair of alternatives can help to determine whether one of the alternatives is superior to the other or if they are equal in "goodness."

3.3. INCOMPARABILITY INDEX

Because median pre-orders are used in computing the rank acceptability indices, it is no longer possible to model incomparability. As some DMs might be accustomed to make decisions that take incomparabilities into account, another index is introduced. Incomparability index σ_{ik} measures the share of feasible parameter values that cause alternatives a_i and a_k to be incomparable. For this reason, we define the incomparability function:

$$R(i,k,\xi,\tau) = \begin{cases} 1, & \text{if the alternatives } a_i \text{ and } a_k \text{ are incomparable,} \\ 0, & \text{if not.} \end{cases}$$

This function corresponds to running ELECTRE III with the given parameter values and checking if the alternatives are judged incomparable in the final partial pre-order. In practice we do not compute the final partial pre-order, because this information can be extracted from the two partial pre-orders Z_1 and Z_2 as shown (1). By using the incomparability function, the incomparability index is computed numerically as a multidimensional integral over the feasible parameter spaces as:

$$\sigma_{ik} = \int_W f_W(w) \int_X f_X(\xi) \int_T f_T(\tau) R(i,j,\xi,\tau) dT dw d\xi.$$

4. Computation

All of the indices mentioned above are computed with Monte Carlo simulation. The procedure is similar to that presented and analyzed by Tervonen and Lahdelma [16]. SMAA-III differs in the sense that it applies the ELECTRE III procedure to derive the descriptive values instead of a utility function.

In each simulation iteration, sample parameter values are generated from their corresponding distributions, and ELECTRE III is executed with these values. Then the corresponding hit counters are updated as with the original SMAA. If standard distributions are used for defining the imprecise parameter values, then all sampling operations except weight generation are computationally very light. In the case of weight generation, if tight upper bounds are used, we can have very high weight rejection ratios (up to 99.9%). Nevertheless, even with 99.9% weight rejection, the method is fast enough to use in an interactive decision-making process with problems of reasonable size.

To obtain sufficient accuracy for the indices, we suggest using at least 10,000 simulation iterations. This gives error limits of less than 0.01 with 95% confidence [16].

5. Conclusions

In this paper we introduced a new method, SMAA-III, which allows the parameters and criteria measurements of ELECTRE III to be imprecise and to be defined with various types of constraints: no deterministic values are

required. This has numerous advantages, especially in the context
with multiple DMs, because the parameters can be determined a
that contain the preferences of all DMs. It also allows an easily ɩ
robustness analysis to be performed.

References

1. Belton, V., Stewart, T. J. (2002) *Multiple Criteria Decision Analysis—An Inte* *Approach*, Kluwer, Dordrecht.
2. Butler, J., Dia, J., Dyer, J. (1997) Simulation techniques for the sensitivity analysis of n criteria decision models, *European Journal of Operational Research* 103 (3), 531–545.
3. Figueira, J., Mousseau, V., Roy, B. (2005) ELECTRE methods. In: Figueira, J., Greco, Ehrgott, M. (Eds.), *Multiple Criteria Decision Analysis: State of the Art Surveys*, Ch. Springer Science+Business Media, New York.
4. Figueira, J., Roy, B. (2002) Determining the weights of criteria in the ELECTRE ty methods with a revised Simos' procedure, *European Journal of Operational Research* 139 317–326.
5. Hokkanen, J., Salminen, P. (1997) Choosing a solid waste management system using mul ticriteria decision analysis, *European Journal of Operational Research* 98, 19–36.
6. Keeney, R., Raiffa, H. (1976) *Decisions with Multiple Objectives: Preferences and Value Tradeoffs*, Wiley, New York.
7. Lahdelma, R., Salminen, P. (2001) SMAA-2: Stochastic multicriteria acceptability analy sis for group decision making, *Operations Research* 49 (3), 444–454.
8. Lahdelma, R., Salminen, P. (2002) Pseudo-criteria versus linear utility function in sto chastic multicriteria acceptability analysis, *European Journal of Operational Research* 14, 454–469.
9. Leskinen, P., Viitanen, J., Kangas, A., Kangas, J. (2006) Alternatives to incorporate uncer tainty and risk attitude in multicriteria evaluation of forest plans, *Forest Science* 52 (3), 304–312.
10. Mousseau, V. (1995) Eliciting information concerning the relative importance of criteria. In: Pardalos, Y., Siskos, C., Zopounidis, C. (Eds.), *Advances in Multicriteria Analysis*, pp. 17–43, Kluwer, Dordrecht.
11. Rogers, M., Bruen, M. (1998) A new system for weighting environmental criteria for use within ELECTRE III, *European Journal of Operational Research* 107, 552–563.
12. Rogers, M., Bruen, M., Maystre, L.-Y. (2000) *ELECTRE and Decision Support: Methods and Applications in Engineering and Infrastructure Investment*, Kluwer, Dordrecht.
13. Roy, B., Présent, M., Silhol, D. (1986) A programming method for determining which Paris metro stations should be renovated, *European Journal of Operational Research* 24, 318–334.
14. Smets, P. (1991) Varieties of ignorance and the need for well-founded theories, *Information Sciences* 57–58, 135–144.
15. Tervonen, T., Figueira, J.R. (2007) A survey on stochastic multicriteria acceptability analysis methods, *Journal of Multi-Criteria Decision Analysis*, DOI:10.1002/mcda.407.
16. Tervonen, T., Lahdelma, R. (2007) Implementing stochastic multicriteria acceptability analysis, *European Journal of Operational Research* 178 (2), 500–513.

ATTRACTING ADDITIONAL INFORMATION FOR ENHANCING THE UNCERTAINTY MODEL

Towards Improved Risk Assessments

V.G. KRYMSKY

Industrial Electronics Department, Ufa State Aviation Technical University
12 K. Marx Street
Ufa, Bashkortostan 450000 Russia
kvg@mail.rb.ru

Abstract: The paper outlines a reasonable modification of an approach developed in the framework of imprecise prevision theory and adapted to the available information about some features of probability density functions. This reduces the uncertainty associated with risk analysis operations and as a result leads to obtaining the close interval estimations of statistical characteristics necessary for decision support.

1. Introduction

Reasonable use of available information on factors and phenomena is indeed the root principle of obtaining adequate risk assessments for effective decision support. As risk is typically considered in the form of composed probabilities of events and their consequences, the statistical model of the situation (scenario) is of great importance for achieving correct analytical results. Everybody who deals with risk analysis confirms that the level of uncertainty can be very high (this is caused by the lack of initial statistical data; data collection is poor because the events are rare). The only option is to elicit subjective information from experts [1]. However, we would like to use the most reliable expert judgements to derive a model with acceptable accuracy. This means that suitable but inaccurate assumptions are not allowed.

If the uncertainty is so radical that nothing can be said even about the distribution families related to events or influencing factors, then we face a problem statement in which all the distributions are plausible. This problem statement falls in the scope of the imprecise prevision theory (IPT), established in fundamental publications by Walley and Kuznetsov [2, 3]. IPT is unique in searching for at least some conclusions about the performance of extremely uncertain characteristics. The main advantage of IPT is its capacity to combine both objective statistical and expert information to estimate the lower and upper bounds of probabilities and other relevant data. Such estimates can be obtained without any assumptions of a specific prior distribution law by solving linear programming problems.

As has been demonstrated [4], the impediment to previous IPT methodology is that optimal solutions are defined for a family of degenerated distributions (in other words, distributions composed of δ-functions). The existence of solutions for degenerated distributions often leads to high imprecision, negating the pragmatic value of the assessments of interest (especially for risk analysis applications).

The negative issues associated with attempts to quantify uncertainty via IPT algorithms can be reduced by incorporating some additional information on model features. This paper discusses a strategy of enhancing the estimation technique by means of 'economic' addition of available information, which allows computing more precise bounds of the intervals for the resulting assessments.

2. Imprecise Previsions: Traditional Problem Statement

Let $\rho(x)$ be unknown probability density function of a continuous random variable X distributed in the interval $[0, T]$. Traditional IPT problem formulation (one-dimensional case) [2–4] considers the following constraints:

$$\rho(x) \geq 0, \int_0^T \rho(x)dx = 1, \text{ and } \underline{a}_i \leq \int_0^T f_i(x)\rho(x)dx \leq \overline{a}_i, \quad i = 1, 2, \dots, n. \quad (1)$$

Here $f_i(x)$ are the given real-valued positive functions ("gambles") and $\underline{a}_i, \overline{a}_i \in R_+$ are the given numbers.

Computing the coherent lower and upper previsions $\underline{M}(g)$ and $\overline{M}(g)$ for expectation $M(g)$ of any function $g(x)$, which is also a gamble, requires estimating

$$\inf_{\rho(x)} \int_0^T g(x)\rho(x)dx, \text{ as well as } \sup_{\rho(x)} \int_0^T g(x)\rho(x)dx, \quad (2)$$

subject to constraints (1).

As is known [2, 3], optimization problem (1), (2) is of the linear programming type. So the main approach to searching for a corresponding solution involves forming a dual of initial problem statement. In turn such a dual can be easily solved in many practical cases.

The dual for optimization problem (1), (2) follows:

$$\underline{M}(g) = \sup_{c_0, c_i, d_i} \left(c_0 + \sum_{i=1}^{n} (c_i \underline{a}_i - d_i \overline{a}_i) \right)$$

(3)

subject to $c_0 \in \mathbf{R}$, c_i, $d_i \in \mathbf{R}_+$ and for any $x \geq 0$, $i = 1, 2, \ldots, n$,

$$c_0 + \sum_{i=1}^{n} (c_i - d_i) f_i(x) \leq g(x).$$

(4)

and

$$\overline{M}(g) = \inf_{c_0, c_i, d_i} \left(c_0 + \sum_{i=1}^{n} (c_i \overline{a}_i - d_i \underline{a}_i) \right),$$

(5)

subject to $c_0 \in \mathbf{R}$, $c_i, d_i \in \mathbf{R}_+$ and for any $x \geq 0$, $i = 1, 2, \ldots, n$,

$$c_0 + \sum_{i=1}^{n} (c_i - d_i) f_i(x) \geq g(x).$$

(6)

Investigation [4] shows that function $\rho(x)$ for which $M(g)$ attains the values of $\underline{M}(g)$ or $\overline{M}(g)$ belongs to a family of degenerated distributions (this density is composed of δ-functions). This undesirable fact is like a "payment" for reasoning under too high a level of uncertainty. Very often we may incorporate some limited additional information (typically elicited from experts), which has the capacity to provide more valuable analytical results. One possible method has been described previously [5].

3. The Case of Bounded Densities

The first portion of additional information which allows achieving improvement when solving the optimization problem (1), (2) is presented in the form of the bounded probability densities. To get these data we have to ask an expert questions like "What is the largest possible percentage of accidents per year/decade for a given plant with definite age?" The resulting judgement is reflected by inequality:

$$\rho(x) \leq K = const,$$

(7)

where K is a real positive number satisfying the condition $KT \geq 1$.

New problem formulation requires optimizing the objective function (2) subject to constraints (1), (7). This problem can be solved via the methods of the calculus of variations [5]. The resulting optimal density function becomes a member of a family of step-functions equal to either zero or K (so degenerated solutions are eliminated). This leads to much more precise previsions (numerical examples confirm improvement of 50% in estimating the upper and the lower bounds of $M(g)$).

The knowledge of the solution type creates an opportunity for reducing the initial problem that belongs to scope of the calculus of variations to the easier-to-solve problem of optimizing a multivariable function subject to algebraic constraints.

Indeed, denote the intervals $[x_0, x_1), [x_2, x_3), [x_4, x_5), \ldots$, where $\rho(x) = K \neq 0$. Also denote

$$G(x_j, x_{j+1}) = \int_{x_j}^{x_{j+1}} g(x) dx; \tag{8}$$

$$\Phi_i(x_j, x_{j+1}) = \int_{x_j}^{x_{j+1}} f_i(x) dx, \quad i = 1, 2, \ldots, n. \tag{9}$$

Then we can reformulate our optimization problem:

$$\min_{x_0, x_1 \cdots} \left\{ K \cdot \sum_{j=0}^{m} G(x_{2j}, x_{2j+1}) \right\} \text{and} \max_{x_0, x_1 \cdots} \left\{ K \cdot \sum_{j=0}^{m} G(x_{2j}, x_{2j+1}) \right\} \tag{10}$$

$$K \cdot \sum_{j=0}^{m} (x_{2j+1} - x_{2j}) = 1; \tag{11}$$

$$\underline{a_i} \leq K \cdot \sum_{j=0}^{m} \Phi_i(x_{2j}, x_{2j+1}) \leq \overline{a_i}, \quad i = 1, 2, \ldots, n. \tag{12}$$

To solve such multivariable optimization problems in the general case, we can apply a lot of numerical methods like gradient algorithms, simplex-planning search, and genetic algorithms. In some simple situations, a solution can be reached in analytical form.

The remaining question is, how to choose the value of m? Very often we don't know this value $a\ priori$.

The recommendation for these situations is as follows: start from small values of m (e.g., set $m = 0$) to solve the optimization problem. The value of m can be increased ($m = 1$), continuing to solve the problem. The process can be stopped if the step-function for $\rho(x)$ begins retaining its form (this means that newly introduced intervals become the same as for the previous value of m). This finalizes the process of seeking the resulting assessment.

4. The Case of Bounded Modules of Density Derivatives

The next additional portion of information can be represented by constraints related to the maximum values of the density derivatives [6].
Sometimes it is realistic to elicit these data from experts by asking them a
question like "What is the largest possible difference between the percentages of accidents computed for two neighboring years/decades for a given
plant with definite age?"

Let us denote $M \in \mathbf{R}_+$ an upper bound on the values of the probability
density derivative module; i.e., for $\forall x$

$$|d\rho(x)/dx| \le M = const. \tag{13}$$

Now we have to optimize the objective function (2) subject to constraints (1),
(7) and (13). This is also a problem that can be solved via the methods of the
calculus of variations (very similar to the approach described in [5]). This
shows that optimal density functions belong to a family of trapezoid—or
triangular—functions (Figure 1). Correspondingly the intervals for the final
assessments are expected to be closer as the speed of density change is constrained. Another effect of recognizing the form of the optimal solution is the
possibility of reducing the initial problem to an easier-to-solve optimization of
a multivariable function subject to algebraic constraints (as was done above).

Indeed, let $[x_0,x_1),[x_2,x_3),[x_4,x_5),\ldots,[x_{2m},x_{2m+1})$ be the intervals that play
the role of the trapezoid lower bases. It is easy to see that the trapezoid upper
bases for which $\rho(x) = K$ are located within the intervals

$$[x_0 + K/M, x_1 - K/M),[x_2 + K/M, x_3 - K/M),\ldots,[x_{2m} + K/M, x_{2m+1} - K/M)$$

Let $[x_1, x_2), [x_3, x_4), [x_5, x_6),\ldots,[x_{2m+1},x_{2m+2})$ be the intervals on which $\rho(x) = 0$
(Figure 1).

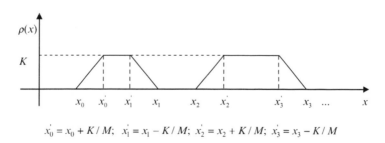

$$x_0' = x_0 + K/M; \quad x_1' = x_1 - K/M; \quad x_2' = x_2 + K/M; \quad x_3' = x_3 - K/M$$

Figure 1. The Plot of Optimal Bounded Density with Bounded Module of Derivative.

Then the problem statement can be easily reformulated in relation to optimizing the multivariate function, which depends on the variables $x_0, x_1, \ldots, x_{2m+1}$.

The choice of a value for m can be accomplished as was done above. Specifically, we can start from minimum values of m (e.g., $m = 0$) and then try to increase it with testing in parallel if change to m is followed by a change to optimal density.

The numerical examples confirm the improvement of up to 75% in resulting accuracy. However, the incentive to find a more promising model for uncertainty remains, as the bounded densities and bounded density derivatives lead to partially unrealistic solutions: the densities are equal to zero for some argument intervals, which means that the 'probabilistic mass' is concentrated in separate zones. This explains the desire to restrict ourselves by considering only the class of smooth differentiable density functions.

5. Application of Generalized Distribution Family

Let us introduce a family of distribution densities described by smooth differentiable functions as

$$\rho(x) = \left(\sum_{k=1}^{n} C_k \exp(-\alpha_k x) \right)^2 = \sum_{k=1}^{n} C_k^2 \exp(-2\alpha_k x) + 2 \sum_{\substack{l \neq r \\ l=1, r=1}}^{n} C_l C_r \left(\exp{-(\alpha_l + \alpha_r)x} \right), \quad (14)$$

in which C_k, $\alpha_k \geq 0$, $k = 1, 2, \ldots, n$, are real numbers satisfying the condition

$$\sum_{k=1}^{n} C_k^2 / (2\alpha_k) + 2 \sum_{\substack{l \neq r \\ l=1, r=1}}^{n} C_l C_r / (\alpha_l + \alpha_r) = 1. \quad (15)$$

It is easy to verify that $\rho(x) \geq 0$ and $\int_0^\infty \rho(x) dx = 1$.

Now consider the previsions that may be given for $\int_0^\infty f_i(x) \rho(x) dx, i = 1, 2, \ldots, m$. In the case where $\rho(x)$ satisfies (14) we obtain

$$\int_0^\infty f_i(x) \rho(x) dx = \sum_{k=1}^{n} C_k^2 F_i(s) \Big|_{s=2\alpha_k} + 2 \sum_{\substack{l \neq r \\ l=1, r=1}}^{n} C_l C_r F_i(s) \Big|_{s=\alpha_l + \beta_r}, i = 1, 2, \ldots, m, \quad (16)$$

in which $F_i(s)$, $i = 1, 2, \ldots, m$, are Laplace transformed functions $f_i(x)$.

Note that Laplace transform $F_i(s)$ for any continuous function $f_i(x)$ is introduced by the expression

$$F_i(s) = \int_0^\infty f_i(x)\exp(-sx)dx,$$

in which s is the Laplace variable (which may take complex values in general case: $s = \mathrm{Re}(s) + j \cdot \mathrm{Im}(s)$; here $\mathrm{Re}(s)$, $\mathrm{Im}(s)$ denote real and imaginary parts of s respectively).

Meanwhile it is proven by D.V. Widder [7] that if we know the performances of $F_i(s)$ for real positive values of s then we have a unique expansion of its behavior to the whole complex plane of s values.

The tables containing results of the Laplace transformation for different functions are widely presented in relevant literature.

For instance, if $f_i(x) = x$ then $F_i(s) = 1/s^2$; if $f_i(x) = x^2$ then $F_i(s) = 2/s^3$, etc.

Hence, we can write the following general formulae for expectation and variation:

$$E[X] = \int_0^\infty x\rho(x)dx = \sum_{k-1}^n C_k^2/(4\alpha_k^2) + 2\sum_{\substack{l\neq r \\ l=1,r=1}}^n C_l C_r/(\alpha_l + \alpha_r)^2, \qquad (17)$$

$$Var[X] = \int_0^\infty x^2\rho(x)dx = \sum_{k=1}^n C_k^2/(8\alpha_k^3) + 2\sum_{\substack{l\neq r \\ l=1,r=1}}^n C_l C_r/(\alpha_l + \alpha_r)^3. \qquad (18)$$

In turn probability of the event $X \leq x$ can be found as

$$P(X \leq x) = \int_0^x \rho(x)dx = \sum_{k=1}^n \frac{C_k^2}{2\alpha_k}(1 - \exp(-2\alpha_k x))$$

$$+ 2\sum_{\substack{l\neq r \\ l=1,r=1}}^n \frac{C_l C_r}{\alpha_l + \alpha_r}\left(1 - \exp\left(-(\alpha_l + \alpha_r)x\right)\right) \qquad (19)$$

Formulae (17)–(19) allow the reduction of typical IPT problems to easier-to-solve standard problems that belong to the scope of optimizing nonlinear multivariable functions depending on the values of C_k, α_k, $k = 1,2,\ldots,n$.

For instance, if we would like to estimate $P(X \leq x_0)$ for given x_0 on the basis of interval previsions for moments (17), (18), then we have to substitute x_0 instead of x into objective function (6) and search for $\max_{C_k,\alpha_k} P(X \leq x_0)$ and $\min_{C_k,\alpha_k} P(X \leq x_0)$ subject to constraints (1) as well as

$$\underline{E[X]} \leq E[X] \leq \overline{E[X]}, \underline{Var[X]} \leq Var[X] \leq \overline{Var[X]}. \qquad (20)$$

Here $\underline{E}[X]$, $\underline{Var}[X]$, $\overline{E}[X]$, $\overline{Var}[X]$ are the lower and the upper bounds of the intervals for the values of the moments respectively.

An important particular case of the introduced distributions which can be obtained if we consider only two exponential terms in the sum for $\sqrt{\rho(x)}$ in equality (16) is analyzed below.

Consider the case in which

$$\rho(x) = \left(C_1 \exp(-\alpha x) + C_2 \exp(-\beta x)\right)^2 = C_1^2 \exp(-2\alpha x) + C_2^2 \exp(-2\beta x)$$
$$+ 2C_1 C_2 \exp\left(-(\alpha + \beta)x\right). \tag{21}$$

Here $C_1, C_2, \alpha \geq 0$, $\beta \geq 0$ are the distribution parameters.

First, analyze which type of statistical characteristic behavior can be presented by Expression (21).

If C_1 and C_2 are of the same sign (i.e., $C_1 C_2 \geq 0$), then Expression (21) corresponds to monotonic density functions (Figure 2).

Note that function behavior like of $\rho^{\bullet}(x)$ is more typical for different non-zero values of C_1, C_2 and $C_1 C_2 \geq 0$; the behavior reflected by $\rho^{\bullet\bullet}(x)$ ('pure' exponential type) takes place if $C_1 = 0$ or $C_2 = 0$. The last situation appears also if $\alpha = \beta$.

If C_1 and C_2 have different signs (i.e. $C_1 C_2 \leq 0$) then Expression (21) may correspond to nonmonotonic density functions (Figure 3).

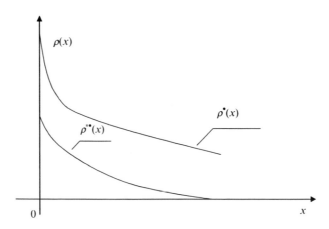

Figure 2. Types of Density Functions Presented by Expression (21) if $C_1 C_2 \geq 0$.

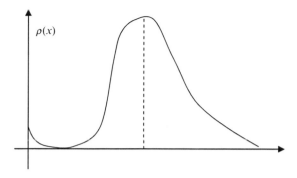

Figure 3. General form of Density Functions Presented by Expression (21) if $C_1 C_2 < 0$.

It becomes clear that the introduced family of density functions covers a wide class of practically important distribution types (unimodal and even antimodal).

Using this generalized distribution family allows reduction of the problem to optimization of the objective function (2) as a multivariable function depending on the initially unknown parameters C_k, $\alpha_k \geq 0$, $k = 1,2,\dots,n$.

The improvement achieved for numerical assessments becomes somewhat higher than for cases of nonsmooth density functions.

6. Concluding Remarks

Some reasonable data elicited from experts and accessible for verification can significantly improve the decisions made under the conditions of uncertainty. Adding information on density bounds, density derivative bounds, or any generalized form of distribution function with unknown parameters is, on the one hand, a kind of reasonable enhancement and, on other hand, does not actually restrict us in taking into account possible (probable) scenarios. Meanwhile, this technique provides promising modification of traditional approaches based on imprecise previsions and creates a bridge between the strict concepts of the corresponding theory and practical needs for assessment accuracy.

The proposed methodology opens the door for next steps associated with incorporating additional information elicited from experts. Thus it makes sense sometimes to ask an expert if s/he is ready to give preferences about some kinds of data. By presenting these preferences in the form of subjective probabilities, we have an opportunity to compute the expectations of the upper and the lower bounds derived for previsions. In turn this strengthens support for responsible decisions in the framework of risk analysis.

Acknowledgements

Dr. Igor Kozine of Risø National Laboratory, Denmark, introduced me to the world of interval-valued probabilities and helped me with his valuable comments. Dr. Igor Linkov of Intertox Inc., USA, was among the key organizers of NATO Advanced Research Workshop at Lisbon, so his kind invitation to participate was the necessary condition which allowed me to join the team of this book's authors. The contribution of both Igors to supporting my research activity is gratefully acknowledged.

References

1. Cooke R.M., *Experts in Uncertainty*, Oxford University Press, Oxford, 1991.
2. Walley P., *Statistical Reasoning with Imprecise Probabilities*, Chapman & Hall, New York, 1991.
3. Kuznetsov V., *Interval Statistical Models*, Radio & Sviaz, Moscow, 1991 (in Russian).
4. Utkin L.V., Kozine I. O., Different faces of the natural extension, *Proceedings of the Second International Symposium on Imprecise Probabilities and Their Applications*, ISIPTA '01, 2001, pp. 316–323.
5. Krymsky V.G., Computing Interval Bounds for Statistical Characteristics Under Expert-Provided Bounds on Probability Density Functions, *Applied Parallel Computing, State of the Art in Scientific Computing, Revised Selected Papers of 7th International Workshop PARA'04*, Lecture Notes in Computer Science, Springer, 2006, pp.151–160.
6. Kozine I.O., Krymsky V.G., Enhancement of Natural Extension, *Proceedings of the Fifth International Symposium on Imprecise Probabilities: Theories and Applications (ISIPTA'2007)*, Action M Agency, Prague, 2007, pp.253–262.
7. Widder D.V., *The Laplace Transform*, Princeton University Press, Princeton, NJ, 1972.

MODELING STAKEHOLDER PREFERENCES WITH PROBABILISTIC INVERSION

Application to Prioritizing Marine Ecosystem Vulnerabilities

R. NESLO

Delft University of Technology
Delft, The Netherlands

F. MICHELI

Stanford University
Palo Alto, California, USA

C.V. KAPPEL

National Center for Ecological Analysis and Synthesis
Santa Barbara, California, USA

K.A. SELKOE

University of Hawaii
Manoa, Hawaii, USA

B.S. HALPERN

National Center for Ecological Analysis and Synthesis
Santa Barbara, California, USA

R.M. COOKE

Delft University of Technology and Resources for the Future
Delft, The Netherlands
Cooke@rff.org

I. Linkov et al. (eds.), Real-Time and Deliberative Decision Making.
© Springer Science + Business Media B.V. 2008

Abstract: A panel of 64 experts ranked 30 scenarios of human activities according to their impacts on coastal ecosystems. Experts were asked to rank the five scenarios posing the greatest threats and the five scenarios posing the least threats. The goal of this study was to find weights for criteria that adequately model these stakeholders' preferences and can be used to predict the scores of other scenarios. Probabilistic inversion (PI) techniques were used to quantify a model of ecosystem vulnerability based on five criteria. Distinctive features of this approach are:

1. A model of the stakeholder population as a joint distribution over the criteria weights is obtained. This distribution is found by minimizing relative information with respect to a noninformative starting distribution, but makes no further assumptions about the interactions between the weights for different criteria. Criteria distributions with dependence emerge from the fitting procedure.

2. The multicriteria preference model can be empirically validated with expert preferences not used in fitting the model.

1. Introduction

This article presents an analysis of the 64 experts' rankings of 30 scenarios of human activities and their impacts to coastal ecosystems. The elicitation protocols were designed and executed by researchers at the National Center for Ecological Analysis and Synthesis. Experts were asked to rank the five scenarios posing the greatest threats and the five scenarios posing the least threats. The goal of this study was to find weights for criteria that adequately model these stakeholders' preferences and can be used to predict the scores of other scenarios. Probabilistic inversion (PI) techniques were used to quantify a model of ecosystem vulnerability based on five criteria. Stakeholder preference modeling can also serve as a form of expert elicitation when the stakeholders are domain experts, as in the present case. Their preferences are taken to prioritize threats to marine ecosystems, with a view to optimizing mitigation and abatement actions.

Other multicriteria weighting methods [9, 10, 22] require stakeholders to evaluate the criteria directly. Of course, the weights assigned to a criterion cannot be assessed independently of the scale on which *all* criteria scores are measured—a fact that is sometimes overlooked. The present approach asks the stakeholders to rank scenarios rather than evaluate criteria. Criteria weights are then derived to fit the stakeholder preference rankings as well as possible. This has the significant advantage of allowing us to assess the validity of our fitted model of stakeholder preference.

Probabilistic inversion denotes the operation of inverting a function over a probability distribution, rather than at a point. Such problems arise in quantifying uncertainty in physical models [8, 13, 14, 15, 23]. One has uncertainty distributions on observable phenomena, either from data or from expert judgment, and one wishes to find a distribution over the parameters of a predictive model, such that one recovers the observed distributions when the parameter distributions are "pushed through" the model. PI algorithms used in the past were computationally intensive, involving sophisticated interior point optimization techniques and duality theory as well as ad hoc steering [16]. Recent computational advances [26, 34] clarify the mathematical foundations for PI and yield simple algorithms with proven convergence behavior, suitable for use by nonspecialists. The results depend on a variant of the classical Iterative Proportional Fitting algorithm [6–8, 12, 17, 19, 20, 26].

In stakeholder preference modeling, the data is discrete-choice preference data elicited from a set of stakeholders. The distributions to be inverted are those of indicator variables such as:

- Alternative i is better than alternative j.
- Alternative i is ranked third in the given set of alternatives.

We are interested in the probability of such variables, taking the values "yes" or "no" for a set of stakeholders. We can measure these probabilities by querying a large representative set of stakeholders. Existing discrete-choice—or random-utility—techniques construct a value or utility function from discrete-choice data [1, 3, 23, 24, 27, 28, 30–33], and they strongly restrict the form of the utility functions. Using PI, this form can be inferred from choice data.

We first discuss the data, then address model adequacy and model fit. Summary statistics for the 30 scenarios are then given. The conclusion of this analysis is that the data are broadly consistent with a linear model of stakeholder preferences.

2. Data

The 30 threat scenarios were scored on five criteria:

- C1 Spatial scale
- C2 Frequency
- C3 Trophic (functional) impact
- C4 Recovery time
- C5 Resistance

These criteria were developed and tested elsewhere [11, see also 5, 25, 29].

The stakeholders' preference data is represented with a linear model:

Score for scenario $S = \Sigma_{i=1...5}$ (score of S on C_i × weight for C_i)   (1)

The weights are random variables that are nonnegative and sum to 1. The (joint) distribution for the weights is modeled to represent the distribution of weights in a population of stakeholders, of which the 64 elicited experts are a random sample. Since the weights are normalized, the scores are transformed so that the product *score* × *weight* is positive and falls within the same range. Spatial scale is given in square kilometers, and the values for spatial scale range from 0.1 to 50,000 km². These values are transformed to $\ln(100\,m^2)$, whose values thus range from 2.3 to 15.4. Frequency was scored as $\ln(360^*\#$ /year). Trophic or functional impact is the number of trophic layers affected. Resistance is scored as the percent of species affected per trophic layer. These transformations are chosen for mathematical convenience.

A salient feature of these data is dominance. Scenario A dominates Scenario B from above if A's scores on all five criteria are greater or equal to the scores of B. A dominates B from below if A's scores on all five criteria are less than or equal to those of B. If A dominates B from above, then B can never be ranked above A in any model that computes the scenario score as a monotonic function of the five criteria scores. The presence of dominated scenarios enables us to analyze whether the experts' rankings are broadly consistent with a monotonic model of criteria scores.

3. Model Adequacy

Of the 30 scenarios, only seven were nondominated. This means that none of the 23 scenarios dominated from above could be ranked 1 by a stakeholder whose preferences were consistent with the model. In fact, 22.4% of the top rankings were inconsistent in this sense: 77.6% of the top rankings went to four of the seven nondominated scenarios. A scenario dominated from above by two or more scenarios could not consistently be ranked second; in fact, 23.7% of the second rankings were inconsistent in this sense. Dominance from below was much less prevalent than dominance from above.

In view of the large number of dominated scenarios, we view the percentages of inconsistent rankings as indicating that the stakeholders' preferences were broadly, though not wholly, consistent with a monotonic model.[1] We therefore proceeded to fit the linear model (1).

The 30 scenarios and their criteria scores are shown in Table 1. The nondominated scenarios are shaded.

---

[1] If the 64 experts had chosen their top-ranked scenario at random, the probability that 14 or fewer would chose one of the 23 dominated scenarios is in the order of 10^{-20}.

TABLE 1. Scenarios and Criteria Scores.

Nr	Code	Scenario	Scale ln((km*10)²)	Freq ln (360°#/ year)	Func (# trophic layers)	Recov (years)	Resist
1	am	Aquaculture: marine plant	5.30	11.77	1	1	0.2
2	as	Aquaculture: shellfish	6.21	11.77	1	0.1	0.05
3	cl	Climate change: sea level rise	13.82	5.19	2	5	0.2
4	ct	Climate change: sea temp	15.42	5.89	3	50	0.25
5	cu	Climate change: UV	13.82	3.58	1	1	0.05
6	ca	Coastal engineering: habitat alteration	4.61	5.89	4	25	0.75
7	dh	Direct human impact: trampling	9.62	11.77	2	25	0.35
8	fd	Fishing: demersal destructive	6.68	2.89	4	0.5	0.1
9	fn	Fishing: demersal nondestructive low bycatch	2.30	2.89	1	0.5	0.1
10	fa	Fishing: nondestructive artisanal	4.61	2.89	1	1	0.5
11	fp	Fishing: pelagic high bycatch	6.21	1.28	1	0.5	0.05
12	fr	Fishing: recreational	6.68	9.84	2	5	0.2
13	fu	Freshwater input: increase	6.91	4.28	2	1	0.1
14	is	Invasive species	14.51	11.77	1	20	0.25
15	ma	Military activity	6.91	8.37	1	5	0.1
16	nh	Nutrient input: causing harmful algal blooms	9.21	4.28	2	1	0.1
17	nz	Nutrient input: causing hypoxic zones	6.68	4.28	3	1	0.05
18	no	Nutrient input: into oligotrophic waters	8.29	4.97	1	0.5	0.3
19	og	Ocean dumping: lost fishing gear	2.30	5.89	3	3	0.15
20	os	Ocean dumping: ship wrecks	3.91	2.89	4	10	0.5
21	ox	Ocean dumping: toxic materials	6.91	2.89	1	1	0.1
22	po	Ocean pollution	6.91	6.58	1	3	0.2
23	pa	Pollution input: atmospheric	9.62	3.58	1	0.5	0.2
24	pi	Pollution input: inorganic	8.29	4.28	2	3	0.2
25	pr	Pollution input: organic	8.52	5.19	2	5	0.2
26	ps	Power, desalination plants	4.61	11.77	3	10	0.5
27	sr	Scientific research: collecting	2.30	8.37	1	2	0.15
28	sd	Sediment input: decrease	3.91	1.28	1	0.5	0.05
29	si	Sediment input: increase	10.82	5.19	2	10	0.3
30	ts	Tourism: surfing	2.30	10.49	1	1	0.05

4. Model Fitting: Criteria Weights

We fit the linear model by finding a distribution over criteria weights which fit as well as possible the probabilities of rankings given by the stakeholders. The fitting is done by probabilistic inversion. We start with a noninformative distribution over criteria weights (which however are constrained to add to 1). We then adapt this distribution to optimally recover the stakeholders' rankings. That is, if we sample randomly from the adapted distribution, the probability of drawing a set of weights with which Scenario A is ranked first equals, to the extent possible, the percentage of experts who ranked A first, and so on. The fitting based on first ranks applies only to the percentages for the scenarios that were ranked first. Similarly, the fitting based on the first two ranks applies only to the percentages for the scenarios ranked 1 or 2.

We are interested in finding a fitting that can be validated by predicting rankings *not* used in the fitting. Since the goal is to prioritize threats, the top rankings are most important. Satisfactory results were found by fitting the model based on the first four rankings; this model could then and used to predict the fifth rankings. Table 2 and Figure 4 compare the predicted and observed percentages of rankings. The model is first used to "retrodict" or "recover" the first four rankings. These are the data actually used to fit the model, so this comparison is a check of model fit rather than model prediction. Using the model, we can predict the percentages of experts ranking the various scenarios in the fifth position (Figure 5). These percentages were not used in fitting the model and test the ability of the model to predict preferences of the population of stakeholders. Of course, we should hope that the predictions and retrodictions show similar agreement with the observed rankings.

Because we are fitting a linear model, the expected score of any scenario may be computed by using the expected values of the criteria weights in the adapted distribution. A new scenario, not among the original 30, can be scored by multiplying its (transformed) criteria scores by the expected weight of each criterion. This of course is the great advantage of a linear model, and explains the preference for this model above more complex models, even though the latter might yield a better fit. Figure 1 shows the expected criteria weights based on fitting only the first ranks, the first two ranks, the first three ranks, and the first four ranks, and finally, based on fitting all ranks. We observe that these expected weights do not change significantly between the two-, three-, and four-rank options. Using all ranks causes changes, and also causes greater variance in the criteria scores (see Table 4).

TABLE 2. Model Predictions and Stakeholder Probabilities for Top Five Rankings.

Constraint	Prediction I	Prediction I,II	Prediction I,II,III	Prediction I,II,III,IV	Stakeholders
#S3 = 1	0.0000	0.0000	0.0000	0.0000	0.0597
#S4 = 1	0.3424	0.3428	0.3420	0.4359	0.3433
#S6 = 1	0.2695	0.2687	0.4164	0.3008	0.2687
#S7 = 1	0.0296	0.0299	0.0453	0.0329	0.0299
#S8 = 1	0.0000	0.0000	0.0114	0.0000	0.0149
#S9 = 1	0.0000	0.0000	0.0000	0.0000	0.0149
#S11 = 1	0.0000	0.0000	0.0000	0.0000	0.0149
#S12 = 1	0.0000	0.0000	0.0000	0.0000	0.0149
#S14 = 1	0.0744	0.0748	0.0580	0.0800	0.0746
#S16 = 1	0.0000	0.0000	0.0000	0.0000	0.0299
#S19 = 1	0.0000	0.0000	0.0000	0.0000	0.0149
#S22 = 1	0.0000	0.0000	0.0000	0.0000	0.0448
#S25 = 1	0.0000	0.0000	0.0000	0.0000	0.0149
#S28 = 1	0.0000	0.0000	0.0000	0.0000	0.0299
#S29 = 1	0.0000	0.0000	0.0000	0.0000	0.0299
#S2 = 2	0.0000	0.0000	0.0000	0.0000	0.0339
#S3 = 2	0.0001	0.0339	0.0442	0.0392	0.0339
#S4 = 2	0.2295	0.2213	0.1713	0.2218	0.2203
#S5 = 2	0.0000	0.0000	0.0000	0.0000	0.0169
#S6 = 2	0.4753	0.0511	0.0663	0.0661	0.0508
#S7 = 2	0.1557	0.0676	0.0825	0.0681	0.0678
#S8 = 2	0.0000	0.0679	0.0432	0.0725	0.0678
#S9 = 2	0.0000	0.0000	0.0000	0.0000	0.0169
#S11 = 2	0.0000	0.0000	0.0000	0.0000	0.0169
#S14 = 2	0.0275	0.2700	0.1855	0.2829	0.2712
#S16 = 2	0.0000	0.0000	0.0000	0.0000	0.0508
#S18 = 2	0.0000	0.0000	0.0000	0.0000	0.0169
#S20 = 2	0.0238	0.0170	0.0214	0.0174	0.0169
#S22 = 2	0.0000	0.0000	0.0000	0.0000	0.0508
#S23 = 2	0.0000	0.0000	0.0000	0.0000	0.0169
#S24 = 2	0.0000	0.0000	0.0000	0.0000	0.0169
#S29 = 2	0.0000	0.0000	0.0000	0.0000	0.0508
#S2 = 3	0.0000	0.0000	0.0000	0.0000	0.0317

(continued)

TABLE 2. (continued)

Constraint	Prediction I	Prediction I,II	Prediction I,II,III	Prediction I,II,III,IV	Stakeholders
#S3 = 3	0.0015	0.0084	0.1924	0.3305	0.1587
#S4 = 3	0.0798	0.0656	0.0769	0.1486	0.0635
#S6 = 3	0.0707	0.2063	0.0713	0.1131	0.0635
#S7 = 3	0.5732	0.4615	0.0816	0.1401	0.0794
#S8 = 3	0.0005	0.0053	0.0328	0.0514	0.0317
#S9 = 3	0.0000	0.0000	0.0000	0.0000	0.0159
#S12 = 3	0.0000	0.0000	0.0000	0.0000	0.0635
#S14 = 3	0.0730	0.0649	0.1276	0.1616	0.1270
#S16 = 3	0.0000	0.0000	0.0000	0.0000	0.0317
#S17 = 3	0.0000	0.0000	0.0000	0.0000	0.0635
#S18 = 3	0.0000	0.0000	0.0000	0.0000	0.0317
#S20 = 3	0.0968	0.1582	0.0158	0.0189	0.0159
#S21 = 3	0.0000	0.0000	0.0000	0.0000	0.0159
#S22 = 3	0.0000	0.0000	0.0000	0.0000	0.0159
#S24 = 3	0.0000	0.0000	0.0000	0.0000	0.0159
#S25 = 3	0.0000	0.0000	0.0000	0.0000	0.1111
#S26 = 3	0.1044	0.0293	0.0160	0.0181	0.0159
#S29 = 3	0.0000	0.0000	0.0000	0.0000	0.0794
#S3 = 4	0.0137	0.0687	0.0137	0.2174	0.1864
#S4 = 4	0.1508	0.1125	0.0150	0.0392	0.0339
#S5 = 4	0.0001	0.0036	0.0372	0.0392	0.0339
#S6 = 4	0.0889	0.3417	0.2099	0.0958	0.0847
#S7 = 4	0.1196	0.1948	0.2580	0.1091	0.1017
#S8 = 4	0.0031	0.0074	0.0028	0.0737	0.0678
#S11 = 4	0.0000	0.0000	0.0000	0.0000	0.0339
#S12 = 4	0.0017	0.0000	0.0000	0.0000	0.0847
#S14 = 4	0.2851	0.0784	0.0235	0.0910	0.0847
#S16 = 4	0.0000	0.0000	0.0000	0.0000	0.0339
#S17 = 4	0.0000	0.0000	0.0000	0.0000	0.0169
#S18 = 4	0.0000	0.0000	0.0000	0.0000	0.0169
#S20 = 4	0.0990	0.0795	0.3630	0.0176	0.0169
#S22 = 4	0.0000	0.0000	0.0000	0.0000	0.0678
#S24 = 4	0.0000	0.0000	0.0000	0.0000	0.0169

(continued)

TABLE 2. (continued)

Constraint	Prediction I	Prediction I,II	Prediction I,II,III	Prediction I,II,III,IV	Stakeholders
#S25 = 4	0.0000	0.0000	0.0000	0.0000	0.0508
#S29 = 4	0.0000	0.0001	0.0000	0.1024	0.1017
#S2 = 5	0.0122	0.0024	0.0035	0.0705	0.0333
#S3 = 5	0.0140	0.0845	0.0276	0.0000	0.0500
#S4 = 5	0.1189	0.1671	0.0182	0.0270	0.0500
#S6 = 5	0.0544	0.0789	0.0974	0.0969	0.0667
#S7 = 5	0.1121	0.1460	0.4151	0.2279	0.1667
#S8 = 5	0.0034	0.0091	0.0057	0.0000	0.0167
#S12 = 5	0.0094	0.0005	0.0000	0.0004	0.0167
#S13 = 5	0.0000	0.0000	0.0000	0.0000	0.0167
#S14 = 5	0.1848	0.1641	0.1137	0.1036	0.1167
#S16 = 5	0.0000	0.0000	0.0000	0.0000	0.0833
#S17 = 5	0.0000	0.0000	0.0114	0.0000	0.0167
#S19 = 5	0.0000	0.0000	0.0000	0.0000	0.0167
#S20 = 5	0.1714	0.0864	0.0883	0.2331	0.0333
#S22 = 5	0.0000	0.0000	0.0000	0.0000	0.0333
#S24 = 5	0.0000	0.0000	0.0000	0.0000	0.0167
#S25 = 5	0.0000	0.0000	0.0000	0.0000	0.0500
#S26 = 5	0.2552	0.1646	0.1003	0.0493	0.0667

Although the expected weights are most important in using the model, it is also of interest to examine the distributions of weights. Figure 2 shows the cumulative distribution functions of the five weights in the four cases shown in Figure 1. The joint distributions for one rank, four ranks, and all ranks are shown in Figure 3.

The rightmost cumulative distributions indicate greatest importance. The picture from Figure 2 echoes that in Figure 1 for the first two ranks: resistance is most important, followed by trophic impact. Of course, we must bear in mind that these results are relative to the scaling chosen to represent the criteria scores.

Figures 1 and 2 show that the mean values and marginal distributions are somewhat similar in all fitting situations. The joint distributions, however, are quite different. One sample of weights represents one virtual stakeholder. If we plot these five weights on five vertical lines, we get a jagged line representing

Weights First Rank

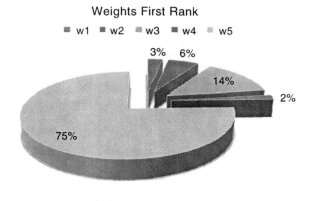

Weights First Two Ranks

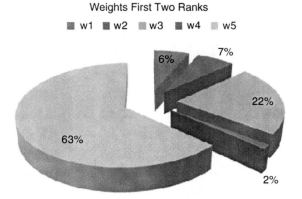

Weights First Three Ranks

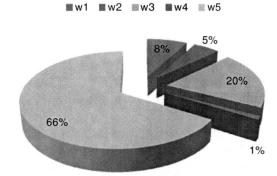

Figure 1. Expected criteria weights based on ranks 1, 1&2, 1&2&3, 1&2&3&4, and all ranks.

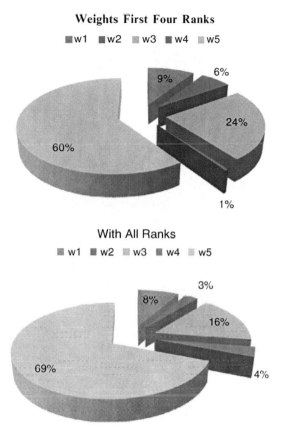

Weights First Four Ranks

■w1 ■w2 ▨w3 ■w4 ▨w5

With All Ranks

▨ w1 ■ w2 ▨ w3 ■ w4 ▨ w5

Figure 1. (continued)

one virtual stakeholder. If we plot 16,000 such lines we get a picture of the population of stakeholders. We say that the stakeholder weights have *interactions* if, for example, knowledge that a stakeholder assigns high weight to the "frequency" criterion gives significant information regarding weights for other criteria. A quick visual impression of the joint distributions is given by the "percentile cobweb plots" shown in Figure 3. Instead of the weights themselves, Figure 3 plots the weights' percentiles, as this makes the dependence structure more visible. Evidently the joint distributions are complex, and are different for the different fitting situations. A detailed analysis of interactions is not undertaken here. It is worth noting that the probabilistic inversion infers the dependence structure from the stakeholder data; it does not assume or impose any structure. We note that as we use more ranks in the fitting, the fitting becomes less smooth. The departure from the starting distribution grows more pronounced as the number of constraints that the fitting tries to satisfy increases.

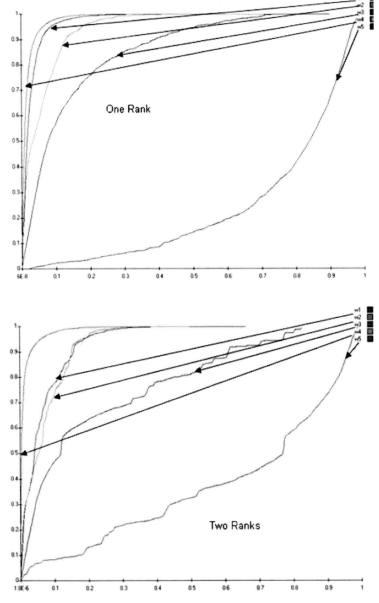

Figure 2. Cumulative weight distributions based on rank 1, 1&2, 1&2&3, 1&2&3&4, and all ranks.

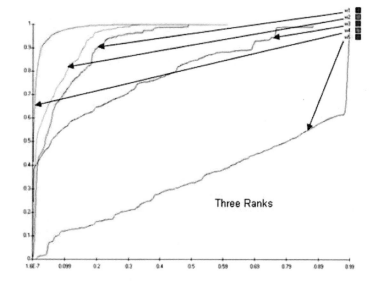

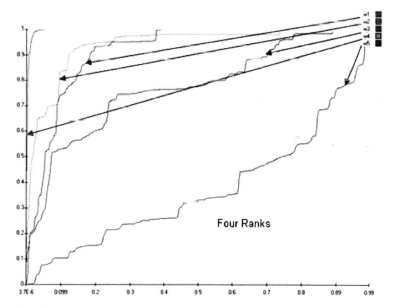

Figure 2. (continued)

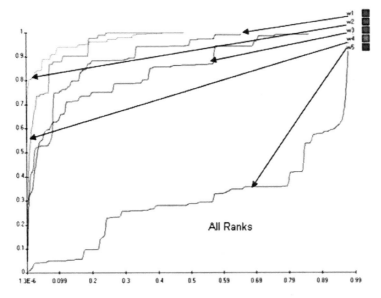

Figure 2. (continued)

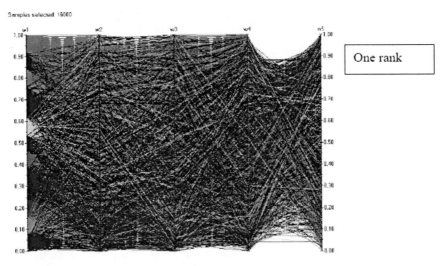

Figure 3. Percentile cobweb plots for criteria weights fitting one rank, four ranks, and all ranks.

Table 2 shows the predicted probabilities of rankings based on the fitting in the four cases discussed above. Thus "prediction I" indicates the prediction based on fitting only the first-ranked scenarios. The first column gives the constraints. "#S4=1" denotes the constraint that Scenario 4 was ranked 1. The last

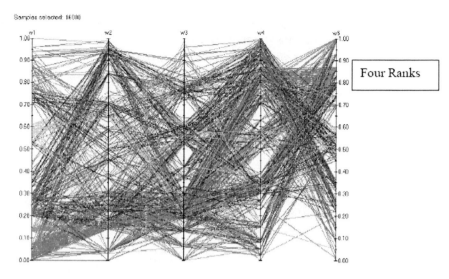

Figure 3. (continued)

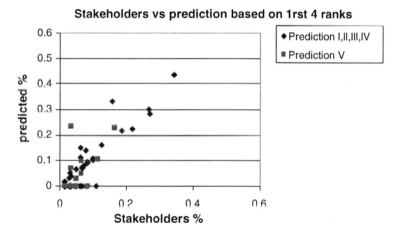

Figure 4. Predictions based on ranks 1–4 of stakeholder percentages for the first four ranks (diamonds), and for the fifth ranks (squares)

column shows that 34.33% of the stakeholders ranked Scenario 4 as 1. Using the fitting based only on the first ranks predicts that 34.24% of the population of stakeholders would rank Scenario 4 as 1. Similarly, using the fitting based on the first four ranks, 43.59% of the population would rank Scenario 4 first. Of course, owing to the presence of inconsistent rankings, the fitting can never be perfect. Indeed, 22.4% of the first ranks were inconsistent with the model; as we fit 77.6% of the consistent rankings, the remaining probability mass must be distributed over the other feasible rankings. Some of the discrepancies are sizeable, as in the case of #S20 = 5 for the prediction based in the top four ranks. On the whole, however, the predictions do capture the drift of stakeholder preferences. Fitting all ranks is numerically quite burdensome and conflates issues that determine the most serious and least serious threats. The fitting based on the top four rankings presents the best compromise.

Figure 4 shows the information in Table 2 graphically. On the horizontal axis are stakeholders' percentages for rankings of scenarios; on the vertical axis are the predicted percentages based on the fitted model. The diamonds are scenarios which were ranked first, second, third, or fourth. These percentages were used to fit the model. The squares are scenarios that were ranked fifth. We see that these percentages are reasonably well predicted by the model. Scenarios plotted on the horizontal axis correspond to rankings that are inconsistent with the model.

5. Scenario Scores

Figure 5 shows the densities of the scores of the top four scenarios, ranked according to their mean values. These densities are generated by the distribution of criteria weights, which models the distribution of participants. It is interesting to note that the modes of these densities are all

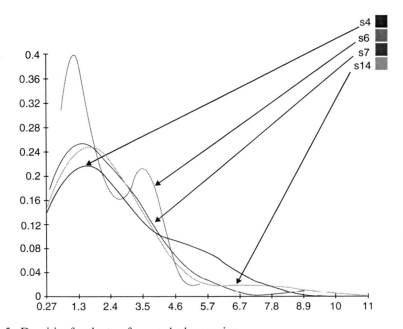

Figure 5. Densities for the top four ranked scenarios.

similar, but the shapes are different. The top-ranked scenario, Scenario 4 (Sea level rise), is distinguished by a large right tail. Scenario 6 (Coastal engineering) shows a bimodal form, suggesting that there are two distinct subgroups of participants. The remaining two scenarios, Scenario 14 (Invasive species) and Scenario 7 (Direct human impact) are quite similar in distribution.

Table 3 shows the mean, variance, and standard deviation of the five criteria weights and the 30 scenarios, based on the first four ranks. Table 4 gives the same information based on all ranks. Note that the variances in Table 4 tend to be larger, sometimes much larger. The top-ranked Scenario 4 has a

TABLE 3. Scenario scores using the first four ranks.

Using first four ranks			
Variable	Mean	Variance	SD
S1	1.572	1.362	1.167
S2	1.561	1.523	1.234
S3	2.214	2.103	1.450
S4	2.901	3.702	1.924
S5	1.763	1.649	1.284
S6	2.328	1.464	1.210
S7	2.420	2.441	1.562
S8	1.825	1.732	1.316
S9	0.693	0.193	0.439
S10	1.146	0.214	0.462
S11	0.923	0.412	0.642
S12	1.844	1.547	1.244
S13	1.446	0.920	0.959
S14	2.540	2.953	1.719
S15	1.472	1.075	1.037
S16	1.657	1.226	1.107
S17	1.639	1.352	1.163
S18	1.446	0.688	0.829
S19	1.405	1.035	1.017
S20	1.858	1.165	1.079
S21	1.118	0.533	0.730
S22	1.412	0.794	0.891
S23	1.465	0.856	0.925
S24	1.642	1.050	1.025
S25	1.729	1.178	1.085
S26	2.220	1.944	1.394
S27	1.065	0.677	0.823
S28	0.712	0.228	0.477
S29	2.024	1.513	1.230
S30	1.129	1.023	1.012

TABLE 4. Scenario scores using all ranks.

Variable	Using all ranks		
	Mean	Variance	SD
S1	1.118	1.725	1.313
S2	1.065	2.115	1.454
S3	2.066	4.997	2.235
S4	4.196	17.187	4.146
S5	1.613	4.405	2.099
S6	2.689	4.173	2.043
S7	2.726	6.270	2.504
S8	1.396	2.081	1.442
S9	0.528	0.255	0.505
S10	1.041	0.482	0.694
S11	0.822	0.932	0.965
S12	1.510	2.294	1.514
S13	1.167	1.585	1.259
S14	2.768	7.954	2.820
S15	1.274	1.975	1.405
S16	1.384	2.433	1.560
S17	1.263	1.874	1.369
S18	1.264	1.688	1.299
S19	1.039	0.961	0.980
S20	1.786	1.714	1.309
S21	0.981	1.232	1.110
S22	1.220	1.596	1.263
S23	1.302	2.185	1.478
S24	1.445	2.066	1.437
S25	1.568	2.362	1.537
S26	1.918	2.361	1.537
S27	0.757	0.651	0.807
S28	0.605	0.431	0.656
S29	2.051	3.663	1.914
S30	0.701	0.902	0.950

variance of 3.7 based on four ranks, and 17.2 based on all ranks. This suggests that trying to fit the top *and* bottom ranks just muddies the water—it does not give more insight into the factors determining high-threat scenarios.

6. Conclusion

By design, this study involved many dominated scenarios. This enabled us to test the extent to which the stakeholder preferences were consistent with a model for scenario scores based on a monotonic function of the five criteria scores. A stakeholder who prefers a dominated to a nondominated scenario is not consistent with any such model. Of course, this does not mean that such a stakeholder is inconsistent, it simply means that his/her preferences are not consistent with this type of model. In view of the large number of dominated scenarios, we may conclude that these stakeholders are broadly, though not wholly, consistent with such a monotonic model. A more complex model—possibly involving other criteria or interactions of criteria—might produce a better fit, but such models would be much more cumbersome in practice.

The linear model (1) is one type of monotonic model. Owing to the inconsistencies noted above it can never yield a perfect fit, but it does seem to capture the main drift of the stakeholder preferences. This means that the expected weights (Figure 1) can be used to score coastal ecosystem threat scenarios, provided their scores on the five criteria are given and scaled appropriately.

References

1. Anderson, S.P., de Palma, A., and Thissen, J-F., 1996. *Discrete Choice Theory of Product Differentiation.* MIT Press, Cambridge.
2. Bradley, R., 1953. Some statistical methods in taste testing and quality evaluation. *Biometrika* 9:22–38.
3. Bradley, R., and Terry, M., 1952. Rank analysis of incomplete block designs. *Biometrika* 39:324–345.
4. Cooke, R. M., and Misiewicz, J., 2007. Discrete choice with probabilistic inversion: application to energy policy choice and wiring failure. Presented at *Mathematical Methods in Reliability*, July.
5. Covich, A. P., Austen, M. C., Barlocher, F., Chauvet, E., Cardinale, B. J., Biles, C. L., Inchausti, P., Dangles, O., Solan, M., Gessner, M. O., Statzner, B., and Moss, B., 2004. The role of biodiversity in the functioning of freshwater and marine benthic ecosystems. *Bioscience* 54:767–775.
6. Csiszar, I., 1975. I divergence geometry of probability distributions and minimization problems. *Annals of Probability* 3:146–158.
7. Deming, W. E., and Stephan, F. F., 1944. On a least squares adjustment to sample frequency tables when the expected marginal totals are known. *Annals of Mathematical Statistics* 40(11):427–444.
8. Du, C., Kurowicka, D., and Cooke, R. M., 2006. Techniques for generic probabilistic inversion. *Computational Statistics & Data Analysis* 50:1164–1187.
9. Linkov, I., Kiker, G. A., and Wenning, R. J. (Eds.) 2007. *Environmental Security in Harbors and Coastal Areas.* Springer, Dordrecht.
10. French, S., 1988. *Decision Theory; An Introduction to the Mathematics of Rationality.* Ellis Horwood, Chichester.
11. Halpern, B. S., Selkoe, K. A., Micheli, F., and Cappel, C. V. 2007. Evaluating and ranking global and regional threats to marine ecosystems. *Conservation Biology* 21:1301–1315.
12. Ireland, C. T., and Kullback, S., 1968. Contingency tables with given marginals. *Biometrika* 55:179–188.

13. Kraan, B. C. P. and Cooke, R. M. 2000. Processing expert judgments in accident consequence modeling. *Radiation Protection Dosimetry* 90(3):311–315.
14. Kraan, B.C.P. and Cooke, R. M., 2000. Uncertainty in compartmental models for hazardous materials - a case study. *Journal of Hazardous Materials* 71:253–268.
15. Kraan, B.C.P., and Bedford. T. J. 2005. Probabilistic inversion of expert judgments in the quantification of model uncertainty. *Management Science* 51(6):995–1006.
16. Kraan, B. C. P., 2002. *Probabilistic Inversion in Uncertainty Analysis and Related Topics*. Ph.D. dissertation, TU Delft, Dept. Mathematics.
17. Kruithof, J., 1937. Telefoonverkeersrekening. *De Ingenieur* 52(8):E15–E25.
18. Kullback, S., 1959. *Information Theory and Statistics*. Wiley, New York.
19. Kullback, S., 1968. Probability densities with given marginals. *The Annals of Mathematical Statistics* 39(4):1236–1243.
20. Kullback, S., 1971. Marginal homogeneity of multidimensional contingency tables. *The Annals of Mathematical Statistics* 42(2):594–606.
21. Kurowicka, D., and Cooke, R. M., 2006. *Uncertainty Analysis with High Dimensional Dependence Modelling*. Wiley, New York.
22. Linkov, I., Sahay, S., Kiker, G., Bridges, T., Belluck, D., and Meyer, A., 2006. Multicriteria decision analysis; comprehensive decision analysis tool for risk management of contaminated sediments. *Risk Analysis* 26(1):61–78.
23. Luce, R. D., and Suppes, P., 1965. Preference, utility, and subjective probability. In *Handbook of Mathematical Psychology*, vol. 3, eds. Luce, R. D., Bush, R., and Calanter, E. Wiley, New York.
24. Luce, R. D., 1959. *Individual Choice Behavior; A Theoretical Analysis*. Wiley, New York.
25. M.E.A. (Millennium Ecosystem Assessment), 2005. *Ecosystems and Human Well-Being: Synthesis Report*. Island Press, Washington, DC.
26. Matus, F., 2007. On iterated averages of I-projections, *Statistiek und Informatik*, Universität Bielefeld, Bielefeld, Germany.
27. McFadden, D., 1974. Conditional logic analysis of qualitative choice behavior. In *Frontiers in Econometrics*, ed. Zarembka, P., 105–142. New York Academic Press, New York.
28. Siikamäki, J., and Layton, D. F., 2007. Discrete choice survey experiments: a comparison using flexible methods. *Journal of Environmental Economics and Management* 53:127–139.
29. Stringer, L. C., Dougill, A. J., Fraser, E., Hubacek, K., Prell, C., and Reed, M. S., 2006. Unpacking "participation" in the adaptive management of social ecological systems: a critical review. *Ecology and Society* 11(2):39.
30. Thurstone, L., 1927. A law of comparative judgment. *Psychological Review* 34:273–286.
31. Torgerson, W., 1958. *Theory and Methods of Scaling*. Wiley, New York.
32. Train, K. E., 2003. *Discrete Choice Methods with Simulation*. Cambridge University Press, New York.
33. Train, K. E., 1998. Recreation demand models with taste differences over people. *Land Economics* 74: 230–239.
34. Vomlel, J., 1999. *Methods of Probabilistic Knowledge Integration*. Ph.D. thesis, Czech Technical University, Faculty of Electrical Engineering.

OPERATIONS RESEARCH AND DECISION ANALYSIS

Software Tools and Applications

D. STRIMLING

DecisionAiding Ltd.
20/10 Ostrovski
Ra'anana 4337, Israel
david.strimling@decisionaiding.com

Abstract: This chapter demonstrates operations research and decision and risk analysis tools and their application in the context of a hypothetical case study of groundwater cleanup at a toxic waste site [1]. The following issues are addressed:

1. Population risk assessment
2. Remediation effectiveness
3. Optimal treatment method
4. Multiple objectives
5. Reducing uncertainty
6. Linked decisions

The chapter concludes with a brief discussion integrating the six parts of the case study into a process for implementation of adaptive management.

1. Population Risk Assessment

A small community gets its water from wells that tap into an old, large aquifer. Recently, an environmental impact study found toxic contamination in the groundwater due to improperly disposed chemicals from a nearby manufacturing plant.

The environmental impact study provided estimates of the following risk factors for each chemical:

- Cancer potency factor (CPF)
- Contamination level (CL)

The study further recommended that a population risk assessment be conducted to determine if any action needs to be taken to correct the situation.

I. Linkov et al. (eds.), Real-Time and Deliberative Decision Making.
© Springer Science + Business Media B.V. 2008

The study said that the risk assessment must account for the variability of
body weights (BW) and volume of water consumed (VWC) by individuals in
the community.

We will use influence diagrams [2] to frame each of the issues to be
addressed. An influence diagram is a compact graphical representation of
a decision scenario that shows the interactions of uncertainties and decisions to be made. The influence diagram in Figure 1 frames the population
risk assessment issue using only uncertainty. We will add decisions later.
The objective at this point is to determine the risk to the population due
to the toxic contamination in the groundwater. The equation for this risk
is given by

$$\text{Population Risk} = \frac{\left(\sum \text{CPF}_i * \text{CL}_i \right) * \text{VWC}}{\text{BW}}$$

The risk factors CPF and CL, which are inputs to this equation for each of
the improperly disposed chemicals, are shown in the influence diagram as
uncertainties.

The computations displayed in the influence diagram can be easily
modeled in an Excel spreadsheet [2] as shown in Figure 2.

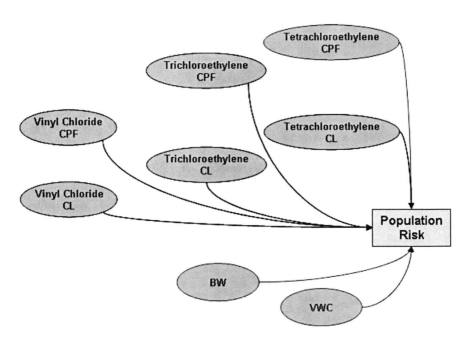

Figure 1. Population Risk Assessment Influence Diagram.

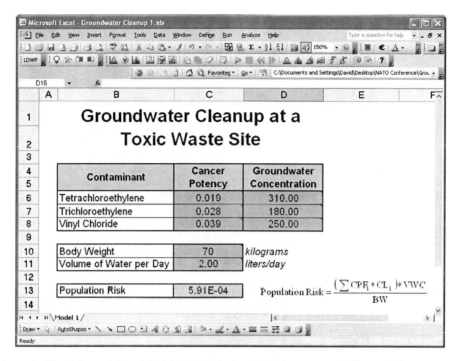

Figure 2. Excel Spreadsheet Model of Population Risk Assessment Influence Diagram.

The values of the parameters shown in the spreadsheet model are mean value estimates provided by the environmental impact study team. The study team, however, also provided distribution data. These distributions can be integrated into the spreadsheet model using the Oracle Crystal Ball [3] add-in to Excel. Crystal Ball allows the point estimates in each cell to be represented by probability distributions. For example, the means value estimates shown in the tetrachloroethylene CPF and CL cells can be replaced with their full distributions as shown in Figure 3.

These probability distributions can then be used to perform a Monte Carlo simulation to generate a complete risk profile for the population risk from toxic contamination in the groundwater. The risk profile is shown in Figure 4.

Acceptable cancer risk levels are on the order of 1 in 10,000 (0.0001) with 95% certainty. The mean value of the population risk from the toxic CL of this aquifer is approximately 0.0006 with a 25% chance that the risk could be greater than 0.0008. This is an exceptionally high population risk level for which remediation is required.

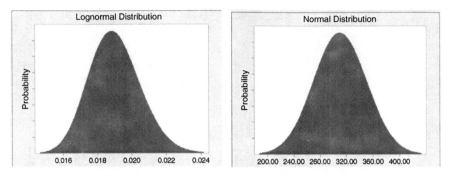

Figure 3. Probability Distributions for Tetrachloroethylene CPF and CL.

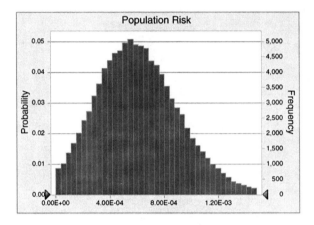

Figure 4. Risk Profile for Population Risk to Toxic Contamination in Groundwater.

2. Remediation Effectiveness

Since this is the community's only source of potable water and the population risk is unacceptable, the task force recommended the following three treatment methods:

1. Air stripping

2. Carbon filter

3. Photo-oxidation

Before proceeding, the task force wants to know the effectiveness of remediation on reduction of the population risk.

We have some limited insight into the effectiveness of the three alternatives. This insight was provided by task force experts as probability distributions of treatment cleanup efficiency. This is represented as a modification to the original influence diagram shown in Figure 5.

The task force assumed in establishing the efficiency probability for cleanup treatments that it applied equally to each alternative method. Since their knowledge of the treatment efficiency was limited, they established the uniform distribution shown in Figure 6. This uniform distribution represents the degree of uncertainty they had in treatment efficiency.

The addition of the treatment efficiency factor to the influence diagram and its associated probability distribution can be incorporated in the spreadsheet model as shown in Figure 7.

Running the Monte Carlo simulation with Crystal Ball provides the results shown in Figure 8. This figure compares the original population risk assessment without remediation with the results of remediation. It can be seen from the figure that remediation can have a significant effect on the population risk.

3. Optimal Treatment Method

Having proven the potential effectiveness of remediation, the task force wants to reduce the level of contamination to recommended standards, using one of the three remediation methods proposed.

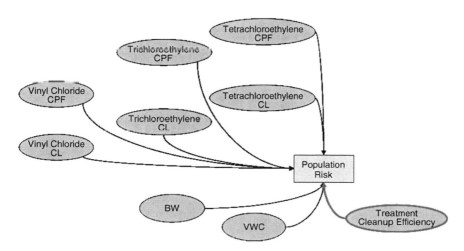

Figure 5. Influence Diagram for Remediation Effectiveness.

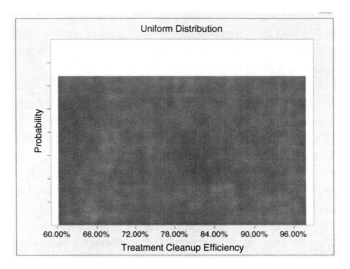

Figure 6. Uniform Distribution of Treatment Efficiency.

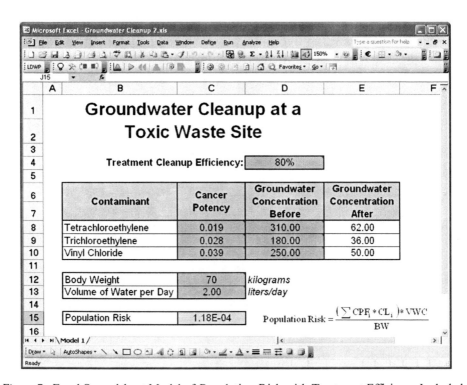

Figure 7. Excel Spreadsheet Model of Population Risk with Treatment Efficiency Included.

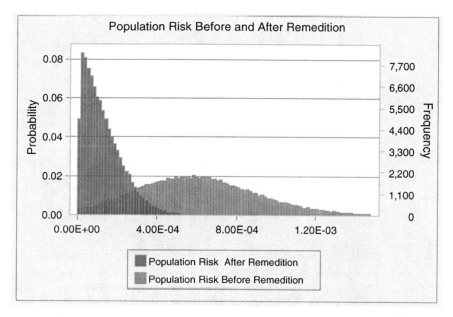

Figure 8. Effect of Remediation on Population Risk.

The costs of the different cleanup methods vary according to the resources and time required for each (cleanup efficiency). With the historical and site-specific data available, the task force wants to find the best process and efficiency level that minimizes cost and still meets the study's recommended standards with 95% certainty. Figure 9 shows the influence diagram modified to include total remediation cost as a function of fixed and variable costs for each contaminant, and two classes of decision variables. The decision variables are the things we can control; in this case, the choice of remediation method and level of cleanup efficiency.

Once again, it is an easy task to represent this influence diagram in the spreadsheet model. Figure 10 shows the modified model.

The professional version of Crystal Ball has a stochastic optimization tool called OptQuest [4] that can be used to solve this problem. The minimum cost treatment model is a mixed integer stochastic mathematical programming model with the following form (Figure 11):

The optimum solution found by OptQuest is to use the photo-oxidation remediation method at 91% cleanup efficiency. This remediation method and cleanup efficiency level costs $10,902 ± $380, and provides an average population risk level of 0.0000516 with 95% confidence (see Figure 12).

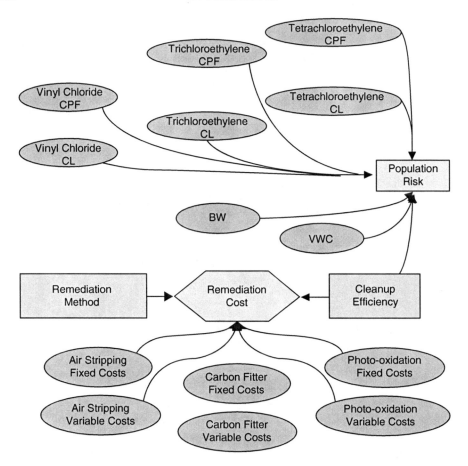

Figure 9. Influence Diagram for Determining the Minimum Cost Treatment Method.

4. Multiple Objectives

Several conservation, political, industrial, and community groups have come forward to raise issues related to their individual agendas. An extensive community consultation was conducted. All groups felt that the selection of the optimal treatment method (if remediation is done at all) must consider objectives other than just cost and population risk. The consensus was to add the following objectives:

- Incremental Health Risk
- Remediation Impacts
  - Contamination

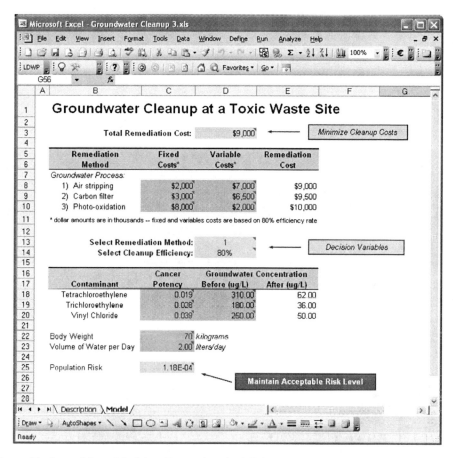

Figure 10. Spreadsheet Model to Determine the Minimum Cost Treatment Method.

- Emissions
- Community
- Safety

These objectives were incorporated into the influence diagram under the category of "Remediation Impacts" as shown in Figure 13.

To incorporate the multiple objectives into the spreadsheet model, a decision analysis using value-focused thinking (VFT) [5] was conducted to assess the overall value of remediation. This analysis was conducted as a community forum where the objectives were modeled as a goals hierarchy from which value functions and objective importance weights were elicited using standard decision analysis procedures. The software package Logical

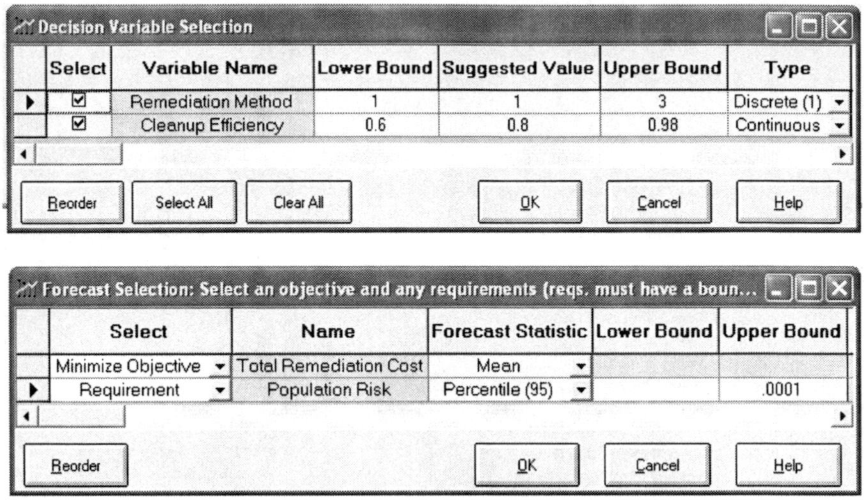

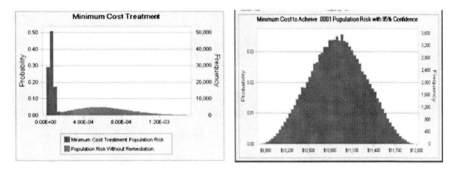

Figure 11. Minimum Cost Treatment Model, Decision Variables (above) and Objective Function and Constraint (below).

Figure 12. Optimal Treatment Method Performance and Cost Risk Profiles.

Decisions for Windows (LDW) [6] was used for the analysis. Figure 14 shows the goals hierarchy that was developed during the community forum.

Data were gathered for each of the objectives shown in the goals hierarchy. These data are shown in Figure 15.

The "Incremental Health Risk" and "Cost" data were the target population risk level and the mean of the cost data used in the previous analyses. The remediation impact objectives were all subjective estimates from the task force using qualitative scales from "Low" to "High." These data were used in the value and importance weight elicitation processes during the community forum. Value functions and importance weights are shown in Figure 16.

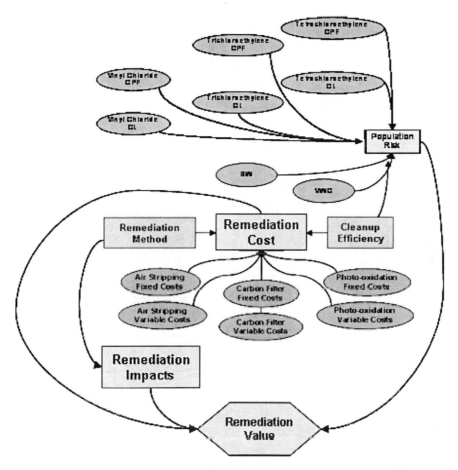

Figure 13. Multiple Objective Extension of Remediation Evaluation.

Using LDW's value function export capability, we incorporated the goals, hierarchy structure, and value functions into the spreadsheet model. The modified spreadsheet model is shown in Figure 17.

The Crystal Ball OptQuest stochastic optimization feature was then used to determine the optimum value remediation method. The result was to use carbon filter remediation at 66% cleanup efficiency with a total value of 0.908. This remediation method and cleanup efficiency level costs $6,909 ± $287, provides an average population risk level of 0.000199 with 95% confidence that the population risk level will be below 0.000392, and satisfies all of the remediation impact concerns (see Figure 18).

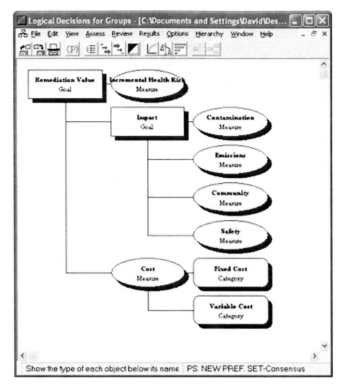

Figure 14. Community Forum Goals Hierarchy.

	Incremental Health Risk	Contamination	Emissions	Community	Safety	Cost	Fixed Cost	Variable Cost
Category Multipliers							1	1
No Remediation	0.0006	Low	Low	Low	Low	0	0	0
Air Stripping	0.0001	High	Medium	High	High	9000	2000	7000
Carbon Filter	0.0001	Low	Low	Low	Low	9500	3000	6500
Photo-oxidation	0.0001	Medium	High	Medium	High	10000	8000	2000

Figure 15. Data for Each Objective in the Goals Hierarchy.

5. Reducing Uncertainty

Having determined an optimum remediation method based on the multiple objectives of the stakeholders, the task force now wants to investigate the value of additional information so that mitigation plans can be developed to manage the population risk.

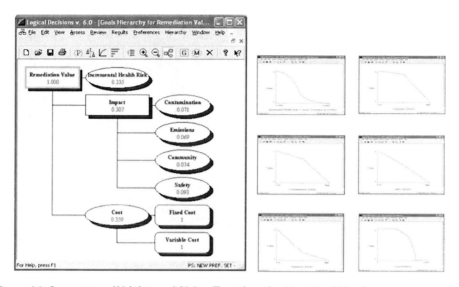

Figure 16. Importance Weights and Value Functions for Decision Criteria.

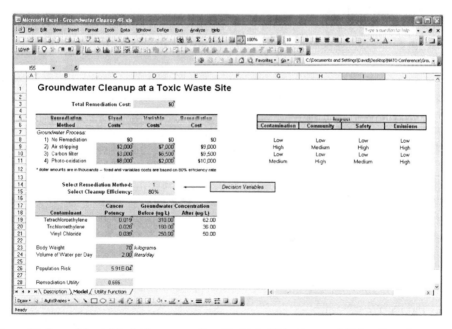

Figure 17. Spreadsheet Model Incorporating Multiple Objectives and Value Functions.

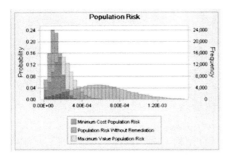
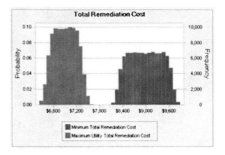

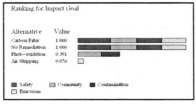

Figure 18. Maximum Value Remediation with Comparison to Minimum Cost Remediation.

Two possible cases of perfect information have been identified:

- Cancer potency and groundwater concentration
- VWC per day

The BW distribution was not considered for perfect information analysis because it represents the community population and is not an uncertainty in the same sense as the cancer potency/groundwater concentration and VWC per day.

To determine the value of perfect information for cancer potency and groundwater concentration, we remove the uncertainty and use the mean value. Figure 19 shows the modified influence diagram and spreadsheet model.

Results of the Monte Carlo simulations using the Crystal Ball are shown in Figure 20. The result shown is an overlay of the maximum-value remediation with and without perfect information for cancer potency and groundwater concentration. The overlay indicates no benefit from perfect information for cancer potency and groundwater concentration.

To determine the value of perfect information for VWC per day, we remove the uncertainty and use the mean value. Figure 21 shows the modified influence diagram and spreadsheet model.

Results of the Monte Carlo simulations using the Crystal Ball add-in to Excel are shown in Figure 22. The result shown is an overlay of the maximum-value remediation with and without perfect information for VWC. The overlay indicates a significant benefit for the perfect information about VWC.

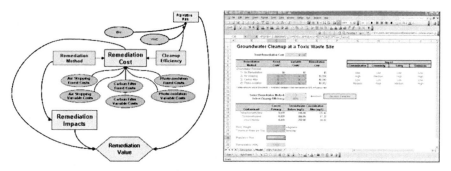

Figure 19. Influence Diagram and Modified Spreadsheet Model for Cancer Potency and Groundwater Concentration Perfect Information.

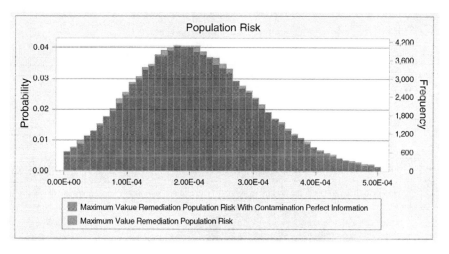

Figure 20. Overlay of Maximum Value Population Risk with and without Cancer Potency and Groundwater Concentration; Perfect Information.

6. Linked Decisions

Having done all of these analyses, the task force believes that there are so many unknowns that it would be wise to tread slowly. As a first step, they agree to proceed with the maximum-value plan to do carbon filter remediation at 66% cleanup efficiency. In addition, they agree to institute a parallel population risk mitigation plan that would supply bottled water above a daily intake of 2 L free of charge to anybody within 2 miles of the contaminated site.

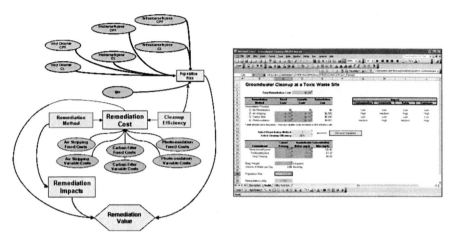

Figure 21. Influence Diagram and Modified Spreadsheet Model for VWC per Day; Perfect Information.

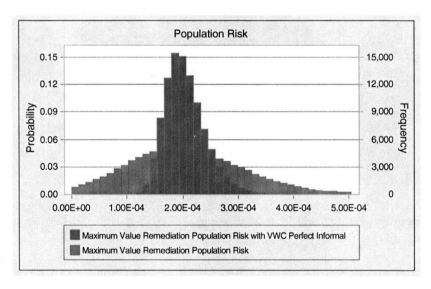

Figure 22. Overlay of Maximum Value Population Risk with and without Cancer Potency and Groundwater Concentration; Perfect Information.

Since the remediation will take many months and it would be too costly to continue this mitigation plan for a long period of time, the task force would like to periodically review the progress of the remediation to determine when to change courses of action. If during the course of the carbon filter remediation the population risk drops appreciably, the task force may decide to reduce, modify, or eliminate the bottled water mitigation plan.

The decision analysis concepts of Real Options and Options Thinking [7] provide the means for such a periodic review. Real Options provides the ability to delay and revise decisions over time as uncertainty is resolved. Options Thinking decomposes a decision into a sequence of decisions over time and reduces risk by delaying resource commitment and reducing uncertainty. Options Thinking also increases value by preserving options to proceed at lower cost, and permitting creation of new possibilities.

Figure 23 shows the flow of the Real Options process. In each phase, multiple objectives are identified together with uncertainties and decision opportunities. A maximum value decision is taken based on this "snapshot" in time, and the risks and their current levels are recorded. Risk tracking and handling plans are developed, and the project is begun.

7. Integrating the Parts: Adaptive Management

This simple case study of groundwater cleanup at a toxic waste sight has demonstrated the effective use of multiple operations research and decision analysis methods. Influence diagrams were used to frame the problem. Monte Carlo simulation was used to quantify the population risk. Stochastic optimization was used to determine a minimum cost solution at an acceptable population risk level. VFT coupled with stochastic optimization was used

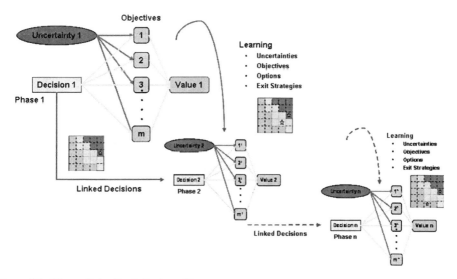

Figure 23. Flow of the Real Options Process.

to determine a maximum value solution with multiple competing objectives in addition to cost. Monte Carlo simulation was again used on the final solution to determine the value of perfect information for purposes of identifying possible mitigations plans. Finally the concept of Real Options thinking was used to provide a management procedure to insure effective completion of the remediation project. All of these operations research and decision analysis methods were implemented with available commercial software that is effective and user friendly.

The demonstrated process has all the tenets of adaptive management. It recognizes that uncertainty is inherent in any natural system, it seeks to minimize the uncertainty by learning about the system being managed over time, and it chooses a management approach and monitors the effects of that approach, making required adjustments based on the monitored results.

The benefits derived from this implementation of adaptive management are many. It provides the ability to delay and revise decisions over time as uncertainty is resolved; it reduces risk by delaying resource commitment until real progress has been made and uncertainty reduced; it increases value by preserving options to proceed more efficiently and effectively and permitting creation of new possibilities as we move from phase to phase.

Figure 24 shows a how the process demonstrated here maps to the Elements of Adaptive Management sanctioned by the National Research Council of the U.S. National Academy of Sciences.

National Research Council US National Academy of Sciences			Issue 1 - Population Risk Assessment	Issue 2 - Remediation Affectivity	Issue 3 - Optimal Treatment Method	Issue 4 - Multiple Objectives	Issue 5 - Reducing Uncertainty	Issue 6 - Linked Decisions
Elements of Adaptive Management	1	Management objectives which are regularly revisited and accordingly revised				X	X	X
	2	A model of the system(s) being managed	X	X	X	X	X	X
	3	A range of management choices			X	X	X	X
	4	Monitoring and evaluation of outcomes						X
	5	A mechanism(s) for incorporating learning into future decisions						X
	6	A collaborative structure for stakeholder participation and learning					X	X

Figure 24. Demonstrated Process Map to NRC Elements of Adaptive Management.

References

1. "Groundwater Cleanup at a Toxic Waste Site," 2007. *Crystal Ball Examples Guide*, Oracle Corporation, Redwood Shores, CA.
2. Powell, S. G., and Baker, K. R., 2004. *The Art of Modeling with Spreadsheets*. Wiley, Hoboken, NJ.
3. Crystal Ball Premium add-in to Excel, version 7.3.1, 2007. Oracle Corporation, Redwood Shores, CA.
4. OptQuest for Crystal Ball, version 2.4), 2007. OptTek Systems, Boulder, CO.
5. Keeney, R. L., 1992. *Value Focused Thinking*. Harvard University Press, Cambridge, MA.
6. Logical Decisions for Windows, Group Version 6.00.22, 2007. Logical Decisions, Fairfax, VA.
7. Mun, J., 2002. *Real Options Analysis*. Wiley, Hoboken, NJ.

MULTICRITERIA DECISION ANALYSIS AND LIFE CYCLE ASSESSMENT

Applications Under High Uncertainty

K. ROGERS

USDA Ecological Science and Engineering Research Fellow
Purdue University
West Lafayette, Indiana, USA

T. SEAGER

Rochester Institute of Technology
Rochester, New York, USA
Thomas.Seager@rit.edu

I. LINKOV

U.S. Army Engineer Research and Development Center
83 Winchester Street, Suite 1
Brookline, Massachusetts 02446, USA
Igor.Linkov@usace.army.mil

Abstract: Assessment of environmental impact is one of the crucial steps in life-cycle assessment (LCA). Current LCA tools typically compute an overall environmental score using a linear-weighted aggregation of normalized inventory data relating to relative performance in impact categories such as global warming, stratospheric ozone depletion, or eutrophication. However, uncertainty associated with quantification of weights is, in general, very high. Moreover, where multiple stakeholder groups are engaged in a particular problem, there may be several different sets of weights that result in disparate scores or ranking. In some cases, the final results may seem entirely dependent upon the relative importance of weights and/or level of data uncertainty. Therefore, we propose to couple life-cycle impact assessment tools with stochastic multiattribute acceptability analysis (SMAA), which is a multicriteria decision analysis (MCDA) technique for exploring uncertain weight spaces. This paper briefly reviews the current state of the art for impact assessment in LCA and compares results using the U.S. Environmental Protection Agency's TRACI model with the SMAA approach for transportation energy

I. Linkov et al. (eds.), Real-Time and Deliberative Decision Making.
© Springer Science + Business Media B.V. 2008

alternatives with uncertain preference information. In both cases, life-cycle inventories are compiled from Argonne National Labs' GREET model. In the typical life-cycle impact assessment (LCIA), case results are based on the total environmental score, allowing dissimilar impacts to be added together, which correlates rank to the highest normalized impact. However, the SMAA approach balances the criteria more evenly, resulting in a different preference ordering. The difference between the two methods is partly due to stochastic versus point representation of weights. Data normalization, which converts incommensurate impact units to dimensionless quantities for the purpose of aggregation, greatly influences the results.

1. Introduction

Life-cycle impact assessment (LCIA) is an important tool for analyzing the environmental impacts of a product or process. It is applied to a wide range of problems, from product design and improvement and selecting alternatives to larger-scale sustainability initiatives such as the choice of industry-wide products (packaging, for example) or a national green construction program. The advantage of the cradle-to-grave approach is the comprehensive environmental profile that brings impacts from various stages in the life cycle into a consistent framework for analysis. This helps avoid the issue of problem-shifting, where an improvement in one stage of the life cycle results in degradation of another stage [1].

While the holistic nature of LCA gives the method its strengths, the scope and scale results in aggregated data that is highly variable and uncertain. Modeling choices, technological and regional variability, data quality, and environmental fate and transport uncertainty are some examples of the sources of uncertainty and variability in LCA. Where information is available, Monte Carlo analysis is sometimes used to propagate uncertainty and/or variability through several LCIA phases [2]. However, few studies incorporate uncertainty through a complete impact assessment [3].

The first two steps of impact assessment are inventory analysis and characterization, which are required by ISO 14040 standards. In inventory analysis, emissions (or consumption) data are categorized into different potential impacts (e.g., global warming potential, acidification potential, or ozone depletion potential). In characterization, data is mathematically aggregated into a single unit of equivalency, such as kg-CO_2-eq. for global warming, representing unique impact categories. Characterization factors are based on several different types of analyses, including environmental fate and transport of chemicals and toxicological and ecotoxicological studies. In many LCAs the characterized data is sufficient for the purpose of the study; however, in situations where environmental tradeoffs occur and there are a large number of stakeholders involved who have varying degrees of experience in LCA and additional concerns beyond environmental impacts, further aggregation of characterized inventory data is desirable to aid decision-making processes.

The critical challenge is to represent disparate or incommensurate impacts on a scale that allows direct comparison. LCIA methods address this challenge with the last two steps in impact assessment, which are normalization and weighting. In LCIA normalization, the characterized inventory data pertaining to the process or product at hand is divided by the total equivalent inventory for an entire region or industry. The result is a dimensionless number representing the fraction of total emissions attributable to the studied alternative, relative to the region (e.g., U.S. or Europe). Many LCIA software packages, such as SimaPro, include normalization factors for multiple regions [4]. Finally, a set of relative importance weights may be applied to each normalized impact category and the weighted categories may be summed to a single environmental score, facilitating comparison of different alternatives.

Because the normalization benchmark and the weights represent value judgments rather than an objective standard, normalizing and weighting are considered optional by ISO 14040 standards [5]. Consequently, many LCAs are left as a set of characterized data, leaving decision or policy makers to confront multicriteria, multi-stakeholder problems unaided. To address this issue, some LCA practitioners have borrowed tools from multicriteria decision analysis (MCDA) to facilitate understanding of tradeoffs and multiple perspectives in the final weighting phase of LCIA [6].

There are two basic categories of MCDA methods: compensatory methods, such as multiattribute value theory (MAVT) or the Analytic Hierarchy Process (AHP), and non-compensatory methods such as ELECTRE or PROMETHEE.

Compensatory methods rely on the concept that criteria can be normalized into the same scale and can ultimately be compared using weights as a relative importance measure when the criteria are aggregated into a single score [7]. MAVT has been applied to LCIA problems because the aggregation method in LCIA is similar to a linear utility function with the objective of minimizing the overall environmental impact [8, 9]. However, MAVT requires a level of certainty in preferences that is typically lacking in environmental decision making. AHP facilitates value elicitation by querying decision makers about their strength of preference between only two assessment criteria at a time. Relative weights are then calculated indirectly. Although intransitivities may occur, in general decision makers find the process more tractable than other approaches. For examples of AHP in LCA, see [10–12].

Non-compensatory, outranking methods have generally not been used in LCA. However, they are particularly applicable to large-scale multiple-stake-holder problems with incommensurate criteria and uncertain preference information, such as LCIA. Outranking methods were developed for situations where there are a large number of alternatives, strong heterogeneity between criteria makes aggregation difficult, and compensation of loss in a given criterion by gain in another is unacceptable [13]. One of the requirements for impact categories in LCIA is that the environmental effects are incomparable or that important effects may be masked or lost upon further aggregation [14]. Nevertheless, there remains considerable uncertainty with regard to value elicitation (i.e., weighting) for the purposes of decision making in the context of LCA.

This paper presents an approach to modeling uncertainty in LCA that extends to stochastic exploration of the weighting or environmental scoring step. The general approach, called Stochastic Multiattribute Acceptability Analysis (SMAA), is appropriate for situations in which the weights may be only partially or even completely unknown due to the large number of decision makers (i.e., variability) or uncertainty associated with value elicitation. In these cases, which are typical of environmental decision-making problems, reducing uncertainty or describing variability may be prohibitively expensive. SMAA methods are capable of determining the sensitivity of ranking alternatives by exploring the weight space in which one alternative may be preferred over others.

2. Method

SMAA is a family of decision analysis methods that facilitate group decision-making problems with uncertain data in both criteria measurement and preference information. SMAA methods work by exploring stochastic weight spaces to describe a probabilistic preference ordering of alternatives. An inverse weight space analysis is used in conjunction with Monte Carlo simulation to obtain rank acceptability indices (the probability that makes an alternative ranked first, second, third, and so on), a central weight vector (the centroid of the weights applied that make an alternative preferable over all other alternatives) and confidence factors [15]. In this case study, only the rank acceptability indices will be analyzed.

The original SMAA was based on a utility function to obtain a weight vector and acceptability index that makes an alternative rank first [16]. SMAA-2 extended the original method by adding rank acceptability indices for all ranks, 1 to n, for each alternative [17]. While SMAA and SMAA-2 are based on a utility model, SMAA-3 and SMAA-III are based on the ELECTRE- III outranking methods, using pseudo-criteria (preference and indifference thresholds) to express criteria preference uncertainty. Alternatives are compared in pairs for each criterion to determine the number of times an alternative outranks other alternatives. Preference is based on threshold values which relate to the difference between criteria measurements. For example, if the difference between two alternatives in a specific criterion is less than the indifference threshold, then neither alternative is preferred, if the difference is greater than the indifference threshold, one alternative is weakly or partially preferred to the other one, and if the difference is greater than the preference threshold, one alternative is completely preferred to the other.

This case study compares the ranks from the traditional LCIA method of adding weighted, normalized impacts into single environmental scores with the SMAA rank acceptability indices of seven transportation fuel options: three petroleum-based low sulfur diesel alternatives, three soy-biodiesel alternatives, and an electricity-powered vehicle (EV). Alternatives and abbreviations are listed in Table 1.

Inventory data was obtained using the GREET v1.8 model from Argonne National Labs [18], which predicts the fuel life cycle (well-to-wheels) energy use and emissions of nine air pollutants including: CO_2, CH_4, N_2O, NO_x, CO, VOC, SO_x, PM2.5, and PM10. The target year for simulation was 2015, and the vehicle type modeled was a passenger car. All GREET default assumptions were used except the biodiesel blend, which was changed from 20% to 100%. The well-to-wheels energy and emissions inventory data was characterized into six environmental impact categories: Fossil fuel depletion (FFD), global warming potential (GW), eutrophication potential (EUT), acidification potential (ACID), photochemical ozone formation potential (SMOG) and human health criteria (HHCR). These particular categories were chosen because the pollutants provided in the GREET model do not have impacts in other categories.

The U.S. Environmental Protection Agency's Tool for the Reduction and Assessment of Chemical and Other Environmental Impacts (TRACI) was

TABLE 1. Alternatives.

Abbreviation	Alternative
LS Diesel	Low sulfur diesel (compression ignition vehicle)
GI LS Diesel HEV	Grid-independent low sulfur diesel hybrid electric vehicle
GC LS Diesel HEV	Grid-connected (plug in) low sulfur diesel hybrid electric vehicle
BD100	100% biodiesel blend (compression ignition vehicle)
GI BD100 HEV	Grid-independent biodiesel hybrid electric vehicle
GC BD100 HEV	Grid-connected (plug in) biodiesel hybrid electric vehicle
EV	Electric vehicle

TABLE 2. TRACI Characterization and Normalization Factors.

Emission	FFD (MJ/BTU)	GW (kg/CO$_2$-eq)	EUT (kg/N-eq)	SMOG (kg/Nox-eq)	ACID (kg/H + eq)	HHCR (kg/milli-dalys)
FFD	0.001055					
CO$_2$		1				
CH$_4$		23		0.002964		
N$_2$O		296				
VOC				0.780645		
CO				0.013387		
NO$_x$			0.00429	1	40.04	0.002213
PM10						0.083448
PM2.5						0.13908
SO$_x$					50.79	
Normalization factors for US emissions	8.53E + 13 MJ/year	6.85E + 12 CO2 eq/ year	5.02E + 09 N-eq/ year	3.38E + 10 NOX eq/year	2.08E + 12 H + eq/ year	1.71E + 11 milli-daly' s/year

used to characterize the inventory data. TRACI characterization factors are determined using fate and transport modeling as well as toxicology assessments for each pollutant in question. For detailed information on TRACI, see [14, 19]. Crystal Ball v5.5 [20] was used to run stochastic simulations for all uncertainty propagation. TRACI characterization and normalization factors are provided in Table 2.

Average characterized data is presented in Table 3. Note that no alternative dominates all categories; bold numbers are highest values; and shaded cells are lowest values. Lower environmental impacts are desirable.

The characterized inventory data was normalized by dividing it by the total US emissions for each TRACI category using the US normalization database in [21], reported in Table 2. This method is commonly used in cases where there is a drastic change in technology that is too broad to be normalized to a specific industry. The goal of normalization is to have comparable units for the different categories so that they can be weighted and added together to obtain a single environmental score.

Table 4 shows the normalized inventory. Note that the highest and lowest impact category values for the alternatives remain proportional to the characterized inventory. However, the scale of impact categories in comparison with one another has changed—in some case by several orders of magnitude.

In the absence of weight information, equal weights are applied to the normalized inventory. Figure 1 shows the overall environmental score of each alternative and the resulting rank. For the majority of alternatives,

TABLE 3. Mean Characterized Inventory (kg or MJ/1,000 Miles).

	FFD (MJ)	GW (CO2-eq)	EUT (N-eq)	SMOG (Nox-eq)	ACID (H + eq)	HHCR (milli-dalys)
LS Diesel	4,819	371	$9.7 \times 10-4$	0.31	13	$1.0 \times 10-2$
GI LS Diesel HEV	3,613	279	$7.7 \times 10-4$	0.24	10	$8.9 \times 10-3$
GC LS Diesel HEV	3,637	299	$9.6 \times 10-4$	0.27	22	$2.4 \times 10-2$
BD100	2,065	149	$1.6 \times 10-3$	0.90	25	$1.4 \times 10-2$
GI BD100 HEV	1,550	113	$1.2 \times 10-3$	0.69	19	$1.2 \times 10-2$
GC BD100 HEV	2,248	187	$1.3 \times 10-3$	0.57	28	$2.6 \times 10-2$
EV	3,146	289	$1.2 \times 10-3$	0.31	42	$4.8 \times 10-2$

TABLE 4. Mean Normalized Inventory (Fraction of US Emissions per 1,000 Miles).

	FFD	GW	EUT	SMOG	ACID	HHCR
LS Diesel	5.6E-11	5.4E-11	6.7E-13	9.0E-12	6.2E-12	6.0E-14
GI LS Diesel HEV	4.2E-11	4.1E-11	5.3E-13	7.2E-12	4.8E-12	5.2E-14
GC LS Diesel HEV	4.3E-11	4.4E-11	6.6E-13	8.1E-12	1.1E-11	1.4E-13
BD100	2.4E-11	2.2E-11	1.1E-12	2.7E-11	1.2E-11	8.1E-14
GI BD100 HEV	1.8E-11	1.6E-11	8.4E-13	2.0E-11	9.3E-12	6.8E-14
GC BD100 HEV	2.6E-11	2.7E-11	8.7E-13	1.7E-11	1.4E-11	1.5E-13
EV	3.7E-11	4.2E-11	8.5E-13	9.0E-12	2.0E-11	2.8E-13

combined performance for FFD and GWP overshadows all other environmental impacts and overperformance can compensate for lagging performance in all other categories. Uncertainty information is provided with error bars in the chart; the alternatives are ordered based on the mean environmental score; however, there is crossover and ambiguity in the ranks due to uncertainty.

CSMAA1.0 [22] software was used to calculate rank acceptability indices for each transportation alternative. The characterized data was used directly in the analysis. The preference threshold was the average standard deviation of each impact category, and the indifference threshold was half the average standard deviation. The rank acceptability indices are shown in Figure 2. The ranking of several alternatives has changed from the LCIA approach and the uncertainty and variability information represented by SMAA is more robust.

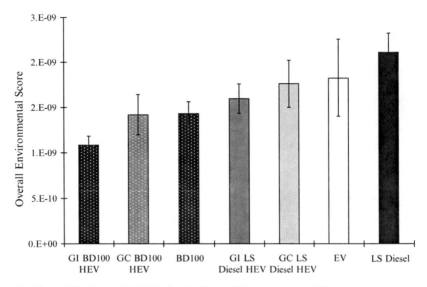

Figure 1. Linear Weights Added Method—Overall Environmental Score.

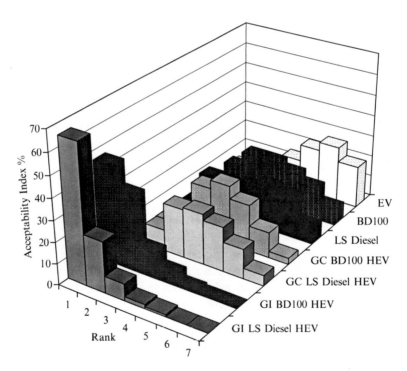

Figure 2. SMAA-III Rank Acceptability Indices.

3. Discussion

Preference ranking of the alternatives based on characterized data is difficult because impact categories are incommensurate and tradeoffs are present. However, the ELECTRE-III outranking approach avoids the issue of compensation for poor performance in some impact categories and SMAA allows the decision maker to explore all weighting possibilities at once to determine the likelihood that an alternative will be preferred—providing an organized method that avoids subjective and often time-consuming weight elicitations to prioritize alternatives.

When the "best" alternative is unavailable, it is important to understand how the remaining alternatives will rank. In the transportation study, the most preferable alternatives are HEVs, determined by the higher acceptability indices for ranks 1 through 4—although there is a chance that any alternative can have a favorable rank depending on the weights used. GI LS Diesel HEV had the lowest criteria measurements in four categories. Thus it is not surprising that this alternative has a high acceptability index for first rank.

Intuitively, good performance in several different categories should yield a higher acceptability, while poor performance in several categories should yield a lower rank. However, in the LCIA linear weighting method, compensation is allowed regardless of knowledge of the actual severity of environmental effects of different categories. Since results in state-of-the-art LCIA are normalized based on total regional emissions, the analysis is skewed to underemphasize the importance of a pollutant with high current regional emissions levels. As a result, potentially important impacts may be masked. Therefore, we conclude that SMAA is best applied in combination with an outranking approach and that both the normalization and stochastic weighting approaches are important to explain the difference in performance ordering between state-of-the-art LCA tools (e.g., TRACI) and SMAA approaches.

Acknowledgements

This work was supported in part by the Environmental Quality Technology Program of the U.S. Army Engineer Research and Development Center (Dr. John Cullinane, Technical Director). Permission was granted by the Chief of Engineers to publish this information.

References

1. Guinee, J. B. (Ed.), 2002. *Handbook on Life Cycle Assessment: Operational Guide to the ISO Standards*. Kluwer, Dordrecht, Boston, MA/London.

2. Lo, S.-C., Ma, H.-W., and Lo, S.-L., 2005. Quantifying and Reducing Uncertainty in Life Cycle Assessment Using Bayesian Monte Carlo Method. *Science of the Total Environment* 340:23–33.

3. Lloyd, S. M., and Ries, R., 2007. Characterizing, Propagating, and Analyzing Uncertainty in Life-Cycle Assessment. *Journal of Industrial Ecology* 11:161–179.

4. Goedkoop, M., De Schryver, A., and Oele, M., 2007. *Introduction to LCA with SimaPro 7.* PRé Consultants, The Netherlands.

5. Steen, B. A., 2006. Describing Values in Relation to Choices in LCA. *International Journal of Life Cycle Assessment* 11:277–283.

6. Miettinen, P., and Hamalainen, R. P., 1997. How to Benefit from Decision Analysis in Environmental Life Cycle Assessment (LCA). *European Journal of Operational Research* 102:279–294.

7. Simpson, L., 1996. Do Decision Makers Know What They Prefer? MAVT and ELECTRE II. *Journal of the Operational Research Society* 47:919–929.

8. Seppala, J., Basson, L., and Norris, G., 2002. Decision Analysis Frameworks for Life-Cycle Impact Assessment. *Journal of Industrial Ecology* 5:45–68.

9. Seppala, J., and Hamalainen, R. P., 2001. On the Meaning of Distance-to-Target Weighting Method and Normalization in Life Cycle Impact Assessment. *International Journal of Life Cycle Assessment* 6:211–218.

10. Noh, J., and Lee, K. M., 2003. Application of Multiattribute Decision-Making Methods for the Determination of Relative Significance Factor of Impact Categories. *Environmental Management* 31:633–641.

11. Daniel, S. E., Tsoulfas, G. T., Pappis, C. P., and Rachaniotis, N. P., 2004. Aggregating and Evaluating the Results of Different Environmental Impact Assessment Methods. *Ecological Indicators* 4:125–138.

12. Pineda-Henson, R., Culaba, A. B., and Mendoza, G. A., Evaluating Environmental Performance of Pulp and Paper Manufacturing Using the Analytic Hierarchy Process. *Journal of Industrial Ecology* 6:15–28.

13. Figueira, J., Mousseau, V., and Roy, B., 2005. ELECTRE Methods. In Figueira, J., Greco, S., and Ehrogott, M. (Eds.). *Multiple Criteria Decision Analysis: State of the Art Surveys*, vol. 78, pp. 133–153. Springer, New York.

14. Bare, J. C., Norris, G. A., Pennington, D. W., and McKone, T., 2003. TRACI: The Tool for the Reduction and Assessment of Chemical and Other Environmental Impacts. *Journal of Industrial Ecology* 6:49–78.

15. Tervonen, T., and Lahdelma, R., 2007. Implementing Stochastic Multicriteria Acceptability Analysis. *European Journal of Operational Research* 178:500–513.

16. Lahdelma, R., Hokkanen, J., and Salminen, P., 1998. SMAA: Stochastic Multiobjective Acceptability Analysis. *European Journal of Operational Research* 106:137–143.

17. Lahdelma, R., and Salminen, P., 2001. SMAA-2: Stochastic Multicriteria Acceptability Analysis for Group Decision Making. *Operations Research* 49:444–454.

18. Wang, M., 2007. GREET: Greenhouse Gases, Regulated Emissions, and Energy Use in Transportation, 1.7 ed. Argonne National Laboratory, Argonne, IL.

19. Norris, G. A., 2003. Impact Characterization in the Tool for the Reduction and Assessment of Chemical and Other Environmental Impacts. *Journal of Industrial Ecology* 6:79–101.

20. Crystal Ball 5.5, 2005. Decisioneering Inc., Littleton, CO.

21. Bare, J. C., Gloria, T., and Norris, G. A., 2006. Development of the Method and U.S. Normalization Database for Life Cycle Impact Assessment and Sustainability Metrics. *Environmental Science and Technology* 40:5108–5115.

22. Tervonen, T., 2007. CSMAA 1.0.

DECISION EVALUATION FOR COMPLEX ENVIRONMENTAL RISK NETWORKED SYSTEMS (DECERNS)

Cost/Benefit Module and Application to Wildlife Reserve Management

A. GREBENKOV, A. YAKUSHAU, A. LUKASHEVICH

Joint Institute for Power and Nuclear Research-Sosny
Minsk, Belarus

T. SULLIVAN

Brookhaven National Laboratory
Upton, New York, USA
tsullivan@bnl.gov

I. LINKOV

US Army Engineer Research and Development Center
Concord, Massachusetts, USA
ilinkov@yahoo.com

Abstract: Decision Evaluation in Complex Risk Network Systems (DECERNS) is a computerized decision support system with the objective of providing a methodology, computer models and software tools that facilitate decision-making for sustainable land use planning, sediment management, and related issues.

Cost-benefit analysis (CBA) is an essential component of environmental assessments. This approach introduces costs (i.e., monetary equivalents) of ecosystem services, but management decisions often require decision makers to integrate heterogeneous technical information with values and judgments.

This paper introduces a CBA module developed as part of DECERNS and describes its linkages with a multicriteria decision analysis (MCDA) module that allows integration of monetized parameters with expert-driven estimates and ecological values of stakeholders that may be difficult to monetize. We describe the model's application to selection of land use alternatives for the Dikoye and Zvanets wildlife sanctuaries in Belarus.

I. Linkov et al. (eds.), Real-Time and Deliberative Decision Making.
© Springer Science + Business Media B.V. 2008

1. Introduction

Selection of land use alternatives requires choosing an optimal remediation strategy, taking into consideration the further use and protection
of natural resources. A simultaneous consideration of wildlife value,
complex contamination patterns, and significant physical disturbances
to natural habitats is required. Cost-benefit analysis (CBA) or cost-effectiveness analysis (CEA) aims at assessment of all the costs and benefits of
alternative options, using monetary values of the main criteria for each
alternative. CBA monetizes the cost of ecosystem services and biological
resources (value of biodiversity, including genetic biodiversity, the value
of life for ecological components, and the value of natural production)
[2] but many criteria cannot be expressed as monetary values. In such
cases, conversion factors that transform benefits and risks into monetary
values are used. CEA is often used to extend CBA when it is not possible
to provide acceptable quantitative cost data for all components. CEA
selects the option that achieves the objective at the minimum cost per unit
of positive effect/benefit. CEA may be considered a secondary (auxiliary)
method; it is often used if the full CBA cannot be correctly realized (e.g.,
when not all advantages and disadvantages can be effectively presented
in monetary terms).

Multicriteria decision analysis (MCDA) allows integration of easily mon-
etizable criteria with value judgments and metrics that cannot be monetized.
Decision Evaluation for Complex Environmental Risk Networked Systems
(DECERNS) is a software tool being developed as a comprehensive decision
support system for land use management. It will integrate a suite of MCDA
tools, a geographic information system (GIS), and process models to support environmental management. This paper presents a modeling approach
that calculates cost (i.e., monetary equivalents) of ecosystem services based
on metrics for biotic and abiotic (non-living) biological resources. Ecological
values are difficult to quantify because they result in long-term benefits
that cannot be compared with short-term economic returns associated with
human infrastructure development. To address this problem, our approach
uses a variable discounting of all costs and benefits of environmental protection measures. Traditional CBA is not directly applicable to land use alternatives when values of non-economic origin dominate monetized costs in the
analysis. Examples of non-economic costs include protection of habitat of
endangered species, long-term value of natural environments, and value of
ecological health. The model still involves a least-cost analysis but it takes
into account variable discount rate values. In addition, the approach contains some relative indices dependent on national GDP to tailor the analysis
to the appropriate conditions for a specific country.

2. Economic Aggregates Associated with Remediation Scenarios

Several socioeconomic parameters may be important for evaluating land use management alternatives [1]. Some of them can be translated into monetary equivalents (e.g., capital and operation cost of remediation technologies, direct and side benefits of utilization of natural resources (fossils, biomass, hunting, fishing, and other products of the natural environment, including ecological goods and services)), but others may be difficult to translate into monetary equivalents because they reflect values not primarily related to utilization (e.g., biodiversity, ecosystem functioning, aesthetics) and value of inheritance (Figure 1).

Any damage to the environment may cause some loss of value and become a cost of economic origin assuming that environmental losses are relevant to the economy and human society (Figure 2).

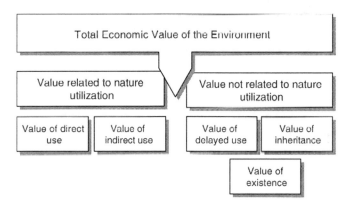

Figure 1. Total Economic Value of the Environment.

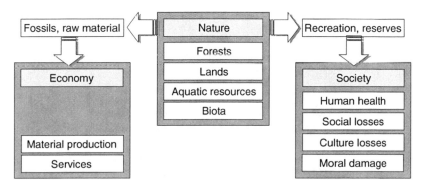

Figure 2. Economic Damage Resulting from Environmental Damage.

In practice, when evaluating economic losses caused by anthropogenic contamination (disturbance) of the environment, one evaluates the cost as a sum of expenses connected with prevention of both: (a) environmental risks for a wildlife representative or entire population/community within the damaged ecosystem and (b) socioeconomic risks within the contaminated (damaged) territory. Usually, the most uncertain step in economic analysis related to environmental damage is to attribute a monetary equivalent to biota not directly utilized by the economy and without other direct links to human society. The principal anthropogenic causes of reduced biodiversity include:

- Destruction, disintegration, and contamination of habitat
- Excessive reduction or extirpation of flora and fauna
- Introduction of invasive species (including that caused by climate change)
- Introduction of communicable diseases to wildlife

A methodological approach for assessing economic losses due to reduced biodiversity can help to project the overall cost of (i) compensating for negative effects to biota, (ii) restoring affected habitat, and (iii) averting possible further negative anthropogenic impacts. The approach should be consistent with other approaches for cost estimates.

Any remediation countermeasure may diminish these losses and result in profits and/or cause other losses. The countermeasure profit consists of renewed (or newly generated) values and/or averted losses, while the countermeasure loss consists of lump-sum costs, operational costs and possible collateral losses. If all the components mentioned in Figure 2 can be expressed through a monetary equivalent, a conventional least-cost evaluation may be used to compare alternative remediation scenarios (Figure 3).

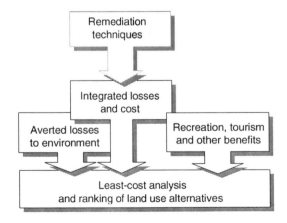

Figure 3. Remediation Benefits and Costs in Least-Cost Analysis.

3. Description of the Model and Algorithm

Several MCDA models are incorporated in a prototype of the DECERNS software [4]. The CBA model described here is under development and has not been completely adapted for environmental restoration. In this version, the algorithm addresses costs associated with biodiversity restoration but does not address all land use scenarios.

The approach is to compare two alternatives, one of which is a do-nothing-option (baseline scenario); the other is an environmental protection scenario (project scenario). Both options (Figure 4) entail environmental risk consisting of residual contamination and exposure or habitat disturbances affecting valuable biota. The risk is assigned a monetary equivalent through specific coefficients established by regulations (e.g., taxation) or through relative indices dependent on national GDP. These quantify the value of environmental losses for both alternatives.

Under the proposed algorithm (Figure 4) in the project scenario, the costs (both capital and operational) of countermeasures applied to restore the affected habitat are evaluated, and the result is summed with environmental losses that might remain after remediation. The resulting value represents the overall cost of the project scenario.

The economic benefit of the project is evaluated as averted environmental losses B_e and, in fact, is the difference between environmental losses of the baseline scenario Y_0 and those of the project scenario Y_p:

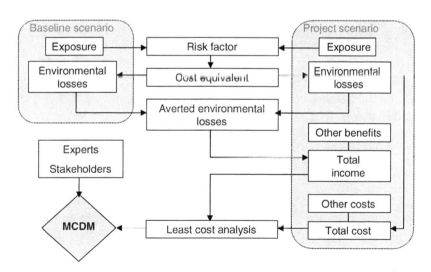

Figure 4. Model Algorithm.

$$B_e = Y_0 - Y_p$$

If B_e is negative, there is no ground for the proposed countermeasures based on environmental losses. If it is positive, the sum of B_e and other direct and side benefits of the countermeasures represents the overall income of the project scenario.

Using the economic indices discussed above, a traditional least-cost analysis can be applied to assess project feasibility. In this approach, the most important and crucial challenges when doing the least-cost analysis are evaluation of the averted environmental losses and selection and use of a proper discount rate.

4. Discount Rate and Averted Environmental Losses

Products of a natural ecosystem have at least three special features. First, most of them cannot be completely reproduced by human technology. Second, natural "technologies" are very conservative as compared to anthropogenic ones, since they are limited by the environment (e.g., incidental solar energy, temperature). Third, most ecosystem products represent common, nonexclusive, and noncompetitive services. The first two features are the argument for the present approach with regard to a specific discount rate for naturally occurring products and services [3]. The third one is useful when determining offsets for environmental protection actions.

If the value of the contaminated territory is primarily from non-economic factors (e.g., biodiversity) and countermeasures to restore damaged ecosystems are applied, the discount rate should correspond with the restoration timing cycle for the ecosystem in order to reflect the alternative worth of natural capital. One of the options for determining such a timing cycle can be the relative time of ecosystem restoration T calculated as a ratio between total ecosystem biomass M and the weighted average value of annual productivity P of its principal structural elements, so that the discount rate r will be a reciprocal value of T:

$$T = \frac{M}{P}, \qquad r = \frac{1}{T}$$

When commercial economic values dominate, the discount rate is calculated as a financial index with the conventional formula using a normative discount rate value r_n (usually a bank interest rate) and an inflation rate i:

$$r = \frac{r_n - i}{1 + i}$$

Under our approach, it is suggested that the environmental losses in both the project and baseline scenarios be calculated as monetary equivalents of physical loss of specified biota expressed in the annual decrease in individuals of a particular species, multiplied by the rate of payment for such loss. The latter value is derived from national regulations or can be a replacement value calculated as a GDP-related function. The resulting expression for loss, applicable to most cases, is given below in a simplified form:

$$Y = \sum_{i=1}^{N} (N_{Yi} - N_{0i}) H_i K_p$$

where:

$i = 1...N$ = Number of biota representing species

N_{Yi} = Number of representatives of species i lost annually as a result of contamination/damage of its habitat

N_{0i} = Number of representatives of species i exempted annually by legal permits

H_{it} = Rate of payment for loss of a single representative of species i established by law

K_p = Biodiversity coefficient established for specified territory

5. Income from Environmental Protection Countermeasures and NPV

In general, countermeasures introduced to remediate contaminated or damaged environmental objects can result in multiple benefits. In addition to averted loss of biota, countermeasures can convert useless, dangerous land into a valuable area suitable for commercial activity; e.g., recreation, tourism, hunting, fishing.

In addition to the use of biological resources, the restored area can provide additional or improved ecological services; e.g., sanitary, carbon deposit, water-purifying, water-protection, and erosion-preventive functions. This type of income can be generalized with the following expression:

$$B = \sum_{k}^{K} \left[R_k^{ind} \cdot \sum_{j=1}^{L} (S_j \cdot \frac{\lambda_j^{nat}}{\lambda_j^{ind}}) \right]$$

where:
$k = 1 ... K$ = Number of ecosystem functions
$j = 1 ... L$ = Number of ecosystems

R_k^{ind} = Annual *present worth cost per unit area of an artificial (industrial)*
      *installation perf*orming the same function k as a natural system

λ_j^{nat} = Annual *specific productivity of ecosystem j performing function k*

λ_j^{ind} = *Annual specific productivity of an industrial installation replacing*
      function k of ecosystem j

S_j = *Area* of ecosystem j

The net present value (NPV) can be calculated as usual when the efficiency of an investment project is estimated:

$$NPV = \sum_{t=0}^{T} \frac{(B_t + B_{et}) - (K_t + C_t)}{(1+r)^t}$$

where:

t = $1 \ldots T$ = Number of years

B_t = Income received during year t from direct and indirect use of ecosystem that is restored as a result of countermeasures

B_{et} = Income received during year t as averted environmental losses in the ecosystem that is restored as a result of countermeasures

K_t = Capital cost during year t of implementing countermeasures

C_t = Operational cost during year t of conducting countermeasures and maintaining the ecosystem that is restored as a result of countermeasures

r = Discount rate

The above expression is used as an indicator of the profitability that can be achieved using countermeasures; i.e., if NPV is a positive figure, the countermeasures are effective and it is worthwhile to implement them.

The suggested model can be applied to rank several possible countermeasures. Under DECERNS the model will address the choice of countermeasures, evaluate the corresponding financial and material requirements, assess and compare the predicted effectiveness of different alternatives, and provide support for decisions on the development of a rehabilitation plan. In order to test the model, two case studies were selected.

6. Dikoye Case Study

The Dikoye area is a unique ornithological territory adjacent to the world-famous Belovezskaya Pushcha National Park. It is located in a watershed of the Baltic Sea and the Black Sea (52° 41' N, 24° 20' E), and is recognized as one of the biggest European eutrophic bogs of the mesotrophic type that remains in its natural condition.

This area provides habitat for about 8% of the European population of the fidgety reedwarbler (*Acrocephalus paludicola*), a globally endangered species. It also provides habitat for 95 other bird species, three of which are also globally endangered: the big columbine (*Aquilla clanga*), the landrail (*Crex crex*), and the great snipe (*Gallinago media*). The Dikoye area is included in the world list of key ornithological territories due to the presence of these birds.

The area also supports and maintains rare plants—e.g., *Betuletum humilis, Caricetum chordorrhizae, Caricetum juncellae,* and *Caricetum limosae*—and two rare mammals, the lynx and the bison.

The case study was undertaken to help select an action plan [5] for converting the entire Dikoye area into conservancy area and integrating it in Belovezskaya Pushcha National Park. The action plan includes several measures, of which the most relevant are:

- Extending the National Park buffer zone to include entire Dikoye area

- Removing several agricultural enterprises and farms

- Restoring the natural hydrological regime to decelerate seral processes by building retaining structures and floodgates in former melioration systems

- Establishing ecotourism services

The results of the case study are presented with the time dynamics of the main indices in Figures 5–7. A discount rate of 8% was used for calculations.

Figure 5 presents the total cost and its components; i.e., the capital cost, operational cost, and cost due to the loss of agriculture products. The capital cost represents lump-sum costs incurred within the first three years when applying all of the measures listed above. The operational cost includes annual fixed charges to maintain and operate the site. The loss of agriculture products is assumed as foregone earnings resulting from extension of the conservancy area by means of removing several agricultural enterprises and farms. The repayment of the bank loan and interest, which come due after ten years, dominates all components of the total cost. The project capital cost and its components were evaluated based on the detailed action plan [5].

Figure 6 shows all expected earnings resulting from utilization of the Dikoye sanctuary as a part of Belovezskaya Pushcha National Park. Project income was calculated as earnings from the following benefits:

- Genetic conservation of endemic species and sustainable development of populations of endangered species

- Protection of the phreatic divide and flood prevention

- Improvement of soil quality

- River basin fishery preservation, timber trade, moorberry products, and ecotourism

- A bank loan that is due for repayment ten years after project start

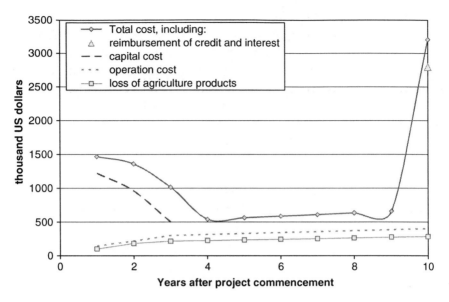

Figure 5. Dynamic of Cost Indices.

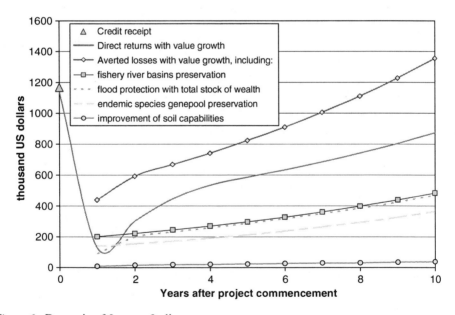

Figure 6. Dynamic of Income Indices.

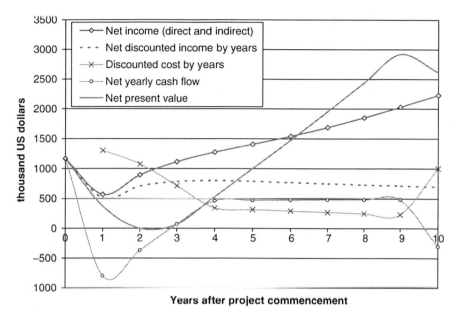

Figure 7. Discounting and NPV.

The net present value calculation (see Figure 7) shows that the action plan designed for protection of the unique Dikoye area is profitable with a payback period of five to six years.

7. Zvanets Case Study

The Zvanets swampland, with almost 16,000 ha of open area, is the biggest natural eutrophic bog in Europe and a world-famous key ornithological territory. Since 1996, the swampland has served as a biological wildlife sanctuary. It is located in the Polesie lowland at the boundary of watershed between Pripiat River and Bug River (52° 05' N, 24° 50' E).

Owing to its size, Zvanets contains a very stable marsh ecosystem, but intensive anthropogenic activities adjacent to and exploiting its hydrological system have negative impacts on regional conservation of biological diversity. Water resources are used without proper control by the Dnieper-Bug canal, melioration systems of surrounding farms, and a fishery co-op (Figure 8). As a result, significant fluctuations in the groundwater table occur frequently. This leads to undesirable plant successions, fires, and underflooding of birds' nests. The biggest population of the fidgety reedwarbler (*Acrocephalus paludicola*), a globally endangered species, and the population of another

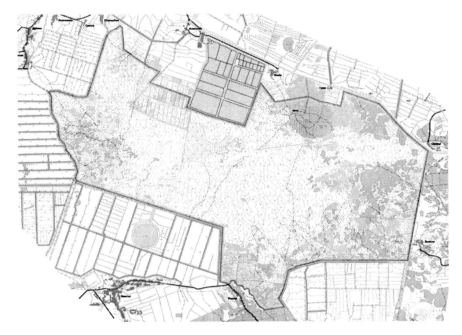

Figure 8. Hydrologic Basin of the Zvanets Wildlife Sanctuary and Adjacent Territory.

endangered bird species, the big columbine (*Aquilla clanga*), are now in real danger of extinction due to habitat disturbance.

Several economic activity factors affect the biodiversity of the entire Zvanets marsh ecosystem:

- Lock activity in the Dnieper-Bug canal
- Water diversion from the Pripiat River through the Beloozersky canal to enhance navigability of the Dnieper-Bug canal
- Operation of melioration systems and sluicing of water resources adjacent to the bog

The case study was undertaken to help select an action plan [6] to provide controlled hydrological regimes for the Zvanets wildlife sanctuary. It is believed that the specified set of countermeasures will protect bird populations. The project includes several measures; the most relevant include:

- Changing the land use of some adjacent territories to incorporate environmentally important areas into the sanctuary
- Limiting some economic activities during large-scale nesting seasons, especially on mineral hillocks
- Installing additional canal pumping stations and providing regulatory regimes

TABLE 1. Results of the Zvanets case study.

Parameters	Units	Data
Investment cost	USD	152,295
Operational cost	USD/year	76,000
Additional losses	USD/year	0
Capital depreciation discount rate		0.000050
Capital depreciation	USD/year	6,096
Averted environmental losses	USD/year	149,531.60
Operating income (sanitary functions)	USD/year	2111.00
Operating income (carbon deposit functions)	USD/year	174.00
Project duration	Years	25
Discount rate for NPV		0.07
NPV	USD	413,816.58
IRR	%	19.358
Payback period	Years	7.0
Dynamic payback period	Years	8.0

- Purging spillway canals of sediments and plants
- Installing an additional pond as a water storage reservoir to buffer rain and snowmelt floods in spring
- Establishing ecotourism services

The project capital cost was evaluated based on the detailed action plan [6] and project income was calculated as earnings from the following benefits:

- Genetic conservation of endemic species and sustainable development of populations of endangered species
- Sanitary and carbon deposit functions of the bog

The results of the case study are presented in Table 1, which shows that the action plan designed for protection of the unique Zvanets area is profitable with a dynamic payback period of about eight years.

References

1. Grebenkov, A., and Yakushau, A., 2007. Ranking of Available Countermeasures Using MCDA Applied to Contaminated Environment. In: *Risk Management Tools for Environmental Security, Critical Infrastructure and Sustainability*. Springer – NATO Science Series, Series IV: Earth and Environmental Series, Vol. 7, p. 131–135.
2. Phillips, A. (series Ed.), 1998. Economic Values of Protected Areas. Guidelines for Protected Area Managers. Best Practice Protected Area Guidelines Series No. 2. IUCN – The World Conservation Union.
3. Kotko, A., 2005. Investment Efficiency Evaluation Methods for Environmental Protection. Izvestiya RAN, Geographical Series, No.4.

4. Sullivan, T., Yatsalo, B., Grebenkov, A., Kiker, G., Kapustka, L., and Linkov, I., 2007. Decision Evaluation for Complex Risk Network Systems (DECERNS). In: Linkov, I., Wenning, R., and Kiker, G. (Eds.), 2008. *Managing Critical Infrastructure Risks*. Springer, The Netherlands.
5. Sheffer, N., Kozulin, A., and Vergeichik, M. (Eds.), 2002. Management Plan for "Dikoye," the Wildlife Sanctuary of Country Significance. The Royal Society for Protection of Birds, UK, and Society for Protection of Birds of Belarus, UNDP, Minsk.
6. Sheffer, N., Kozulin, A., and Vergeichik, M. (Eds.), 2003. Management Plan for "Zvanets", the Wildlife Sanctuary of Country Significance. ... The Royal Society for Protection of Birds, UK, and Society for Protection of Birds of Belarus, UNDP, Minsk.

SUSTAINABLE MANAGEMENT OF WATER RESOURCES AND MINIMIZATION OF ENVIRONMENTAL RISKS

A Multi-Portfolio Optimization Model

E. LEVNER

Holon Institute of Technology
Holon 58102
Bar-Ilan University
Ramat Gan 52900
Israel
levner@hit.ac.il

J. GANOULIS

UNESCO Chair and Network INWEB: International Network
of Water/Environment Centres for the Balkans
Department of Civil Engineering
Aristotle University of Thessaloniki
54124 Thessaloniki, Greece
iganouli@civil.auth.gr

D. ALCAIDE LÓPEZ DE PABLO

University of La Laguna
La Laguna, Tenerife, Spain
dalcaide@ull.es

I. LINKOV

U.S Army Engineer Research and Development Center
83 Winchester Street, Suite 1
Brookline, MA 02446, USA
Igor.Linkov@usace.army.mil

I. Linkov et al. (eds.), Real-Time and Deliberative Decision Making.
© Springer Science + Business Media B.V. 2008

Abstract: This paper aims to analyze options for the sustainable development for the Dead Sea Basin. It presents a model based on the green supply chain paradigm and multi-portfolio selection of strategies for sustainable development and mitigating environmental risks. The underlying assumption is that more sustainable development solutions exist than today's scenario of water treatment and distribution in the basin. The integrated sustainable water management model provides security and coordination of risks of the stakeholders in the basin. The developed model combines the economic, physical, and social conditions of water use in the region.

1. Introduction

The growth of population, rapid industrialization, and extensive use of natural resources (e.g., water, soil, forests, gas, oil) increases the pressure on these resources; in particular, on their availability and the treatment of wastes. There are physical limits to continuing economic growth based on the resource use. Today many countries suffer from permanent water shortages or poor water quality. Growing volumes of municipal and industrial wastes have to be handled. Many industries, like electrochemistry, metal mining, food, and transportation of hazardous materials, use environmentally intensive processing technologies that result in large quantities of waste; contamination of soil, air, and water; destruction of landscape; negative effects on biodiversity and natural water cycles; and high energy consumption. Extraction and transportation of natural resources, including sand, gravel, clay, limestone, and natural stones, cause noise and water/air/soil pollution in addition to most of the problems encountered in the application of industrial processes.

In this paper we will concentrate on sustainable management of water resources in a region as a typical example of an application of the suggested methodology. However, it is important to note that the methodology can find wider applications to sustainable management and environmental risk analysis for other types of natural resources.

One particular large-scale environmental problem generated by nonchemical stressors is the erosion and transformation of land in the Dead Sea beach area caused by shrinking of this water reservoir, and resulting in significant losses of the basic natural functions of the land. The Dead Sea Basin plays a crucial economic, ecological and social role in regional development of industry, tourism, and agriculture in Israel and Jordan. This role is threatened by the shrinking of the Dead Sea. During the last 70 years the water level of the Dead Sea has fallen by about 25 m, about half of this in the last 20 years.

The Dead Sea is the terminal point of the Jordan River watershed. As such, it serves as an indicator for the health of the entire exploited ecosystem. Its rapid decline reflects the present unsustainable water management strategies of the riparian countries. The Dead Sea's contraction and associated soil erosion undermine both the agricultural potential of the basin and its attractiveness as a tourist destination, despite the enormous investment in hotel and resort infrastructure in Israel and in Jordan. The exploitation of water resources of the Jordan River and the Dead Sea by present generations should not be at the expense of its natural and cultural heritage.

Sustainable water resources management and related environmental security not only are crucial at a national level, but also constitute a core element for promoting peace and stability between nations in transboundary regions. This general concept was announced and developed at the World Summit on Sustainable Development (WSSD) held in Johannesburg, August–September 2002 [1]. The summit identified the mutually beneficial management of inter-nation natural resources and environmental security as a key factor for peaceful coexistence and cooperation between countries.

Management of transboundary environmental resources, especially water, involves different preventive and corrective actions and different political and economic instruments; it also calls for integrated plans for water conservation, sharing, and demand management. Environmental risk/cost analysis (ERCA) may be used as an advanced tool and a framework for designing integrated plans for reducing ecological and human risks and implementing cost-effective measures for sustainable use of water resources.

Ganoulis et al. [2] and Tal and Linkov [3] reviewed main environmental issues of integrated water management and presented a conceptual scheme describing how ERCA may be used as a general framework to reduce possible environmental conflicts between different stakeholders. Effective cooperation between different institutions, involvement of numerous national and local stakeholders, and ranking the stakeholders' conflicting interests constitute crucial components for implementing specific plans for integrated water resources management. Within any multiple-agent system, and particularly in large water ecosystems consisting of industrial, agricultural, municipal, recreational, and other participating agents, there are political, organizational, social, psychological, and other barriers between the participants [4]. They may be caused by conflicting values and demands which impede sustainable development and environmental integrity in a region. In recent years it has been well recognized in the management science literature that many of the successes of multicriteria mathematical programming (MCMP) can be explained by breaking down or smoothing out the barriers in large multi-agent systems. This principle underlies the MCMP methodology, and is developed in the present paper.

This paper analyzes options for sustainable development in the Dead Sea Basin. It presents a model based on the green supply chain paradigm and multi-portfolio selection of strategies for sustainable development and mitigating environmental risks. The underlying assumption is that there exist strategies for more sustainable development than today's scenario of water treatment and distribution in the basin. The integrated sustainable water management model provides security and coordination of risks for the stakeholders in the basin. The developed model combines the economic, physical, and social conditions of water use in the region.

2. Definitions and Notation

2.1. ENVIRONMENTAL ("GREEN") SUPPLY CHAIN

The decision-making model presented in this paper translates the problem of sustainable management of natural resources into a nonstandard multi-portfolio selection problem and then to a standard multicriteria optimization formulation, which, in turn, is reduced to a single-criteria mathematical programming problem. The model is an extension and a generalization of the optimization (mathematical programming) models developed in recent years by Fisher et al. [5], Isaac et al. [6], and Cai et al. [7]. The main difference between the present model and the three cited above is that this model considers all the main stakeholders and their ecological and economic interests, and explicitly introduces them into the model. This is achieved by introducing the concepts of an environmental supply chain and multi-portfolio selection. As will be shown below, the multi-portfolio modeling approach permits the introduction of several portfolios of strategies for the separate mitigation of risks caused by chemical, physical, biological and radioactive stressors, and their coordination in a uniform framework.

Bloemhof et al. [8], Carter and Narasimhan [9], and Levner and Proth [10] have independently suggested the concept of an *environmental supply chain*, known also in the management science literature as a *green*, or *closed-loop supply chain*. Recall that the conventional operations management concept of the *supply chain* (SC) refers to a global network of organizations and institutions that cooperate to improve the material and information flows between suppliers and customers at the lowest cost, the highest speed, and with the greatest benefits. The main components of the conventional supply chain are:

- Demand forecasting and planning
- Material requisition and inventory

- Manufacturing and packaging
- Distribution and transportation
- Customer service
- Waste treatment and reuse

Graphically, the supply chain is represented as a graph of a network in which each agent or stakeholder in the system is presented as a node, and each chain link represents an interconnection transforming raw materials into product and services. Thus, the supply chain can be viewed as a visual representation of the technological activities of the agents (stakeholders) of the system during its life cycle. In its simplest form, a supply chain is a linear ordering, or chain.

The definition of the *environmental supply chain* (ESC) integrates the above material, financial, and information flows with flows of natural resources (in particular, water), throughout the product life cycle, and introduces new decisions for suppliers and manufacturers in the SC necessary to decrease waste flows and environmental pollution, even beyond their sale and delivery interests [8–10].

Environmental protection issues are critical in the ESC, and are incorporated into supply chain management strategies; moreover, the environmental dimension should be viewed as an inseparable part of business performance at all stages of supply chain management. Levner and Proth [10] introduced and studied a special class of environmental supply chains called *aquatic logistics supply chains* (ALCSs). These SCs have two specific features:

1. Water flows of different quality constitute not only a final, main product, but also a raw material, a byproduct, a secondary product, and a waste product.

2. Along with costs and benefits, these SCs explicitly incorporate environmental risk as a main objective of design, planning, and management.

The ALSC is a management science paradigm of the concept of the hydrological water flow cycle in hydrological sciences [11] (see Figure 1).

The ALSC is a visual presentation of the technological activities of all the stakeholders of the ecosystem throughout the water life cycle. An example of a simple ALSC is shown in Figure 2.

2.2. DEFINITIONS OF ENVIRONMENTAL RISK

According to the U.S. Environmental Protection Agency (USEPA) [12], risk is the likelihood that a course of actions (or a lack thereof) will result in an undesired event. In this paper we use another, more natural and wider definition suggested by Levner and Proth [10]:

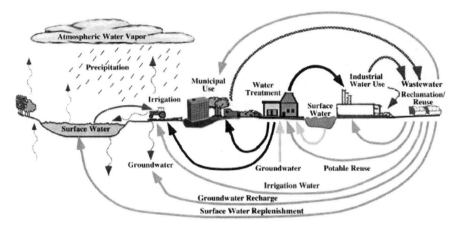

Figure 1. The Hydrological Water Flow Cycle in Hydrological Sciences [11].

Water_Preparation → Water_Transportation → Customers →Wastewater_Sources
(Water_Pollution_Actions) → Wastewater Treatment →Treated_Wastewater_Usage
_(Irrigation) → Wastewater_Disposal

Figure 2. A Schematic Aquatic Logistics Supply Chain [10]

Risk is a two–dimensional entity comprising likelihood **and** magnitude that a course of actions (or a lack thereof) will result in an undesired event. Consistent with this definition, this paper will characterize ecological risk by two parameters: the likelihood of damage and its magnitude (severity).

Two main types of risk assessment are widely used in the literature:

- Discipline-oriented (e.g., engineering, biological, medical, ecological)
- Integrated: integrating space, time, sources of risk, stressors, their pathways, results, and multiple endpoints

Many researchers who have studied integrated risk management problems have noticed that it is comparatively easy to describe and formulate constraints of the problem but it is difficult and troublesome to formulate and quantify the objective function of the risk management problem [5–7].

A plethora of different approaches to characterizing the environmental risk related to water management are known, but none of them are ideal and operational. Known approaches include:

- Daily probability of infection through ingestion of pathogens [12].
- Annual probability of infection through ingestion of pathogens [13].

- Costs of damage to aquifer, soil, and human health [14].
- Product of the probability and magnitude of damage [12] (the deficiency of the latter approach is discussed in Zaidi [15]).
- Two-dimensional risk matrix of probability and magnitude of damages (Figure 3 [10]). This risk matrix is capable of evaluating the effectiveness of risk mitigation measures. In this paper we will adhere to the latter characteristics.

3. A Risk-Based Framework for Integrated Water Resources Management (IWRM)

Effective management of water resources should be based on current best practices, which are grouped under the term Integrated Water Resources Management (IWRM). The term was first used in 1977 at the UN Conference in Mar del Plata and according to the Global Water Partnership (GWP)—an NGO based in Stockholm—IWRM is defined as [16]:

a process which promotes the coordinated development and management of water, land and related resources to maximise the resultant economic and social welfare in an equitable manner without compromising the sustainability of vital ecosystems.

According to IWRM, apart from the technical and economic criteria, at least two more additional general objectives should be considered: environmental security and social equity. In terms of an integrated approach, management issues should be considered at the basin scale.

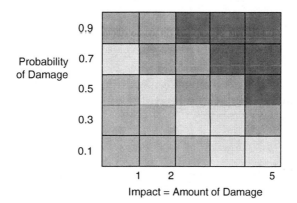

Figure 3. Two-Dimensional Characteristic of Risk. The Matrix Serves to Rank the Risks: the First (South-West) Tier Denotes Low Level, the Next Ones Acceptable, High, and Very High, Respectively.

Natural resources management focuses on five areas: scientific-technical, environmental, socioeconomic, legal, and institutional. The management of water quantity and quality is a complicated multidisciplinary scientific field, requiring good cooperation between various disciplines, such as:

- *Hydrogeology:* geophysical and geological prospecting, drilling techniques, mapping
- Hydrodynamics: quantitative aspects of flows, mathematical modeling, calibration, and prediction scenarios
- Systems analysis: optimization techniques, risk analysis, and multiobjective decision-making methods
- *Hydrochemistry:* chemical composition of the soil and water
- *Hydrobiology:* biological properties of groundwater systems

Modern tools for groundwater development extensively use new information technologies, database development, computer software, mathematical modeling, and remote sensing.

In what follows, the IWRM approach is formulated in terms of a risk-based multidisciplinary methodology called Risk-based Integrated Water Resources Management (RIWRM). Furthermore, Risk-based Multicriteria Multi-portfolio Decision Analysis (RMCDA) is presented as a tool for risk management and conflict resolution in internationally shared groundwater resources.

3.1. A MULTI-PORTFOLIO APPROACH

Generally speaking, risk is a complex function of the probability of failure and its consequences. In the technical literature the product of the probability and its consequences are often taken as the risk function. However, this approach is too rough for our aims, and different risk indices may be introduced for describing economic and social risks.

The RIWRM approach to water resources planning and operation aims to reduce not only technical and economic but also environmental and social risks in order to achieve sustainable development from the following different (and possibly contradictory) points of view:

1. Technical reliability
2. Economic effectiveness
3. Environmental safety
4. Social equity

Each of the above points of view and dimensions of interests can be represented by its own set of risk-mitigating policies, which we will call a *portfolio of strategies.*

In this context the RIWRM problem can be set up as a *multi-portfolio choice problem* which allows a scientifically motivated compromise between the individual utilities (interests) of all stakeholders in the green supply chain, where technological, economic, and social conditions are taken into account in form of constraints in problem formulation. In this approach we follow and extend Markowitz's portfolio choice model [17, 18]. The compromise between the stakeholders is achieved by using the multicriteria mathematical programming (in fact, the quadratic programming) approach. The multi-portfolio choice problem could be formulated as follows.

Given an *n*-dimensional vector budget (amount of money available to invest, along with other tools, such as human and information resources) and a list of management strategies $1,\dots,m$ requiring investment in the main links of the green supply chain, how can the vector budget be optimally divided among the various water resources management strategies? An important feature is that the expected "return" on investment; i.e., the resultant economic and social welfare benefits of environmental protection, is a composite return "paid out" over the life of the considered management strategy. Moreover, it is not necessarily a scalar defining economic welfare in monetary form, but rather a vector characterizing technical, economic, environmental, social, and other dimensions of the expected return from the integrated water resources management strategy in question.

Denote by x_{ij} the amount of the *j*th component of the *n*-dimensional vector budget allocated to management strategy i, for $i = 1,\dots,m, j = 1,\dots,n$. Then the $m \times n$ matrix x, that we call a *multi-portfolio*, is a multidimensional decision variable for the problem. A goal of the optimization process is to characterize and find the optimum portfolio of water resources management strategies.

Let the total return from portfolio x be the random variable matrix v(x), and $\mu(x) = (\mu_{ij})$ the expected value of return v(x) from portfolio x in a specified period. It is a measure of the long-term average return per period from the portfolio. Note that in the present general approach, the return $\mu(x)$ is a matrix whose components reflect separate economic, technical, environmental, and social returns (benefits, welfare) that are quantitatively estimated by using the utility functions for each stakeholder.

Another very important parameter for characterizing an optimum portfolio is the measure of risk associated with portfolio x. Today there is no universal method or formal approach to estimate and precisely measure environmental risk. The challenge for environmental management is to select a suitable unit of measurement for environmental risk. Following a financial risk management approach proposed by Markowitz in 1952 [17, 18], we may recall that the environmental risk of a portfolio can be quantitatively characterized by (is a single function of) the variance of returns from portfolio x. Moreover, there is a plethora of different definitions of environmental risk and environmental safety, which can have a variety of connotations. In this paper we follow two basic risk concepts, the first one being the Markowitzean measure of risk; namely, variance of returns. The second one is a two-dimensional array Rij = (Probability_of_Damage, Amount_of_Damage) developed by the authors [19] and briefly introduced in Section 2.2 (see Figure 3). The first concept basically defines the risk of ineffective (failed) investments in environmental protection projects, whereas the second type of environmental risks are defined as threats to human health, to the natural environment—air, water, and land—upon which life depends, and to health of flora and fauna. In the model presented, we take into account both risk types.

Table 1 below depicts a template showing the relations between strategies and budget components. The cell at the intersection of each row (strategy) and column (budget component) contains three entries: decision variable x_{ij} = *portfolio component*, the expected value of return $\alpha_{ij} = \alpha(x_{ij})$ = *returns*, and the environmental risk R_{ij}, which, in turn, either the variance of returns or the two-dimensional array defined just above.

As mentioned above, in the RIWRM approach four basic risk dimensions—technical, economic, environmental, and social—are considered. Environmental risk is closely related to all other risks, and may be expressed in economic, social, and technical terms. By differentiating environmental risk, we wish to emphasize the environmental aspects; i.e., those aspects related to environmental security and protection from (human-induced) water pollution and disasters.

In this perspective, an optimum portfolio of water resources management strategies should maximize the expected return and minimize the environmental risk; these objectives should be achieved simultaneously. Finding an optimum portfolio of IWRM strategies is therefore a multicriteria

TABLE 1. The Multi-Portfolio Matrix.

Budget components	1	2	...	j	...	n
Strategies						
1						
2						
...						
i				$(x_{ij}, \alpha_{ij}, R_{ij})$		
...						
m						

optimization problem. Our Markowitzean approach is applicable to water resources management and extends the basic Markowitz model [17] in that (1) the variable portfolio x is the $m \times n$ matrix rather than an n-dimensional vector of variable assets, and (2) each objective function (i.e., the return and risk) is in fact a vector of several functions.

3.2. RISK-MITIGATING PROCEDURE

We suggest the following iterative procedure for finding the best compromise solution for minimizing environmental risks in the multi-agent green supply chain described above.

Step 1. *Forming the green supply chain and defining input parameters.* The structure of the supply chain is defined. Expected values of returns as well as estimated individual risk values are defined for all participants (stakeholders) in the supply chain. All problem constraints (hydrological, economic, and technological data) and different objectives of the participants are also obtained in this step.

Step 2. *Finding weights for all objective functions.* Using The fuzzy Borda ranking method [23], the compromised integrated weights w_{ij} for all objectives of the stakeholders are found.

Step 3. *Finding a compromise solution minimizing the total risk for all stakeholders.*

The basic Markowitz portfolio selection model (for the case of a single portfolio) in the vector form is the following quadratic programming problem:

Let $\propto = (\propto_i)^T$ be a vector of expected values of returns (yields), where $\propto_i$ is the expected return for the environmental protection strategy i, and let $\Sigma = (\sigma_{ij})$ be the variance-covariance $m \times m$ matrix. Then the expected return from portfolio x in a period is $(\propto_i)^T x$, and the variance of this return is $x^T \Sigma x$.

Minimize the variance of returns

$$R = x^T \Sigma x \tag{1}$$

Maximize the expected return

$$D = (\mu_i)^T x \qquad [\text{or } (\mu_i)^T x \geq \delta B] \tag{2}$$

(where B is given), subject to the feasibility and resource conditions

$$x \in S \tag{3}$$

where S is the set of feasible solutions.

Notice that even in the case of a single portfolio, our suggested model differs from the above Markowitz model in that we use a different definition of risk which includes one condition corresponding to the impact and another to risk probability:

Minimize the impact (total damage)

$$r = \Sigma_{k=1,...,N}\Sigma_{i=1,...,m}\, r_{ki} \qquad (4)$$

where there are N risk groups, for different risk classes and different stakeholders; r_{ki} is an impact for individual stakeholder k with respect to strategy i,

$$p = \Sigma_{k=1,...,N}\Sigma_{i=1,...,m}p_{ki} \leq p_0, \qquad (5)$$

(this constraint requires that the total probability of damage does not exceed the acceptable risk level p_0. It is assumed here that elementary probabilities p_{ki} (of damage to stakeholder k under strategy i) are independent and sufficiently small.

The suggested multi-portfolio model is an extension of the previous single-portfolio model in the following directions: (1) the variable portfolio x is the $m \times n$ matrix rather than an m-dimensional vector of variable assets, and (2) each objective function (i.e., the return and risk) is in fact a vector of several functions for different strategy portfolios and different stakeholders. The extension for the case of n columns (that is, n portfolios) follows:

For each portfolio j, $j = 1,2,...,n$, minimize the variance of returns

$$R^j = (x^j)^T \Sigma^j x^j \qquad (6)$$

where x^j is the jth column of $m \times n$ matrix x, Σ^j is the variance-covariance $m \times m$ matrix corresponding to portfolio j.

Maximize the expected return

$$D^j = (\mu_{ij})^T x^j \quad [\text{or}\,(\mu_{ij})^T x^j \geq \delta B^j] \qquad (7)$$

(where B^j is given), subject to the feasibility and resource conditions

$$x \in S \qquad (8)$$

where S is the set of feasible solutions.

Similarly to the single-portfolio case, we use a different definition of risk which includes one condition corresponding to the impact and another corresponding to risk probability:

Minimize the impact (the total amount of damage) related to portfolio j

$$r^j = \Sigma_{k=1,...,N}\Sigma_{i=1,...,m}\, r^j_{\,ki} \qquad (9)$$

where there are N risk groups for different risk classes and different stake-holders; $r_{ki}{}^j$ is an impact for portfolio j for individual stakeholder k with response to strategy i.

$$p^j = \Sigma_{k=1,\ldots,N}\Sigma_{i=1,\ldots,m}\, p^j{}_{ki} \leq p^j{}_0 \qquad (10)$$

The feasible solution, which satisfies (6)–(10), is found using one of the standard methods of multicriteria mathematical programming. Then go to Step 2 and, if necessary, change the weights provided by the stakeholders. Iteratively repeat Steps 2 and 3 until a compromise portfolio of environment protection strategies satisfying all stakeholders is found.

Many standard methods are known for solving the obtained multicriteria programming problem: surrogate relaxation (integration of two constraints into one), continuous relaxation, Lagrange relaxation, reduction of variables, approximation schemes, and various heuristics (see e.g., [20–22, 24]).

Notice that our formal scheme of risk computation is much more general than the product of threat probability and damage costs. Despite the fact that the latter definition is computationally simple and widely accepted, especially in the technical literature, it does not offer the same opportunities as the multidimensional approach to risk evaluation. The present multi-portfolio methodology is more complicated and computationally less tractable than the classical Markowitz model. However, it allows powerful mathematical methods of financial risk analysis and multicriteria mathematical programming to be exploited for measuring and minimizing the environmental risks (see e.g., [20–22]).

4. Application: Sustainable Water Management in the Dead Sea Basin

4.1. A BRIEF GEOGRAPHICAL NOTE

The Jordan River (in Hebrew: ‏זדריה רהנ‎ nehar hayarden, in Arabic: ‏ندرأل رهن‎ nahr al-urdun) is a river in Southwest Asia flowing through the Great Rift Valley into the Dead Sea. Historically and religiously it is one of the world's most important rivers, where Christians believe Jesus was baptized. The waters of the Jordan are an extremely important resource to the dry lands of the area belonging to Lebanon, Syria, Jordan, Israel and the Palestinians. It is 251 km (156 miles) long. Its tributaries are the Hasbani, which flows from Lebanon, the Banias, arising from a spring at the foot of Mount Hermon, the Dan, whose source is also at the base of Mount Hermon, and the Iyon, which flows from Lebanon. Two major tributaries enter from the east during the river's last stage before it enters the Dead Sea: the Yarmouk River and Jabbok River.

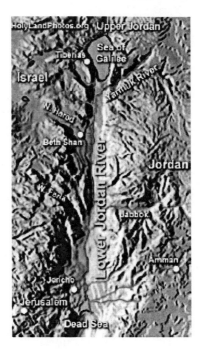

Figure 4. The Jordan River and the Dead Sea Basin.

A map of the region and pictures of the river are presented in Figures 4 through 6.

In modern times the waters are 70–90% used for human purposes and the flow is much reduced. Moreover, the river is heavily polluted; in its lower part, just raw sewage and runoff water from agriculture are flowing into the river. Most polluted is the 60-mile downstream stretch: a meandering stream from the Sea of Galilee to the Dead Sea.

In the early 1960s, the Jordan River moved 1.3 billion cubic meters (46 billion cubic feet) of water every year from the Sea of Galilee to the Dead Sea. Dams, canals and pumping stations built by Israel, Jordan and Syria to divert water for crops and drinking have reduced the flow by more than 90% to about 0.10 billion cubic meters (3.5 billion cubic feet). The practice has almost destroyed the river's ecosystem.

The Dead Sea (in Hebrew: יָם הַמֶּלַח, Yām Ha-Melaẖ, "Sea of Salt"; in Arabic: البَحْرِالمَيّت, al-Barẖl-Mayit, "Dead Sea") is a salt lake between the West Bank and Israel to the west, and Jordan to the east. The Dead Sea is 67 km (42 miles) long and 18 km (11 miles) wide at its widest point. Its main tributary is the Jordan River. At 420 m (1,378 feet) below sea level, its shores are the lowest point on Earth that on dry land. The Dead Sea is the deepest hypersaline lake in the world. At 30% salinity, it is 8.6 times saltier than the

Figure 5. The Upper Jordan River in Spring. [25]

ocean. Today, virtually every major spring and tributary that once flowed into the Jordan River (and then to the Dead Sea) has been dammed or diverted for drinking water and crop irrigation by Israel, Jordan, Lebanon and Syria. The Jordan River now delivers less than 100 million cubic meters of water a year to the Dead Sea, and as much as half of that is raw sewage. Its current surface area is 635 km², 1/3 smaller than before human intervention.

4.2. MAIN THREATS TO SUSTAINABLE MANAGEMENT OF THE DEAD SEA

The main threats to the Dead Sea are:

- Water pumping from Lake Kinneret and the Yarmouk River for water supply has created a water deficit about 800 million cubic meters/year.
- Industrial solar evaporation ponds at chemical works are responsible for about 20% of the total evaporation of Dead Sea waters.
- Additional threats come from the uncoordinated tourism industry, including hotels, transport, and road building.

Impacts of human intervention in the Dead Sea Basin include:

Figure 6. The Lower Jordan River entering the Dead Sea. [25]

- Hundreds of sinkholes caused by decreasing sea water levels; some sinkholes appear suddenly under structures and roads, creating safety risks and economic problems.

- Increased water levels in evaporation ponds due to the accumulation of 20 cm/year of salt on the pond bottoms. The rising water level threatens the foundations of hotels, roads and drainage systems along the shores of the ponds.

- Declining water levels prevent the development of shoreline recreation sites and hotels, and are harmful to seashore agriculture.

The risk-based water resources management problem for the Dead Sea as a multi-portfolio choice problem: Given an m-dimensional vector budget (amount of money available to invest, along with other tools, such as human and information resources) and a list of management strategies 1,..., m requiring investment, how can the vector budget be optimally divided among the various water resources management strategies for the saving of the Dead Sea? Finding an optimum portfolio of water resources management strategies should maximize the expected return and minimize the environmental risk; in other words, these objectives should be achieved simultaneously. The overall goal is to develop a multicriteria optimization model for integrated management of water resources for the Lower Jordan Valley and the Dead Sea Basin.

In accordance with the RIWRM methodology described in Section 3.1, we consider four main portfolios of strategies for the sustainable development of the Dead Sea area:

1. Technological portfolio

2. Economic portfolio

3. Ecological portfolio

4. Social portfolio

Each portfolio represents specific point of view of involved stakeholders. and can be split into smaller sub-portfolios. For instance, the technological portfolio can be hierarchically composed of industrial, agricultural, tourist, transport and other sub-portfolios. The portfolios are not obliged to be disjoint sets as some strategies are naturally related to several different portfolios.

A variety of strategies that are to be integrated for sustainable development of the basin are outlined in Table 2, where the corresponding portfolios are indicated as well.

Expected outputs and results of the suggested model are:

- To increase the understanding of available sources of water savings in the Lower Jordan River Basin.

- To identify the benefits and applicability of different portfolios of ecology-safe strategies, including novel alternative technologies and water pricing policies for sustainable water usage (in particular, the methodology allows comparison of ecology-safe strategies with ecologically risky meta-projects like the Med-Dead or Red-Dead channels).

- Using the results of computer simulation, to estimate the environmental risks at present and in the nearest future.

- Alternative solutions should be reviewed, including the possibility of increasing the flow of freshwater sources to the Dead Sea by limiting diversion from the north and promoting public and private water conservation.

- Finally, to develop recommendations for political decision makers addressing river water challenges and optimizing river waters' contribution to sustainable development.

TABLE 2. Sustainable Management Strategies Examined for the Case Study.

Strategies	Technological portfolio	Economic portfolio	Ecological portfolio	Social portfolio
Construction of desalination stations	X	X	X	
Construction of surface water storage facilities and rainwater harvesting	X	X	X	
Wastewater treatment and reuse	X	X	X	X
Water saving in household use	X	X	X	X
Reduction of water leakage and water losses in pipes and networks	X	X		
New water-saving technologies in industries and process changes	X	X		X
New water-saving technologies in agriculture and new crop types	X	X		X
Improvement of irrigation methods	X	X		X
Importing of water from abroad	X		X	X
Improvement of groundwater exploitation	X			X
Changing of water quotas		X		X
Regional water policy decisions to shift to less-water-intensive sectors		X		X
Changes in water pricing		X		X
Water-related laws and standards		X	X	X
Legal tools: penalties and fines		X	X	X

5. Concluding Remarks

This paper presents a computer-based model and applies it to the sustainable development of the Dead Sea Basin. It presents a tool based on the green supply chain paradigm and the multi-portfolio selection model. The underlying assumption is that solutions for more sustainable development exist than today's unsustainable scenario of water treatment and distribution. The present multi-portfolio methodology is more complicated and computationally less tractable than the classical single-portfolio Markowitz model.

However, it allows powerful mathematical tools developed for multicriteria mathematical programming to be exploited for analyzing, measuring, and minimizing environmental risks.

References

1. *The World Summit on Sustainable Development*, 2002. Johannesburg, South Africa, August–September 2002. Available at: http://www.worldsummit2002.org/.
2. Ganoulis, J., Duckstein, L., Literathy, P., and Bogardi, I. (Eds.), 1996. *Transboundary Water Resources Management: Institutional and Engineering Approaches*, NATO ASI Series, Sub-series 2: Environment, v.7, Springer, Heidelberg.
3. Tal, A., and Linkov, I., 2004. The role of comparative risk assessment in addressing environmental security in the Middle East. *Risk Analysis* 24(5):1243–1248.
4. Levner, E., Zuckerman, D., and Meirovich, G., 1998. Total quality management of a production-maintenance system: a network approach. *International Journal of Production Economics* 56–57(1), 407–421.
5. Fisher, F. M., Arlosoroff, S., Eckstein, Z., Haddadin, M., Hamati, S. G., Huber-Lee, A., Jarrar, A., Jayvousi, A., Shamir, U., and Wesseling, H., 2002. Optimal water management and conflict resolution: the Middle East water project. *Water Resources Research* 38(11), 25.1–25.17.
6. Isaac, J., Kupfersberger, H., Orthoffer, R., and Shuval, H., 2000. *Developing Sustainable Water Management in the Jordan Valley*, 2000. INCO-DC Project JOWA, Applied Research Institute, Bethlehem-Jerusalem, Dec. 2000. Available at: http://systemforschung.arcs.ac.at/jowapubl
7. Ximing, C., McKinney, D. C., and Lasdon, L. S., 2003. Integrated hydrologic-agronomic-economic model for river basin management. *Journal of Water Resources Planning and Management* January/February 2003, 4–17.
8. Bloemhof, J., Van Beek, P., Hordijk, L., and Van Wassenhove, L. N., 1995. Interactions between Operational Research and Environmental Management. *European Journal of Operational Research* 85:229–243.
9. Carter, J. R., and Narasimhan, R., 1998. *Environmental Supply Chain Management*. Center for Advanced Purchasing Studies, Focus study, 1–5.
10. Levner, E., and Proth, J.-M., 2003. Strategic management of ecological systems: a supply chain perspective. Key lecture at the NATO Advanced Study Institute on Strategic Management of Marine Ecosystems, Nice, France, October 2003. In: Levner, E., Linkov, I., and Proth, J.-M. (Eds.), 2005. *Strategic Management of Marine Ecosystems*, Springer, Berlin.
11. Asano, T., and Mills, R. A., 1990. Planning and analysis for water reuse projects. *Journal of the American Water Works Association* 82(1):38–47.
12. United States Environmental Protection Agency,1998. *Guidelines for Ecological Risk Assessment*. Washington, DC, EPA/630/R-96/OO2F.
13. Oron, G., Campos, C., Gillerman, L., and Salgot, M., 1999. Wastewater treatment, renovation and reuse for agricultural irrigation in small communities. *Agricultural Water Management* 38:223–224.
14. Haruvy, N., 2004. Irrigation with treated wastewater in Israel – assessment of environmental aspects. In Linkov, I., and Ramadan, A. (Eds.), *Comparative Risk Assessment and Environmental Decision Making*, Kluwer, Dordrecht.
15. Zaidi, M. (Ed.), 2007. *Wastewater Reuse – Risk Assessment, Decision Making and Environmental Security*. Proceedings of the NATO ARW, Istanbul, 12–15 October 2006, Springer.

16. GWP, 2000. *Integrated Water Resources Management*. Technical Committee Background Paper No. 4. GWP, Stockholm. Available at: http://www.gwpforum.org/gwp/library/Tacno4.pdf

17. Markowitz, H. M., 1952. Portfolio selection. *Journal of Finance* 7(1):77–91.

18. Markowitz, H.M., 1999. The early history of portfolio theory: 1600–1960. *Financial Analysts Journal* 55(4):5–16.

19. Levner, E., Ganoulis, J., Linkov, I., and Benayahu, Y., 2007. Multiobjective risk/cost analysis of artificial marine systems using decision trees and fuzzy expert estimations. In Linkov, I., Kiker, G. A., and Wenning. R. J. (Eds.), *Environmental Security in Harbors and Coastal Areas*, Springer, Berlin, pp. 161–174.

20. Tapiero, C., 2004. *Risk and Financial Management: Mathematical and Computational Methods,* Wiley, London.

21. Cornuejols, G., and Tutuncu, R., 2006. *Optimization Methods in Finance,* Cambridge University Press, Cambridge, UK.

22. Rockafellar, R., and Uryasev, S., 2000. Optimization of conditional value-at-risk. *Journal of Risk* 2:21–42.

23. Levner, E., and Alcaide, D., 2006. Environmental risk ranking: theory and applications for emergency planning. *Scientific Israel – Technological Advantages* 8(1–2):11–21.

24. Elster, K.-H., Gol'shtein, E. G., Levner, E., et al., 1993. *Modern Mathematical Methods of Optimization*, Akademie Verlag, Berlin.

25. Wikipedia. Jordan River. Available at: http://en.wikipedia.org/wiki/Jordan_river

AVOIDING MCDA EVALUATION PITFALLS

J. BARZILAI

Dalhousie University and Scientific Metrics
Halifax, Nova Scotia, Canada
barzilai@scientificmetrics.com

Abstract: Classical decision and measurement theories are founded on errors that have been propagated throughout the literature, leading to a proliferation of tools and methodologies based on flawed mathematical foundations. In this article, incorrect assumptions are addressed and appropriate principles and methodologies are identified.

1. Introduction: the Issues

One is not required to be a mechanical engineer to drive a car and, considering the advanced state of mechanical engineering, most people limit their interest in what lies "under the hood" to finding a competent mechanic. Users of evaluation, risk, and decision analysis tools that are based on classical decision theory should be aware that, as is demonstrated below, classical multicriteria decision analysis (MCDA) has not reached the advanced state of mechanical engineering. Since evaluation and decision tools that are based on flawed mathematical foundations produce meaningless numbers, the purpose of this paper is to give a sample of typical errors and direct the reader to (i) practical tools that are based on sound mathematical foundations and (ii) to these mathematical foundations.

Typically, even the simplest multicriteria evaluation techniques involve numbers and operations such as addition and multiplication. Also typically, it is not recognized that the numbers on which the operations of addition and multiplication are performed represent preference scales and that these are mathematical operations albeit elementary ones. Although the construction of preference scales and the applicability of mathematical operations to preference scale values are of great theoretical and practical importance, the problem of applicability of these operations has been ignored in the literature following the publication of von Neumann and Morgenstern's *Theory of Games and Economic Behavior* [19] and the conditions under which these operations are applicable have not been identified until recently [5–7].

2. A Sample of Typical Problems

2.1. THE APPLICABILITY OF ADDITION AND MULTIPLICATION

Consider the applicability of the operations of addition and multiplication on scale values for a fixed scale; that is, operations that express facts such as "the weight of a given object equals the sum of the weights of two other objects" ($m(a) = m(b) + m(c)$) and "the weight of a given object is two and a half times the weight of another one" ($m(a) = 2.5m(b)$). It may be surprising to learn that these operations are not applicable to any scales in the classical literature, but the correct model for preference scales, which are the scales of interest in evaluation, risk, and decision analysis, is that of a straight line and none of the scales in the classical literature is constructed in accordance with the algebraic and geometric structure of the straight line. In fact, the conditions for applicability of addition and multiplication have not been identified in the classical literature and the issue of applicability of mathematical operations cannot be found in the literature following the publication of von Neumann and Morgenstern's book [19].

A technical note: It is important to emphasize the distinction between the application of the operations of addition and multiplication on scale values for a fixed scale, for example $m(a) = m(b) + m(c)$, as opposed to what *appear to be the same* operations when they are applied to an entire scale whereby an equivalent scale is produced by what amounts to a change of the zero point or unit, for example $t = p + qs$ where s and t are two scales and p, q are numbers [5].

2.2. ON A SCALE OF 1–10, HOW FAR IS LISBON FROM AMSTERDAM?

This is a meaningless question and numbers that are given in answer to similar questions are meaningless as well. Although no mathematical operations are applicable to such numbers, there seems to be nothing in the classical measurement, decision, or evaluation literature to tell marketing experts that there is no basis for the statistical operations that they routinely carry out on numbers received in response to questionnaires that contain such questions [4].

2.3. MEASUREMENT WITHOUT UNITS

Measurement without units produces scales to which addition and multiplication are not applicable [7]. Yet the term *unit* does not appear in Roberts's *Measurement Theory* [21] and there is no formal definition of the term in the literature. Similarly, there is no formal definition of the term *scale* in *Foundations of Measurement* (Krantz et al.) [15].

2.4. GROUP DECISION MAKING

The common view in the classical literature that group decision making cannot be modeled mathematically is an error that is based on a misinterpretation of the implications of Arrow's Impossibility Theorem [1] (cf. [6]). Another approach to group decision making, game theory, cannot serve as a foundation for group decision making either [3, 6].

2.5. UTILITY THEORY

Although utility theory has been the subject of much controversy since its early days, the main flaws in the foundations of this theory have been brought to light only recently. Among other things, the construction phase of utility theory contains a self-contradiction [6].

2.6. THE ANALYTIC HIERARCHY PROCESS

More than 30 years after the publication of Miller's work [16–18], there is still no acknowledgement in the Analytic Hierarchy Process (AHP) literature (or elsewhere) of his contribution to decision theory in general and the AHP in particular. Some of Miller's ideas are valuable while others are mathematically incorrect but almost all of the additions to his original methodology are in error. Many AHP errors are reviewed in Barzilai [8–11] (see also the references there). Not surprisingly, these errors have been misidentified in the literature and some of these errors appear in decision theory. For example, Kirkwood [14, p. 53] relies on Dyer and Sarin [12], which repeats the common error that the coefficients of a linear value function correspond to relative importance [12, p. 820]. Furthermore, "difference measurement," which is the topic of Dyer and Sarin, is not the correct model of preference measurement. As is the case for other preference scales, there is no foundation for the use of the operations of addition and multiplication in the construction of AHP's preference scales (in this case these operations are used to compute the AHP's eigenvector "priorities").

2.7. PAIRWISE COMPARISONS AND PREFERENCE RATIOS

Pairwise comparisons (i.e., comparing two alternatives at a time) and ratios of alternatives cannot be used in the construction of preference scales to which the operations of addition and multiplication are applicable [4]. (Until recently, the use of pairwise comparisons and preference ratios was not related in the literature to the applicability of addition and multiplication.)

3. Preference Function Modeling

Classical evaluation theories, including utility theory, cannot serve as the mathematical foundation of decision theory, game theory, economics, or other scientific disciplines since they do not enable the operations of algebra and calculus, which are needed and widely used in the physical and social sciences and in statistics. A new theory for preference measurement that enables these operations has been developed [6, 7]. Based on this theory, a practical methodology for constructing proper preference scales, Preference Function Modeling (PFM), and a software tool that implements it, Tetra [22], have been developed. Tetra requires only simple and intuitive operations and is a powerful tool for group evaluation and decision making. For future developments of the theory, methodology, and software tools, consult Scientific Metrics [22].

A technical note: In geometrical terms, proper preference scales reflect the objects under measurement to points on a straight line [2, 6, 20].

4. Summary

Classical decision theory (e.g., Keeney and Raiffa [13]) and measurement theory (e.g., Krantz et al. [15]) are founded on errors that go back to early utility theory and which have been propagated throughout the literature and have led to a proliferation of methodologies and software tools that are based on flawed mathematical foundations and produce meaningless numbers. In addition, the common notion in classical decision theory that group decision making cannot be modeled mathematically is incorrect and is based on results that apply to ordinal systems only.

References

1. Arrow, K. J., 1951. *Social Choice and Individual Values*. Wiley, New York.
2. Artzy, R., 1965. *Linear Geometry*. Addison-Wesley, Reading, MA.
3. Barzilai, J., 2007. Game Theory Foundational Errors – Part I, Technical Report, Department of Industrial Engineering, Dalhousie University, pp. 1–2. Available at: http://http://www.ScientificMetrics.com
4. Barzilai, J., 2007. Pairwise Comparisons, in Salkind, N. J. (Ed.), *Encyclopedia of Measurement and Statistics*. Sage Publications, Thousand Oaks, CA, vol. II, pp. 726–727.

5. Barzilai, J., 2006. On the Mathematical Modeling of Measurement, pp. 1–4. Available at: http://www.Scientificmetrics.com/publications.html
6. Barzilai, J., 2006. Preference Modeling in Engineering Design, in Lewis, K. E., Chen, W., and Schmidt, L. C. (Eds.), *Decision Making in Engineering Design*. ASME Press, pp. 43–47.
7. Barzilai, J., 2005. Measurement and Preference Function Modelling. *International Transactions in Operational Research* 12:173–183.
8. Barzilai, J., 2001. Notes on the Analytic Hierarchy Process. Proceedings of the NSF Design and Manufacturing Research Conference, Tampa, FL, pp. 1–6. Available at http://www.ScientificMetrics.com
9. Barzilai, J., 1998. On the Decomposition of Value Functions. *Operations Research Letters* 22(4–5):159–170.
10. Barzilai, J., 1998. Understanding Hierarchical Processes. Proceedings of the 19th Annual Meeting of the American Society for Engineering Management, pp. 1–6.
11. Barzilai, J., 1997. Deriving Weights from Pairwise Comparison Matrices. *Journal of the Operational Research Society* 48(12):1226–1232.
12. Dyer, J. S., and Sarin, R. K., 1979. Measurable Multiattribute Value Functions. *Operations Research* 27(4):810–822.
13. Keeney, R. L., and Raiffa, H., 1976. *Decisions with Multiple Objectives*. Wiley, New York.
14. Kirkwood, C. W., 1996. *Strategic Decision Making*. Duxbury, Belmont, CA.
15. Krantz, D. H., Luce, R. D., Suppes, P., and Tversky, A., 1971. *Foundations of Measurement*, vol. 1. Academic, New York.
16. Miller, J. R., 1966. The Assessment of Worth: A Systematic Procedure and Its Experimental Validation, doctoral dissertation, Massachusetts Institute of Technology.
17. Miller, J. R., 1969. Assessing Alternate Transportation Systems, Memorandum RM-5865-DOT. The RAND Corporation.
18. Miller, J. R., 1970. *Professional Decision-Making*. Praeger, New York.
19. von Neumann, J., and Morgenstern, O., 1944. *Theory of Games and Economic Behavior*. Princeton University Press, Princeton, NJ.
20. Postnikov, M., 1982. *Analytic Geometry* (Lectures in Geometry, Semester I). Mir Publishers, Moscow.
21. Roberts, F. S., 1979. *Measurement Theory*. Addison-Wesley, Reading, MA.
22. Tetra software and PFM publications. Available at: http://www.ScientificMetrics.com

PART 4

**APPLICATIONS OF MULTICRITERIA DECISION ANALYSIS
FOR ENVIRONMENTAL STRESSORS**

APPROACHES USED FOR REMEDY SELECTION AT CONTAMINATED SEDIMENT SITES

Analysis of Two Case Studies

V.S. MAGAR

ENVIRON International Corporation
Chicago, Illinois, USA
vmagar@environcorp.com

K. MERRITT, M. HENNING

ENVIRON International Corporation
Portland, Maine, USA

M. SORENSEN

ENVIRON International Corporation
Atlanta, Georgia, USA

R.J. WENNING

ENVIRON International Corporation
San Francisco, California, USA

Abstract: This paper briefly reviews the complex issues associated with remedy identification, screening, and selection at contaminated sediment sites in North America. We present two case studies illustrating approaches used by stakeholders to arrive at remedy decisions. These approaches include watershed-level thinking and net environmental benefit analysis (NEBA), both of which recognize the influences of chemical and non-chemical, natural and anthropogenic stressors, and their respective influences on the integrity of the aquatic ecosystem.

In the absence of a sitewide human health or ecological risk assessment, and in the absence of a watershed-level approach that balances potential risks and benefits against implementation risks to human health and the environment, site managers typically are ill-equipped to effectively select environmentally appropriate and protective remedies for contaminated sediment sites. The U.S. Environmental Protection Agency (USEPA)

I. Linkov et al. (eds.), Real-Time and Deliberative Decision Making.
© Springer Science + Business Media B.V. 2008

Superfund feasibility study process, and a variety of innovative multi-criteria decision frameworks provide sound frameworks for remedy assessment and selection. Omitting these approaches can result in the selection of a remedy in which the environmental harm caused by the remedy can outweigh its perceived benefits.

Two case studies are presented that involve distinct, but mutually supporting, approaches to remedy decision making that reflect unique outcomes in terms of the goals of ecological and human health risk reduction, environmental protection, and watershed improvement. Combined, these case studies evaluate a range of stressors and compare net environmental benefits of each remedy alternative. Outcomes include meaningful risk reductions and minimal adverse impacts to the environment, workers and the local community residents.

1. Introduction

Sediment contamination affects inland and coastal water resources, and may pose risks to human and ecological health. In aquatic environments affected by chemicals in sediment, risk management strategies focus on either removing affected sediments or interrupting exposure pathways to reduce or eliminate risks over time [1]. Major approaches to sediment remediation include:

- Dredging (removal)
- Capping (isolation)
- Monitored natural recovery (MNR) (natural transformation and isolation)

In the past, conservative overestimates of contaminant risks and underestimates of the effectiveness of isolation and natural recovery processes have led to an assumption that removal of sediment by dredging was the most effective risk reduction approach. While public confidence in dredging remedies is typically high, there is growing evidence that this approach may not achieve desired environmental improvement and risk reduction goals [2].

Effective environmental decision making is a multifaceted activity, involving diverse stakeholders with different priorities and objectives. Advanced assessment tools, such as comparative risk analysis (CRA) and multicriteria decision analysis (MCDA), are emerging as integral components of risk evaluation and decision making [3, 4]. Briefly, these tools combine the probabilities associated with different threats (manmade and natural), the various consequences of the threats, and the risks (mission, asset, human health, and ecological) posed by the threats. Both CRA and MCDA facilitate the assignment of values and probabilities in a consistent manner to diverse decisions and potential outcomes, including manmade events (e.g., terrorism) and natural events (e.g., extreme weather conditions).

CRA and MCDA integrate quantitative and qualitative information from a variety of sources, including environmental modeling and risk assessment, cost-benefit analyses, opinion polls, and ranking activities. These tools offer frameworks for quantifying uncertainties and visualizing tradeoffs, and aid in the comparison of multiple response actions, decision criteria, and preventive measures.

This paper addresses the complex issues associated with sediment remedy identification, screening, and selection. A rationale for a watershed-level decision-making framework is discussed, and two case studies demonstrate the benefits of using this framework in remedy analysis and selection.

2. Sediment Remedy Considerations

The complexity of sediment remedy identification, screening, and selection is increasingly recognized. Relying on a presumptive remedy without careful, unbiased, and comprehensive evaluation of alternatives tends to minimize this complexity by exaggerating the risk of leaving contaminants in place, overestimating the benefits of the presumptive remedy, and understating the potential for risk reduction of alternate remedies. As a result, selected remedies may cause environmental harm that outweighs their perceived benefits.

Chemicals that receive particular attention typically include bioaccumulative compounds such as polychlorinated biphenyls (PCBs), chlorinated pesticides, dioxins and furans, and methyl mercury; petroleum and coal-derived hydrocarbons such as polycyclic aromatic hydrocarbons (PAHs); and metals such as chromium, mercury, lead, copper, cadmium, and zinc. Parsing chemical stressors and their sources from one another and from other types of stressors, such as siltation, overfishing and habitat degradation, remains a significant challenge.

One of the advantages commonly attributed to dredging is greater confidence in the long-term effectiveness of cleanup, assuming risk-based action levels can be attained [5]. In fact, dredging implementation is usually more complex, costly, and energy-intensive than other sediment management approaches, and uncertainties and negative impacts are typically underestimated. For example, the negative impacts of habitat destruction and significant safety concerns associated with heavy construction pose significant short-term risks. Further, post-dredging residual contamination—combined with sediment resuspension and release—presents persistent short- and long-term site risks [5].

Among the goals of the remedy selection process (the feasibility study or remedial alternatives analysis (RAA) in the U.S.) is to achieve cost-effective risk reduction. Because sediment is integral to larger systems that include water bodies, terrestrial features, and diverse processes of natural and human origin, the scope of the remedy selection analysis must expand beyond the immediate area of contamination to include the whole watershed. A watershed-level approach entails development of a conceptual site model (CSM) to support the evaluation of diverse risk sources, including sediment contaminants, other environmental stressors, and remedy implementation. Risks also must be measured in the context of stakeholder interests and diverse values, which may be encapsulated as net environmental benefits.

Watershed-level thinking recognizes the influences of chemical and non-chemical, natural and anthropogenic stressors, and their respective influences on the integrity of a lake, river, coastal, or estuary ecosystem. Intrinsically, a watershed-level approach relies on multiple criteria to evaluate remedy effectiveness and performance. Without a watershed-level perspective, environmental managers and regulators risk misrepresenting known stressors because they lack an understanding of the broad range of potential ecological stressors and the natural and anthropogenic sources that adversely affect the ecosystem [4].

Moreover, absent a watershed-level approach, risk analyses conducted for aquatic environments will continue to overemphasize the risks associated with historical chemical releases, uncertainties associated with in situ remedies, and the benefits of contaminated sediment removal. A narrow focus on the contaminated area tends to undervalue resource conservation and protection, which can lead to the implementation of an intrusive remedy that destroys natural resources (e.g., wetlands, forests, shorelines) under the banner of ecological risk reduction and protection.

In a time of increasing energy conservation and concern for atmospheric release of carbon dioxide (CO_2), remedy approaches that are less energy-intensive and minimize negative impacts to the natural environment, while still achieving risk reduction goals, should be looked upon more favorably. As the regulatory community and the environmental industry adopt a more holistic approach to environmental ecology and protection, the consideration of the full range of stressors is increasingly integral to the remedy selection process. This broadening of the selection process is critical for identifying and evaluating mitigation options to meaningfully reduce or eliminate long-term watershed impacts.

3. Net Environmental Benefits and Risk of Remedy

Net environmental benefits are the gains in environmental services or other ecological properties attained by actions, minus the environmental injuries caused by those actions [7]. A net environmental benefit analysis (NEBA) is used to compare and rank the net environmental benefit associated with multiple management alternatives. NEBAs can be conducted for a variety of stressors and management options, including chemical contaminant mitigation, hydropower mitigation, and global climate change mitigation (e.g., carbon sequestration).

NEBA for contaminated sites typically involves the comparison of the following management alternatives [7]: (1) leaving chemicals in place; (2) physically, chemically, or biologically remediating the site through traditional means; (3) improving ecological value through onsite and offsite restoration alternatives that do not directly focus on removal of contamination; or (4) a combination of those alternatives. Increasingly, ecological restoration is being integrated into remedy analyses and decision making, with the goal of using sediment remedies to enhance ecological value, beyond simply controlling contaminant transport and exposure.

One type of NEBA is the risk-of-remedy analysis, the objective of which is to provide detailed quantitative and qualitative information relevant for the evaluation of short-term and long-term human and environmental risks, overall risk reduction, and comparison of different remedy alternatives. In accordance with USEPA guidelines and National Contingency Plan (NCP) criteria, evaluation of long-term remedy effectiveness focuses on minimizing risks to human health and the environment and remedy permanence. Short-term effectiveness is evaluated to "minimize short-term impacts to the extent practicable" [8], including short-term risks to the community, workers, and ecosystem associated with remedy implementation.

In the context of human health, failure to adequately evaluate implementation risks during the remedy selection process can result in unanticipated injuries (or even fatalities) to workers and nearby residents during cleanup [2]. Consequences may also include costly delays associated with substantial remedy modifications or abandonment of an incomplete remedy [9].

Post-remedy risks are determined primarily by changes in exposure resulting from residuals and redistributed sediment. These often are referred to as longer-term risk considerations. The ability to predict changes in bioaccumulation and other risks depends not only on the ability to describe post-remedy exposure, but also the degree to which pre-remedy characterization and modeling accurately reflect the relationship between sediment, water, and food chain exposures and actual bioaccumulation. Because many sediment management projects identify both short-term and long-term risk reduction goals, it is important for risk evaluations to consider both short- and long- term environmental changes and corresponding risks. Although it may be reasonable to assume negligible short-term residual risks after sediment cap placement, for example, due to the creation of a clean sediment surface via placement, background chemicals common to the site, such as PAHs or metals, are likely to recontaminate the sediment surface after remediation.

The potential for recontamination is particularly acute in urbanized watersheds. At a minimum, long-term residual risks are likely to resemble background risks associated with background chemicals, and may be greater than expected due to the presence of uncontrolled chemical sources.

The comparison of relative risk reduction, relative risk increase, or static risk conditions, provides additional information for decision making that might not otherwise be available. Risk-of-remedy analyses often are used in conjunction with the evaluation of costs to identify the appropriate and logical remedy for a given site. The goal is to meet risk-reduction goals at the most reasonable cost, while minimizing negative impacts to the natural environment, and minimizing short-term risks to human health associated with remedy implementation.

4. Case Studies

The following two case studies involved watershed-scale approaches to remedy decision making at contaminated sediment sites. Each case resulted in a unique outcome in terms of the goals of ecological and human health risk reduction, environmental protection, and watershed improvement. By evaluating a broad range of stressors and comparing the net environmental benefits of each remedy alternative, reductions in ecological and human health risk were achieved, while minimizing adverse impacts to the environment and risks to workers and nearby community residents.

4.1. CASE STUDY 1: HACKENSACK RIVER, NEW JERSEY, USA

The Hackensack River, New Jersey, is one of two large tributaries that flow into the northern portion of Newark Bay. Newark Bay is commonly included as part of the larger New York/New Jersey Harbor Estuary. Sediment along the eastern shore of the river near the confluence with Newark Bay is known to contain trivalent chromium (Cr(III)), due—in part—to a 0.14-km^2 former waterfront commercial property where approximately 800,000 m^3 of Chromium Ore Processing Residue was disposed from 1905 to 1954 [10–12]. Sediment sampling conducted along the property revealed the presence of elevated concentrations of total chromium and a wide variety of other metals and organic chemicals [10–12]. Sediment investigations and other remedy feasibility studies conducted between 2003 and 2006 contributed to the preparation of a remedy alternatives report generated in compliance with an order from the United States District Court of New Jersey.

The Court Order required a sediment remedy in the Hackensack River in the vicinity of the site containing total chromium at levels at or exceeding the effects-range median (ERM) marine/estuarine sediment screening value of 370 milligrams per kilogram (mg/kg) [16]. Figure 1 shows the area of the site delineated to this concentration.

4.1.1. Risk-of-Remedy Approach

The sediment remedy analysis incorporated a risk-of-remedy analysis to evaluate and compare net risk reduction. This analysis involved quantification of the short-term risks associated with implementation as well as the long-term risk reduction expected for each remedy. The analysis was conducted in the course of meeting established remedial action objectives; namely, to maintain low baseline chromium risks and minimize short-term implementation risks to community residents, workers, and the environment.

4.1.2. Remedy Alternatives Analysis

Among the sediment remedies evaluated, in situ remedies included MNR, thin-layer capping, and isolation capping and ex situ alternatives included surface dredging and deep sediment dredging. In addition, various combinations of these alternatives were considered. The analysis of remedy alternatives was supported by an understanding of chromium geochemistry,

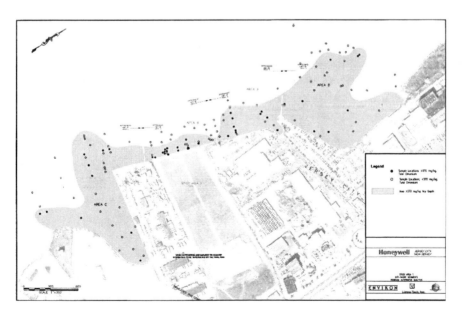

Figure 1. Delineation of Total Chromium Concentrations to the ERM of 370 mg/kg.

risk assessment, hydrological and sediment conditions, and the range of industry-available remediation technologies relevant to sediment.

The absence of measurable Cr(VI) in sediment pore water attested to the dominance of Cr(III) in sediment. These findings were particularly germane to the risk assessment, as Cr(III) is much less bioavailable and much less toxic than Cr(VI) [10, 11]. These conclusions were also consistent with available surface water data. Surface water Cr(VI) measurements were also below the Cr(VI) USEPA Ambient Water Quality Criterion of 50 μg/L.

Sediment suspension and oxidation tests demonstrated the geochemical stability of Cr(III) [12]. Repeated tests showed no sign of chemical Cr(III) oxidation when sediment samples were aggressively mixed and aerated in suspension for 24 h [12]. These findings were consistent with the understanding that natural reducing conditions in sediment preclude the presence of Cr(VI) in favor of the less toxic, less soluble Cr(III) [13–16]. Once reduced, Cr(III) is stable in aquatic environments and unlikely to oxidize to Cr(VI), even in the presence of oxygen [11, 12].

The results of the baseline risk assessment showed that the presence of total chromium in sediment does not pose unacceptable risks to human health or ecological receptors, even at locations with total chromium concentrations greater than the ERM sediment benchmark value of 370 mg/kg [16].

Water current velocities, geochronological studies, and sediment shear strength measurements were collected during field investigations to characterize sedimentation and sediment cohesive strength of the lower Hackensack River in the vicinity of the site. The site was characterized as net depositional, with well consolidated and cohesive sediment. Results of sediment shear strength testing and hydrodynamic monitoring indicated negligible potential for sediment scouring during normal flow conditions, and only moderate sediment scouring during extreme high-flow conditions.

4.1.3. Engineering Support of Relevant Sediment Remediation Technologies

In accordance with USEPA [6], sediment remedies were engineered to achieve risk reduction with respect to total chromium in sediment. Further, in accordance with the Court Order, the 370 mg/kg ERM guided the boundaries of the sediment remedies. Consideration also was given to sediment remedies based on delineation of sediment containing total chromium above 2,000 mg/kg, which represented the lower-bound estimate of a site-specific sediment quality value calculated from data collected during the sediment investigations. Remedial alternative boundaries were established on the basis of whether surface (upper 30 cm) or buried sediment had total chromium concentrations above 370 or 2,000 mg/kg. Each remedial alternative also included source control as a component of the final remedy.

Results of the remedy screening analysis [17] established MNR as readily implementable and highly effective for a low-risk site such as this one, with multiple lines of evidence supporting ongoing natural recovery processes. Sediment capping was found to be an effective, mature technology that has been implemented routinely at low-risk, low-energy, net depositional sites like this one. The removal response action (dredging) was found to be implementable, though ineffective for addressing total chromium risks in sediment. The primary advantage of dredging was that dredging would remove contaminated sediment from the aquatic environment. However, removal alone did not necessarily reduce risks, and in fact threatened to increase short- and long-term risks, particularly risks associated with chemicals other than chromium. The potential for increased short- and long-term risks was due to uncontrolled dredge residuals, sediment resuspension and offsite transport and deposition during dredging, potential sediment spills during dredging and dredge materials management (e.g., dewatering, transfer to and from barges and trucks, and offsite transportation), and construction/transportation risks to the local community and workers.

The risk-of-remedy evaluation, and the low baseline risks associated with chromium, strongly favored in situ remedies (i.e., MNR, thin-layer capping, or isolation capping) over ex situ remedies that involve removal and disposal of contaminated sediment. An in situ remedy was expected to achieve the requirements embodied in the NCP remedy evaluation criteria of overall protection of human health and the environment, short- and long-term effectiveness, and implementability. In contrast, an ex situ remedy posed significant construction related risks (Figure 2) to neighboring residential communities, businesses, and workers while providing little, if any, added short- or long-term environmental improvements to current sediment and ecological conditions. In fact, post-remedy long-term risks for both capping and dredging remedies were expected to return to current levels, particularly with respect to contaminants other than chromium (for which exposures were already declining because of natural recovery processes), due to natural deposition of contaminated sediment in the lower Hackensack River, in general, and in the vicinity of the site, in particular.

The recommended remedy for addressing total chromium concentrations above 370 mg/kg in sediment entailed a combination of source control with isolation capping where the total chromium concentration was greater than 2,000 mg/kg in surface (0–30 cm) sediment, plus MNR of remaining areas. This remedy provided significant risk reduction while limiting remedy-imposed risks. However, despite the low risk, a more aggressive remedy was established to include the following components (Figure 3):

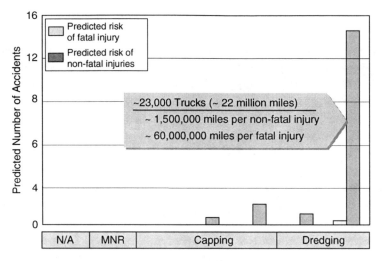

Figure 2. Construction (i.e., transportation) Related Risks Calculated for the Range of Remedies Considered for the Lower Hackensack River site.

- Dredging of approximately 2,000 cubic yards (CY) (1,540 m³) of material to a depth of 2 ft (60 cm) in a source area where surface sediment total chromium concentrations exceeded 2,000 mg/kg, followed by an 18-in. (45 cm) armored cap.

- Placement of 30 cm cap in areas where surface (upper 15 cm) sediment total chromium concentrations exceeded 370 mg/kg, and placement of a 15 cm cap in remaining areas where surface sediment (upper 30 cm) total chromium concentrations exceeded 370 mg/kg.

- MNR for remaining areas where surface (upper 30 cm) sediment total chromium concentrations were less than 370 mg/kg, but total chromium concentrations at sediment depths greater than 30 cm depth exceeded 370 mg/kg.

4.1.4. Determining Success

The success of the risk-of-remedy process rested in the evaluation and unbiased comparison of long-term potential benefits associated with alternative remedies and their relative potential to reduce human health and ecological risk, and the consideration of short-term risks associated with remedy implementation across the watershed. Comparison of short-term implementation risks and long-term potential risk reduction with associated costs made it possible to evaluate the relative cost-benefit of the remedies (Figure 4), pointing toward relatively low-impact, in situ remedies that minimized the impact and costs of unnecessarily removing sediment that posed no unacceptable risk to human health or the environment.

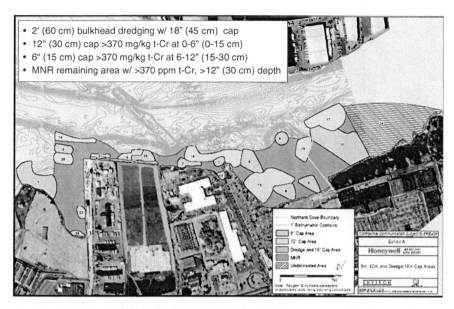

- 2' (60 cm) bulkhead dredging w/ 18" (45 cm) cap
- 12" (30 cm) cap >370 mg/kg t-Cr at 0-6" (0-15 cm)
- 6" (15 cm) cap >370 mg/kg t-Cr at 6-12" (15-30 cm)
- MNR remaining area w/ >370 ppm t-Cr, >12" (30 cm) depth

Figure 3. Negotiated Remedy for the Lower Hackensack River, Including 2,000 CY (1,540 m³) Dredging Over 0.5 Acres (0.2 ha); a 14-Acre (6.7 ha), 12-in. (30 cm) Cap; a 15-Acre (6.1 ha), 6-in. (15 cm) Cap; and MNR Over 20 Acres (8.1 ha).

While the evaluation successfully resulted in a recommendation that significantly reduced risk while avoiding imposing further risk on the watershed, environmental remedies are not selected, engineered, and implemented in a vacuum. In this instance, the final remedy avoided large-scale dredging (involving the removal of at least 530,000 CY (400,000 m³) of sediment—the estimated in situ volume of sediment with total chromium concentrations greater than 370 mg/kg). The watershed-scale approach to the remedy analysis (including consideration of remedy implementation risks) was instrumental in achieving an outcome that substantially limited risks to nearby residential communities, workers, and ecological receptors, as may have originated from sediment resuspension and offsite contaminant transport, construction hazards, aquatic and upland spills of contaminated sediment material, and the unnecessary expenditure of energy and corresponding CO_2 release.

4.2. CASE STUDY 2: PENINSULA HARBOUR SEDIMENT, LAKE SUPERIOR, CANADA

Peninsula Harbour is located in northeastern Lake Superior (Figure 5). Approximately 3 km wide and 4 km long, the harbor is sheltered from the open waters of Lake Superior by two islands (Hawkins Island and Blondin Island)

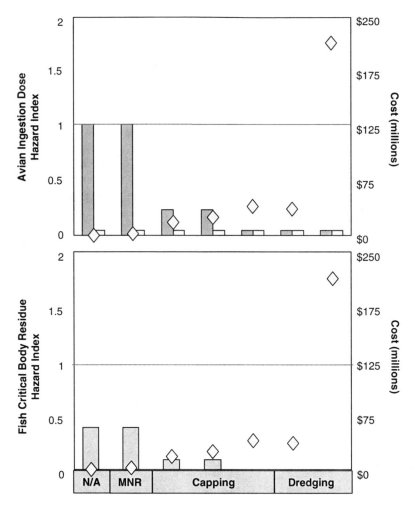

Figure 4. Cost-benefit Analysis of Risk reduction Contributed by MNR, Capping, and Dredging at the Hackensack River Site. Note that None of the Risks Exceeded a HI of 1.0, Suggesting no Further Action was Necessary. (Diamonds refer to Cost on the Right-hand Axes; Bars refer to Hazard Index on the Left-hand axes).

and two peninsulas to the north and south. The harbor contains multiple coves, including Jellicoe, Beatty, and Carden [18].

Peninsula Harbour is listed as a Great Lakes area of concern due to elevated concentrations of mercury and PCBs in sediment, resulting from historic operation of a bleached kraft pulp mill and chlor-alkali facility in Jellicoe Cove. In order to evaluate the need for sediment remediation, human health and ecological risk assessments were undertaken to determine whether anglers, benthic organisms, fish, or piscivorous wildlife were likely to be adversely affected by the presence and/or concentration of PCBs and

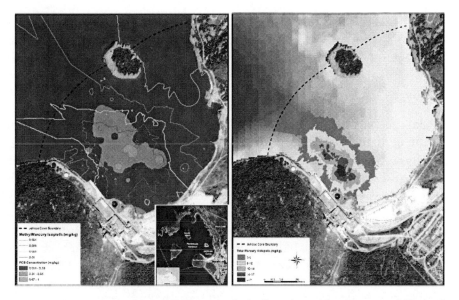

Figure 5. Peninsula Harbor Area of Concern, Lake Superior, with a Focus on Jellicoe Cove. Left Panel Shows Methylmercury Concentrations as Isopleths and PCB Concentrations as Solid Color Contours in Jellicoe Cove. Right Panel Shows Total Mercury Concentrations as Solid Color Contours. The 3 mg/kg Mercury Contour is Truncated at the Outer Perimeter of Jellicoe Cove.

methylmercury in sediment and the food web. Based on the findings of the risk assessments [18, 19], two remedial action objectives were established:

- Minimize the potential for future exposure to sediment-associated PCBs and methylmercury in Jellicoe Cove and the rest of Peninsula Harbor.

- Source control through reducing the potential for migration of chemicals away from Jellicoe Cove.

Sediment in Jellicoe Cove consists primarily of a relatively thin deposit (~30 cm average) on top of bedrock. The ecological risk assessment calculated a sediment management goal for methylmercury of 0.0020 mg/kg, to protect fish populations. By way of comparison, pre-remediation spatially weighted average surface (0–15 cm) sediment concentrations of methylmercury were 0.0051 mg/kg in Jellicoe Cove and 0.0019 mg/kg in the rest of Peninsula Harbour. Thus, the methylmercury concentration in Jellicoe Cove was 2.6 times higher than the sediment management goal; methylmercury concentrations in the rest of Peninsula Harbor did not exceed this goal.

The sediment management goal for PCBs of 0.06 mg/kg was intended to protect mink, with a second goal, 0.19 mg/kg, intended to protect sport anglers. Spatially weighted average surface (0–15 cm) sediment PCB concentrations in Jellicoe Cove and the rest of Peninsula Harbour were 0.14 and 0.12 mg/kg, respectively . Thus, the surface sediment PCB concentration in

Jellicoe Cove and the rest of Peninsula Harbour exceeded the mink sediment management goals by about two-fold, but did not exceed the angler sediment management goal.

4.2.1. Prescreening Sediment Management Options

Sediment management alternatives that were considered for Peninsula Harbour included MNR, capping, or dredging. The scale of the remedy alternatives ranged from complete removal of all sediment in areas where surface sediment concentrations exceeded cleanup goals defined for either PCBs or total mercury to various combined remedies that included MNR, capping, and dredging. The concentration of total mercury (rather than the methylated fraction) was considered in the selection of remedy alternatives for a variety of reasons including cost effectiveness (in assessment and remedy) and a stated focus on the feasibility of source control. The purpose of the preliminary screening was to eliminate from further consideration those technologies or process options that are clearly ineffective, not readily implementable at the Peninsula Harbour Area of Concern, or too costly (provided that less costly remedies can achieve the same degree of effectiveness). The maximum possible area considered for risk-based sediment management was approximately $300,000\,m^2$, corresponding to the area within Peninsula Harbour where surface sediment mercury concentrations exceeded $2\,mg/kg$. The maximum possible volume of sediment considered for risk-based sediment management was approximately $43,000\,m^3$, corresponding to that volume of sediment with mercury concentrations exceeding $2\,mg/kg$. A range of higher concentration mercury thresholds (corresponding to smaller area and volume estimates) were also assessed based on the potential for ecological and human health risk reduction.

Whereas preliminary risk-based assessments of areas and/or volumes of sediment warranting remediation did not explicitly focus on remedial strategies (i.e., MNR, capping, or dredging), it was clear that achieving immediate risk-based sediment management goals would require extensive remedial actions in both Jellicoe Cove and the rest of Peninsula Harbour. Dredging or capping of all sediment to achieve the risk-based sediment management goals were judged not to be cost effective; specifically, the remedy costs and the potential negative environmental impacts of widescale dredging were judged to outweigh the perceived risk-reduction benefits. Further, there were concerns regarding the technical feasibility of widescale dredging in the rest of Peninsula Harbour, where the water depth exceeds $60\,m$. Therefore, localized sediment management received closer consideration for purposes of source control and to enhance MNR by reducing offsite sediment transport potential.

4.2.2. Detailed Evaluation of Sediment Management Options

After screening general response actions in terms of cost, effectiveness, and feasibility, remedial alternatives underwent detailed evaluation. Sediment remedial alternatives evaluated in detail included thin layer capping, hydraulic dredging, and a combination of thin layer capping and hydraulic dredging with dredged materials disposed at an offsite landfill. Onsite landfill disposal also was evaluated but was determined not to be cost effective, and posed greater immediate and long-term onsite risks from buried contaminants. Remedial alternatives were evaluated based on the following criteria:

1. Ability to achieve the sediment management goals identified in the ecological risk assessment

2. Technical feasibility, with respect to reliability, timeline, and construction and operation requirements

3. Community acceptance

4. Environmental impacts and human health implications, as well as the need for measures to control and address residual contamination

5. Requirements for chemical, biological, and/or physical monitoring, in order to achieve both short-term and long-term goals

6. Compliance with regulatory requirements and the effects of these requirements on implementation

7. Overall detailed site-specific costs related to implementation, long-term maintenance and monitoring, and disposal

Although hydraulic dredging is a technically feasible means of achieving mass removal for Jellicoe Cove, there is substantial uncertainty regarding the ability of dredging alone to achieve the risk-based goals for this site. Dredging of Jellicoe Cove would require, for example, removal of sunken pulp logs and removal of thin sediment strata overlying bedrock. These conditions increase the cost and complexity of dredging, likely lead to high post-dredge residual concentrations, and increase the risk of resuspension and offsite contaminant release [5], threatening the rest of Peninsula Harbour. For these reasons, a dredging-only remedy was rejected.

Capping in Jellicoe Cove is a technically feasible means of reducing risk by limiting biotic exposure to contaminated sediment and by reducing the potential for chemical resuspension, release, and transport offsite. As the continued presence of the cap is a precondition of its successful function (either through sequestration or dilution of chemicals), capping involves long-term monitoring to ensure that materials remain in place. Because it is not necessary to remove the sunken logs for successful cap placement, implementation is generally straightforward. The Jellicoe Cove cap would provide the following benefits:

- Immediately reduce elevated surface sediment total mercury concentrations in the cove
- Provide a healthy substrate/habitat for benthic recolonization
- Provide source control and containment by limiting the transport of contaminated sediment from Jellicoe Cove to the rest of Peninsula Harbour
- Minimize dredging and corresponding physical and environmental disturbances associated with dredging, including dredged residuals, resuspension and offsite contaminated sediment transport, and short-term surface water exposures during dredging
- Minimize risks to workers and the community associated with dredging, dewatering, and offsite sediment disposal
- As compared to remedies involving dredging, lower risk of traffic accidents and lower carbon footprint
- Reduce the footprint of the construction staging area, and corresponding impacts to the environment by reducing dewatering and sediment staging requirements

4.2.3. Value-Weighting Method

To further aid in the selection of final remedial alternatives, a simplified MCDA process was used to compare the ability of each remedial alternative to achieve these seven criteria. Each of the seven criteria was ranked qualitatively in terms of relative importance (i.e., low, medium, high). Evaluation of the performance of each remedial alternative in the context of each criterion was then classified as "good," "better," or "best." These terms were not treated as being mutually exclusive, in that multiple remedial alternatives could receive the same qualitative rank for any given assessment criterion. The qualitative responses used as matrix inputs for weighting (w) and evaluation (c) were converted into numeric values, with "low" and "good" receiving scores of 1, "medium" and "better" receiving scores of 2, and "high" and "best" receiving scores of 3. For each remedial alternative, a weighting rank score (S) was then calculated as a weighted average:

$$S = \frac{\Sigma(w_i \times c_i)}{\Sigma w_i}$$

Where:

S = Weighted rank score

w_i = The weighting factor applied to each criterion

c_i = The performance evaluation of each remedial alternative with respect to specific criterion

TABLE 1. Weighting matrix to compare sediment remedial alternatives.

Criteria	Relative importance of criteria	Capping	Dredging	Combined capping and dredging
Effectiveness	High	Better	Good	Best
Technical feasibility	Medium	Best	Good	Good
Community acceptance	Medium	Good	Good	Good
Risk of remedy	High	Better	Good	Better
Monitoring needs	Low	Good	Better	Good
Compliance with regulations	Medium	Better	Good	Good
Cost effectiveness	High	Best	Good	Better
Score		2.1	1.1	1.8

Table 1 summarizes the qualitative conclusions of the detailed evaluation of the sediment management options relative to the seven criteria. The two principal conclusions of the simplified MCDA exercise, as presented in Table 1, are that: (1) the benefits of a capping-only remedy outweigh the benefits of a dredging-only remedy; and (2) the benefits of combined remedy (i.e., capping combined with dredging) do not improve upon the benefits of a capping-only remedy.

This inherently subjective matrix offers a means of explicitly organizing professional judgments, as well as framework for exploring the effects of those judgments on assessment outcome. Reasonable modifications to the matrix to account for differences in opinions among engineers did not change the conclusion that the dredging-only remedy performed the least well of the three alternatives evaluated. However, there was disagreement regarding the usefulness of partial dredging. Although this weighting matrix is useful for technology comparison, considerations not identified here factor in the final remedy selection and more conservative valuations prevailed when disagreement could not be readily resolved. Moreover, as uncertainties associated with the performance and cost of dredging outweigh those associated with capping, the combined capping and dredging remedy did not outperform the capping-only remedy.

4.2.4. Determining Success

Pre-screening determined that Peninsula Harbour would benefit from MNR throughout, with more active sediment management in areas of significantly elevated surface sediment concentrations of total mercury. Further analysis, using a weighting matrix to compare the ability of remedial alternatives to achieve multiple evaluation criteria, led to the understanding that either a capping-only or a combined dredging plus capping remedy could be effective for this site.

The remedy implemented for Jellicoe Cove is expected to involve capping of 200,000 m² of sediment containing mercury concentrations equal to or greater than 3 mg/kg. MNR would be undertaken for the remainder of Peninsula Harbour. A 15-cm cap has been found to provide adequate surface sediment coverage and sufficient thickness for benthic recolonization, while minimizing loss of aquatic habitat and alteration to local bathymetry. The physical stability of the cap is sufficient to withstand typical hydrodynamic flow conditions and erosive forces resulting from 100-year-interval storm events. The area targeted for the thin layer cap may also extend beyond the spatial extent of the 3 mg/kg mercury contour in the sediment, and may cover elevated concentrations of PCBs and, to a lesser extent, methylmercury.

The success of the risk analysis approach applied to Peninsula Harbour rested on a watershed-scale approach that avoided presumptive remedies, recognized implementation risks, and sought to evaluate the cost-effectiveness of a variety of risk-reduction strategies.

5. Conclusions

In aquatic environments affected by sediment contaminants, risk management strategies focus on either removing contaminated material or interrupting exposure pathways to reduce or eliminate risks over time. Remedies that may be applied to manage risk therefore include MNR, capping, and sediment dredging. In the past, conservative overestimates of chemical risks and underestimates of the effectiveness of isolation and natural recovery processes led to an assumption that dredging was the most effective approach to risk reduction. While public confidence in dredging is typically high, there is growing evidence that this approach may not achieve desired environmental improvement and risk reduction goals.

This paper addressed the complex issues associated with sediment remedy identification, screening, and selection. We presented two approaches used for remedy screening, analysis, and selection. These approaches include watershed-level thinking and NEBA approaches that recognize the influences of chemical and non-chemical, natural and anthropogenic stressors, and their respective influences on the integrity of a lake, river, coastal area, or other aquatic ecosystem. Without NEBA and watershed-level perspectives, environmental managers and regulators may misrepresent known stressors because they may lack an understanding of: (1) the broad range of potential ecological stressors; (2) the natural and anthropogenic sources that adversely affect the ecosystem; (3) the full range of potential remedies that exist; and 4) the impact of any and all remedies on chemical risks and ecological integrity.

The two case studies presented here involved distinct approaches to remedy decision making and demonstrated unique outcomes in terms of the goals of ecological and human health risk reduction, environmental protection, and watershed improvement. Combined, these case studies evaluated a range of stressors and compared the net environmental benefits of each remedy alternative. Outcomes included reductions in risk through minimizing adverse impacts to the environment and risks to workers and nearby community residents.

The Hackensack River, NJ case study demonstrated the "risk-of-remedy" process that is founded on the evaluation and unbiased comparison of long-term potential benefits associated with alternative remedies and their relative potential to reduce human health and ecological risk, plus the consideration of short-term risks associated with remedy implementation across the watershed. Arguably, the low baseline risks associated with total chromium in sediment at the site supported the selection of a sitewide MNR remedy, originally proposed in the sediment feasibility work [16]. However, overreliance on whole sediment chemistry led to a more invasive remedy than necessary, resulting in the addition of dredging and capping of sediment that had been shown to present no unacceptable risks to human health or the environment. The watershed-scale approach (including consideration of remedy implementation risks) was instrumental in achieving a negotiated remedy that combined MNR, capping, and dredging, and substantially limited risks to nearby residential communities, workers, and ecological receptors. Such risks included resuspension and offsite transport of contaminants, construction hazards, aquatic and upland spills of contaminated sediment material, and the unnecessary expenditure of energy and corresponding CO_2 release.

For the Peninsula Harbour site, prescreening and a risk-analysis weight-of-evidence approach avoided presumptive remedies (such as dredging-only), recognized implementation risks, and sought to evaluate the cost-effectiveness of a variety of risk-reduction strategies. These approaches established that Peninsula Harbour would benefit from active management of sediment containing the most elevated concentrations of mercury and PCBs followed by MNR for the remainder of the harbor. A simple form of MCDA (the weighting matrix) facilitated the comparison of the effectiveness of remedial alternatives to achieve multiple evaluation criteria, and led to the understanding that whereas either a combined dredging plus capping remedy or a capping-only remedy would be effective for remediation of this site, the benefits of combined remedy would not significantly improve upon the benefits of a capping-only remedy.

In the absence of a site wide risk assessment, and in the absence of a watershed-level approach that balances potential benefits against implementation risks to human health and the environment, site managers may be ill-equipped to effectively select environmentally appropriate and protective remedies. The USEPA CERCLA RI/FS process, and a variety of innovative multicriteria decision frameworks such as those presented in this chapter, provide sound foundations for remedy assessment and selection. Omitting these approaches risks environmental harm that can outweigh the perceived benefits of the presumed remedy.

References

1. Magar, VS and RJ Wenning. 2006. "The Role of Monitored Natural Recovery in Sediment Remediation." Integr. Environ. Assess. Manag. April. 2(1):66–74.
2. Wenning, RJ, M Sorensen, and VS Magar. 2006 "Importance of Implementation and Residual Risk Analyses in Sediment Remediation." Integr. Environ. Assess. Manag. 2(1):59–65.
3. Kiker, GA, TS Bridges, A Varghese, TP Seager, and I Linkov. 2005. "Application of Multicriteria Decision Analysis in Environmental Decision Making." Integr. Environ. Assess. Manage. 1(2):95–109.
4. Linkov, I, FK Satterstrom, G Kiker, C Batchelor, T Bridges, and E Ferguson. 2006. "From Comparative Risk Assessment to Multi-Criteria Decision Analysis and Adaptive Management: Recent Developments and Applications." Environ. Int. 32:1072–1093.
5. Bridges, TS, S Ells, D Hayes, D Mount, SC Nadeau, MR Palermo, C Patmont, and P Schroeder. 2008. The Four Rs of Environmental Dredging: Resuspension, Release, Residual, and Risk. US Army Corps of Engineers, Dredging Operations and Environmental Research Program. ERDC/EL TR-08-4.
6. Magar, VS, RJ Wenning, C Menzie, and SE Apitz. 2006. "Parsing Ecological Impacts in Watersheds." J. Environ. Eng. 132(1):1–3.
7. ORNL. 2008. Ecological Risk Analysis: Guidance, Tools, and Applications. Oak Ridge National Laboratory. Available at http://www.esd.ornl.gov/programs/ecorisk/net_environmental.html
8. USEPA. 2005. Contaminated Sediment Remediation Guidance for Hazardous Waste Sites. EPA-540-R-05-012. Office of Solid Waste and Emergency Response OSWER 9355.0-85. December.
9. Church, BW 2001. "Remedial Actions: The Unacknowledged Transfer of Risk." Environ. Sci. Pollut. Res. Special. 1:9–24.
10. Sorensen, MT, JM Conder, PC Fuchsman, LB Martello, and RJ Wenning. 2007. "Using a Sediment Quality Triad Approach to Evaluate Benthic Toxicity in the Lower Hackensack River, New Jersey." Arch. Environ. Contam. Toxicol. 53:36–49.
11. Martello, LB, MT Sorensen, PC Fuchsman, VS Magar, and RJ Wenning. 2007. "Chromium Geochemistry and Bioaccumulation in Sediments from the Lower Hackensack River, New Jersey, USA." Arch. Environ. Contam. Toxicol. 53(3):337–350.
12. Magar, V, L Martello, B Southworth, P Fuchsman, M Sorensen, and RJ Wenning. 2008. "Geochemical Stability of Chromium in Sediments from the Lower Hackensack River, New Jersey." Sci. Total Environ. 394(1):103–111.
13. Berry, WJ, WS Boothman, JR Serbst, and PA Edwards. 2004. "Predicting the Toxicity of Chromium in Sediments." Environ. Toxicol. Chem. 23(12):2981–2992.
14. Besser, JM, WG Brumbaugh, NE Kemble, TW May, and CG Ingersoll. 2004. "Effects of Sediment Characteristics on the Toxicity of Chromium(III) and Chromium(VI) to the Amphipod, *Hyalella azteca*." Environ. Sci. Technol. 38(23):6210–6216.

15. Becker, DS, ER Long, DM Proctor, and TC Ginn. 2006. "Evaluation of Potential Toxicity and Bioavailability of Chromiumin Sediments Associated with Chromite Ore Processing Residue." Environ. Toxicol. Chem. 25(10):2576–2583.

16. Long, ER, DD MacDonald, SL Smith, and FD Calder. 1995. "Incidence of Adverse Biological Effects Within Ranges of Chemical Concentrations in Marine and Estuarine Sediments." Environ. Manage. 19(1):81–97.

17. ENVIRON. 2006. Sediment Remedial Alternatives Analysis Report: Study Area 7, Jersey City, New Jersey. Prepared for Honeywell International Inc., Morristown, NJ. December.

18. Henning, M, K Leigh, K Merritt, V Magar, and R Santiago. 2008. "Assessment and Mitigation of Ecological Risks Posed by Mercury and PCBs in Peninsula Harbour Sediment Lake Superior." In proceedings of the Federal Contaminated Sites National Workshop. Vancouver, BC. April 28–May 1.

19. Magar, VS, K Merritt, R Santiago, and J Anderson. 2008. "Monitored Natural Recovery at Contaminated Sediment Sites in Canada and the U.S." In proceedings of the Federal Contaminated Sites National Workshop. Vancouver, BC. April 28–May 1.

MULTICRITERIA DECISION ANALYSIS FOR CHOOSING THE REMEDIATION METHOD FOR A LANDFILL BASED ON MIXED ORDINAL AND CARDINAL INFORMATION

R. LAHDELMA

University of Turku, Department of Information Technology
Joukahaisenkatu, 3–5
FIN-20520 Turku, Finland
risto.lahdelma@cs.utu.fi

P. SALMINEN

University of Jyväskylä, School of Business and Economics
P.O. Box 35
FIN-40014 Fvinland

Abstract: We describe a real-life application of a multicriteria method in the context of remediating a landfill in Finland. The landfill was used for dumping industrial waste during the years 1950 to 1965. During that time many harmful chemicals were stored in landfills among other waste. Therefore, this—and possibly several other—landfills have to be remediated. In this application, seven different remediation options were evaluated based on 13 criteria. For some criteria, cardinal measures with associated uncertainties were obtained. For other criteria, only ordinal (ranking) information was available. The problem was analyzed using the Stochastic Multicriteria Acceptability Analysis with Ordinal Criteria (SMAA-O) multicriteria method, which is able to handle this kind of mixed data. SMAA-O represents inaccurate or uncertain cardinal criteria values with a joint probability distribution. Ordinal data are converted into stochastic cardinal data by simulating all consistent mappings between ordinal and cardinal scales that preserve the given rankings. Decision makers' (DMs') unknown or partly known preferences are at the same time simulated by choosing weights randomly from appropriate distributions. The main results of the analysis are so-called acceptability indices for alternatives describing how large a variety of DMs' preferences support an alternative for the first rank or any given rank. The method also computes what kinds of preferences favor each alternative, and provides confidence factors measuring if the data are sufficiently accurate for making an informed decision. In this application, the SMAA-O analysis also aided the DMs in forming a new alternative as a combination of two original alternatives. This new alternative was identified as the preferred solution.

1. Introduction

As a result of industrial activities, several so-called risk landfills exist in Finland. The remediation of these landfills is estimated to require hundreds of millions of Euros during the next 20 years. To define the need and level of remediation, local conditions as well as environmental quality and potential negative effects have to be evaluated carefully. In this paper, we describe the multicriteria decision analysis (MCDA) process used to choose a remediation method for the Huuna landfill in Tervakoski, Finland [17]. The industrial waste landfill urgently needed to be remediated. The landfill was in operation from 1950–1965. Until 1962, the main waste treatment method was incineration. Soon after this, the landfill became full. It was then covered with clean soil, and put under cultivation.

The size of the landfill area is 1.5 ha, and its volume is 40,000 m^3 (60,000 t). The waste is covered under a 0.5–5-m-deep layer of clean soil. Essentially, the problem is that the waste brought to incineration at that time included—for example—polychlorinated biphenyls (PCBs). The proportion of PCBs averaged 4.9 mg/kg in soil tests. Polychlorinated dibenzo-para-dioxin (PCDD) and polychlorinated dibenzofuran (PCDF) compounds were measured at somewhat higher concentrations within the landfill than in its surroundings.

Different cleaning options were evaluated based on 13 criteria defined by eight experts representing the most important interest groups on this matter. It was not considered realistic to measure all criteria on interval scales. Instead, seven criteria were measured on ordinal scales based on expert judgment.

The Stochastic Multicriteria Acceptability Analysis with Ordinal Criteria (SMAA-O) multicriteria method [11] was applied to the problem, since this method is able to deal with mixed ordinal and uncertain cardinal criteria. SMAA methods have been developed for discrete multicriteria problems containing uncertain or inaccurate criteria measurements and where it is difficult to obtain accurate weight information from the decision makers (DMs) [7, 16]. The SMAA methods are based on exploring the weight space in order to describe the preferences that would make each alternative the most preferred one, or that would imply a certain rank for a specific alternative. Related ideas have been presented elsewhere [2–4, 9, 19].

The SMAA [10] and SMAA-2 [13] methods apply assumed value/utility functions and stochastic criteria, while SMAA-3 [7, 14] and SMAA-III [23] use the pseudocriteria model and outranking procedure of the ELECTRE III method [18, 20, 26]. SMAA-O [11] extends SMAA-2 for problems with mixed ordinal and cardinal criteria. The SMAA-D method [15] evaluates the alternatives in terms of a DEA-like efficiency measure. Ref-SMAA [12] and SMAA-A methods [5] compare the alternatives by applying Wierzbicki's achievement scalarizing functions. SMAA-TRI is an extension of the ELECTRE TRI ordinal classification method [27] to include robustness analysis [21]. For real-life applications of SMAA-methods, see [25]. The efficient implementation and computational efficiency of SMAA methods have been described [25]. For a survey on different SMAA methods, see [22].

This paper is organized in seven sections. Section 2 describes alternative treatment options for the landfill including the "current state" alternative; Section 3 deals with the choice of the evaluation criteria, and Section 4 describes how the criteria were measured. Next, Section 5 describes the SMAA-O multicriteria method used, and Section 6 presents the results of the analysis. Finally, Section 7 presents some concluding remarks on this real-life multicriteria application.

2. Alternative Options for Dealing with the Landfill

Initially, six alternative techniques (I–VI) for dealing with polluted soil were identified:

I Current solution: Groundwater is pumped away in such quantities that the groundwater surface remains below the contaminated soil. Monitoring of groundwater quality is conducted frequently.

Fifteen years of groundwater monitoring has proved that the induced lowering of the groundwater surface has been an adequate control measure. Groundwater quality remains high and exceeds drinking water quality standards. Laboratory analyses of monthly samples include information on PCBs, oxygen, iron, manganese, nitrite, nitrate, and chloride in the groundwater. An annual monitoring report is sent to the local authorities.

II Current solution with frequent control measurements, but without groundwater pumping.

If the current pumping of groundwater is ceased temporarily, it is possible to monitor the effects that the rising groundwater has on groundwater quality. If no adverse effects to groundwater quality occur, it may be possible to permanently stop pumping, saving approximately EUR 20,000 per year.

III Horizontal isolation capping system for the area. The current soil layer is replaced with a 0.5-m mineral isolation layer and at least 1 m of soil.

The creation of a capping system providing horizontal isolation over the contaminated area would prevent the infiltration of soil water through the contaminated material and consequently the leaching of constituents of concern deeper into the soil profile. If the present groundwater pumping is also stopped, the groundwater will rise and again saturate the contaminated material irrespective of this horizontal capping system. However the cap will prevent the volatilization of PCBs and inhalation of dusts containing D/F compounds. The design of horizontal isolation systems is a technically well known solution and several construction companies are available.

IV Horizontal and vertical isolation of the area.

Vertical isolation combined with the horizontal isolation of the surface prevents the leaching of constituents of concern from the contaminated material via rainfall and prevents their lateral migration via the groundwater.

V In-situ bioremediation techniques.

In-situ treatment methods are very interesting and development is quite fast in this area. A literature survey and contacts with leading research laboratories in Europe and USA provided no evidence for a full-scale method for in-situ treatment of D/F compounds and PCBs in this situation.

VI Partial or complete removal of the substance and replacement with clean soil. The removed substance is transferred to special landfills or to an incineration plant.

Excavation of the contaminated soil and transportation to the national hazardous waste incineration plant in Riihimäki was the solution suggested by local authorities in the first environmental permit application at the beginning of the 1990s; however, the solution is very expensive.

During the process, a seventh alternative was created as a combination of the Alternatives I and III:

VII Horizontal isolation and groundwater pumping.

The first stage of remedial alternative assessment led to the inclusion of a new solution: basically, a combination of the current groundwater pumping (I) and the suggested horizontal isolation (III). This combination can also be readily implemented and the costs are moderate.

3. Choosing and Measuring the Criteria

Our aim was to form a complete but non-overlapping set of independent criteria that can be used with an additive utility function. A supervisory group consisting of experts chose the criteria using the Delphi technique [8].

Two Delphi rounds were required to form the final set of criteria. Eight experts representing the most important interest groups participated in the process from the beginning. Later on, four additional experts joined the supervisory group. These new members accepted the criteria, but expressed differing opinions on the criteria measurements and valuations.

The first Delphi round focused on identifying interest groups that are affected by the polluted land. Also, the exact effects on each group were identified. The following list of interest groups and effects was produced:

- Farmers: health effects, disadvantages for agriculture and forestry
- Land owners: costs, possible revenue
- Inhabitants: health effects, recreation, habitability, safety
- Zoning administration: effects on current and planned land use
- Builders: costs, workers health, profits
- Groundwater use: groundwater quality, health risks
- Environment: biodiversity
- Waste treatment: risks of a new waste treatment facility, investments, releases to air
- Research: new innovations
- Authorities: research costs, permit processing

The second Delphi round did not reveal any new criteria. The effects on agriculture and forestry were considered insignificant in this context. Therefore, these impacts were excluded from the final set of criteria. The following list of 13 criteria was then formed:

- Effects on nature

  C1 Soil effects
  C2 Groundwater risks
  C3 Surface water risks
  C4 Ecological risks
  C5 Effects on air

- Effects on people

  C6 Health risks
  C7 Recreation, habitability

- Business

  C8 Research and new innovations

- Community structure

  C9 Land use

- Economy

  C10 Costs
  C11 Profit expectations

- Technical feasibility

  C12 Feasibility
  C13 Reliability

4. Measuring the Criteria

Some criteria were measured on cardinal scales. Uncertain cardinal measures were expressed as intervals. For other criteria, only ordinal (ranking) information based on expert judgment was available.

- **C1: Soil effects.** This criterion was evaluated on an ordinal scale. Alternative II, where the pumping of groundwater is stopped, was evaluated to have the most negative effects on soil, whereas the removal Alternative VI was evaluated as best under this criterion.

- **C2: Groundwater risks**. The risks to groundwater were evaluated on an interval scale where 0 = no risk, 1 = possible small risk, 2 = small risk.

- **C3: Surface water risks**. This criterion was evaluated on an ordinal scale but no differences between the alternatives could be identified.

- **C4: Ecological risks**. This criterion was evaluated on an ordinal scale. Measurements of the current Alternative I did not reveal any possible ecological risks. It was estimated that the other alternatives would not be worse under this criterion. Initially, no differences could be found for this criterion.

- **C5: Effects on air**. This criterion was evaluated on an interval scale. Among the alternatives the scale became rather degenerated with only two values: 0 = no effect, 1 = possible small effect.

- **C6: Health risks**. Based on the measurements of the current Alternative I, no alternative was expected to exceed the current regulations. Using an ordinal scale, Alternatives IV, V, and VI were given the best rank under this criterion while Alternative II was evaluated as worst.

- **C7: Recreation, habitability**. An interval scale—where 0 = no limitations for recreation and habitability and 1 = some limitations—was used.

- **C8: Research and new innovations**. This criterion was evaluated on an ordinal scale. Obviously, the alternative that requires the most know-how and methodological development, Alternative V, obtained the highest rank.

- **C9: Land use**. If the remediation method did not limit the future use of the land for different purposes, it was given 1 on an interval scale. Values 2 and 3 represent possible and clear land use limitations, respectively.

- **C10: Costs**. Only investment costs were considered for this criterion (in million Euros in Table 1). The uncertainties of cost estimates were represented as ±15% ranges around their expected values (unless otherwise stated).

- **C11: Profit expectations**. The possible profits here are based on increased land value (in million Euros in Table 1). The value of the current alternative corresponds to the price of meadowland. If the quality of the land improves, it can be used for industrial purposes or even for housing. This, of course, increases the land value. Nevertheless, the estimates of the amount of the increase were considered very uncertain. Accordingly, an uncertainty factor of ±25% was applied to profit expectations.

- **C12: Feasibility**. This criterion was measured on an ordinal scale. Simple and traditionally used techniques were considered best under this criterion. Complicated and poorly known techniques received the worst ranks.

- **C13: Reliability**. Alternatives requiring the least maintenance and follow-up were ranked best.

In retrospect, ordinal measurement could have been more natural also for some of the criteria that the experts wanted to treat as cardinal. These include C2, C5, C7, and C9. However, this would not have changed the results significantly. Table 1 summarizes the criteria and presents the initial criteria values for the alternatives.

5. The SMAA-O Method

The SMAA-O method is developed for discrete multicriteria problems, where alternatives are evaluated in terms of mixed ordinal and uncertain cardinal criteria, and where for some reason it is impossible to obtain accurate weight information from the DMs [11]. Compared with other SMAA-family methods, SMAA-O differs in its treatment of ordinal criteria.

As with the earlier versions of SMAA, the results obtained from the SMAA-O analysis are descriptive. The main results are rank acceptability indices, central weight vectors, and confidence factors for different alternatives. The rank acceptability index is a measure of the variety of preferences granting a certain rank for an alternative. The central weight vectors describe the typical preferences favoring each alternative. The confidence factors measure whether the criteria data are accurate enough to discern the alternatives when the central weight vector is used.

TABLE 1. The Criteria Values of Alternatives. Criteria C3 and C4 were not Included in the First Analysis, the alternatives being Equal under These Criteria.

Alternative	C1	C2	C3	C4	C5	C6	C7	C8	C9	C10	C11	C12	C13
I	5	0	1	1	1	3	1	6	3	0.017	0.003	1	3
II	6	2	1	1	1	4	1	4	3	0	0.003	4	1
III	4	2	1	1	0	2	0	3	2	0.25	0.73	2	3
IV	3	1	1	1	0	1	0	2	1	0.42	1.5	3	2
V	2	1	1	1	1	1	0	1	1	3.3–6.7	1.5	6	4
VI	1	0	1	1	0	1	0	5	1	5–10	1.5	5	1
Direction of improvement	Min	Min	Min	Min	Min	Min	Min	Min	Min	Min	Max	Min	Min
Scale Lin/Ord	O	L	O	O	L	O	L	O	L	L	L	O	O
Uncertainty		0.5			–		–		0.5	15%	25%		

In SMAA-O, a multicriteria decision problem with m alternatives $\{x_1,\ldots x_m\}$ is measured in terms of n criteria. An additive utility function

$$u(x_i, w) = \sum_{j=1}^{n} w_j u_j(\xi_{ij}), \qquad w \in W, \tag{1}$$

is assumed. $u_j(.)$ is the partial utility function for criterion j; the vector of weights w represents the preferences of a DM, and ξ_{ij} is the value of the jth criterion for the ith alternative. However, neither the weights nor the criteria values need to be known precisely. Instead, the analysis is based on simultaneous simulation of different weight vectors and criteria values.

Uncertain or imprecise criteria values are represented by stochastic variables ξ_{ij} with a joint probability distribution and density function $f(\xi)$ in the space $X \subseteq R^{m \times n}$. For cardinal criteria, a suitable distribution is assumed or estimated based on statistical methods. For example, imprecise cardinal measurements with the expected values x_{ij} can be modeled using independent uniform distributions within confidence intervals $[x_{ij} \pm \Delta x_{ij}]$. During the simulation, the criteria values ξ_{ij} are generated from their distribution and the partial utility functions are used to compute the partial utilities required in (1). The DMs must jointly accept the partial utility functions for cardinal criteria. In practice, this often means using linear transformations from the full scale to the interval [0,1].

Ordinal criteria values x_{ij} represent the rank of the ith alternative with respect to the jth criterion. The partial utility function $u_j(.)$ is a mapping from ordinal criteria values to a linear scale. To know this mapping is not necessary. Instead, the SMAA-O analysis simulates different mappings consistent with the ordinal information; i.e., all decreasing mappings $u_j(x_{ij}) \rightarrow [0,1]$. This is done by generating random numbers from a uniform distribution in the interval $[0,1]$, sorting the numbers into descending order, and assigning each rank the corresponding number. It is also easy to consider additional constraints on the ordinal scale intervals by rejecting mappings not satisfying the constraints.

A weight distribution $f(w)$ is used in the simulation to model the DMs' unknown or partly known preferences. A total lack of knowledge about weights is represented in 'Bayesian' spirit by a uniform weight distribution in the set of non-negative normalized weights $W = \{w \in R^n: w \geq 0 \text{ and } \Sigma_j w_j = 1\}$. Various kinds of preference information can be handled as weight constraints [13]. For example, a complete order of importance for criteria $j^1 > j^2 > \ldots > j^m$ is modeled as inequality constraints between weights $w_j^1 > w_j^2 > \ldots > w_j^m$. Unspecified importance between two or more succeeding criteria in the list is modeled by omitting the inequality constraint (e.g., C1 > C2 ? C3 > C4 $\rightarrow$ $w_1 > (w_2, w_3) > w_4$). Such weights are generated efficiently by sorting uniformly distributed weights into consistent order with a minimal number of exchanges.

After the criteria values and the weight vector have been generated in the simulation, the rank of each alternative is determined as an integer from the best rank (=1) to the worst rank (=m) by means of a ranking function

$$rank(\xi_i, w) = 1 + \sum_{k=1}^{m} \rho(u(\xi_k, w) > u(\xi_i, w)) = 1 + \sum_{k=1}^{m} \rho(\xi_k w > \xi_i w), \quad (2)$$

where $\rho(\text{true}) = 1$ and $\rho(\text{false}) = 0$. The SMAA-O method is then based on analyzing the so-called sets of *favorable rank weights* $W_i^r (\xi)$ defined as

$$W_i^r(\xi) = \{w \in W : rank(\xi_i, w) = r\}. \quad (3)$$

The favorable rank weights are simply the (stochastic) set of weights that result in rank r for an alternative. The rank acceptability indices, central weight vectors, and confidence factors are then computed as properties of W_i^r

The rank acceptability index b_i^r is the expected volume of W_i^r. The rank acceptability index is a measure of the variety of different valuations granting alternative x_i rank r, and is computed as a multidimensional integral over the criteria distributions and the favorable rank weights using

$$b_i^r = \int_X f(\xi) \int_{W_i^r(\xi)} f(w)\, dw\, d\xi. \quad (4)$$

The rank acceptability indices can be examined graphically in order to compare how different varieties of weights support each rank for each alternative. Candidates for most acceptable alternatives should be those with high (clearly non-zero) acceptabilities for the first rank. When seeking compromises, alternatives with large acceptabilities for the worst ranks should be avoided. However, the acceptability indices should not be used for an absolute ranking of alternatives, because the magnitude and mutual order of the acceptability indices depend on the chosen (assumed) distributions and scaling of criteria. Also, a holistic acceptability index

$$a_i^h = \sum_{r=1}^{m} \alpha_r b_i^r, \tag{5}$$

is computed for each alternative. The holistic acceptability index is a weighted sum of the rank acceptability indices using so-called *meta-weights* $1 = \alpha_1 \geq \alpha_2 \geq ... \geq \alpha_m \geq 0$. The holistic acceptability index is thus in the interval $[0,1]$ and aims to measure the overall acceptability of alternatives. There are many ways to choose the meta-weights [13]. In this application we assigned $\alpha_m = 0$ and used m-1 dimensional centroid meta-weights for the remaining α_p; i.e.,

$$\alpha_r = \sum_{i=r}^{m-1} \frac{1}{i} \Big/ \sum_{i=1}^{m-1} \frac{1}{i}, \tag{6}$$

The central weight vector W_i^c is the expected center of gravity of $W_i^r(\xi)$ (favorable first-rank weights). The central weight vector is computed as an integral of the weight vector over the criteria and weight distributions by

$$w_i^c = \int_X f(\xi) \int_{W_i^1(\xi)} f(w) \, w \, dw \, d\xi \Big/ b_i^1. \tag{7}$$

With the assumed weight distribution, the central weight vector represents the preferences of a typical DM supporting alternative i. The central weights can be presented to the DMs in order to help them understand what kinds of preferences correspond to the different choices. Furthermore, the central weights are used for computing the confidence factor. The confidence factor p_i^c is the probability of alternative i obtaining the best rank when the central weight vector is chosen. The confidence factor measures whether the criteria data are accurate enough to discern the alternatives. The confidence factor is computed as an integral over the criteria distributions ξ by

$$p_i^c = \int_{\xi \in X | \mathrm{rank}(i,\xi,w_i^c)=1} f(\xi) \, d\xi. \tag{8}$$

6. Results

Initially the six alternatives were analyzed using the SMAA-O method without preference information. Table 2 presents the holistic acceptability indices, confidence factors, and rank acceptability indices as percentages. The rank acceptability

TABLE 2. Results (in %) of the First SMAA-O Analysis: Holistic Acceptability Index a_i^h, Confidence Factor p_i^c, and Rank Acceptability indices b_i^1 ;,..., b_i^6. Alternatives are sorted by their Holistic Acceptability Index.

Alternative	a_i^h	p_i^c	b_i^1	b_i^2	b_i^3	b_i^4	b_i^5	b_i^6
IV	84	99	66	31	3	0	0	0
VI	62	79	32	41	18	7	1	1
III	31	13	1	16	35	43	5	0
V	29	22	2	10	39	36	9	4
I	10	40	0	1	4	10	63	22
II	3	31	0	0	1	3	22	73

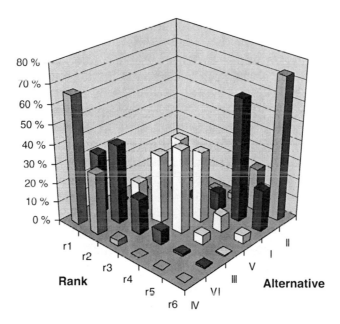

Figure 1. Rank Acceptability Indices from the First SMAA-O Analysis for Each Alternative (I,..., VI) and Rank (r1,..., r6). Alternatives are Sorted by their Holistic Acceptability Index.

indices are also illustrated in Figure 1. It can be observed that only Alternatives IV and VI receive significant acceptability for the first rank. Alternatives I and II are clearly eliminated from consideration on the basis of the model used. The holistic acceptability index (a^h) supports the same conclusions. Alternatives III and V appear very weak, but should not automatically be eliminated without considering preference information. The low confidence factors (13%, 22%) indicate that even if central weight vectors of the alternatives are chosen by the DMs, these alternatives will not undisputedly obtain the first rank.

These results deepened the discussion of the measurement of criteria: initially, Alternative III attracted several participants, but based on Table 2 its choice is difficult to justify. Reevaluation resulted in changes to some criteria values. For Criterion C1, the experts considered that no difference existed between Alternatives V and VI, or between Alternatives I and II. For the previously removed ordinal Criterion C4, some differences in ecological risks between the alternatives could be identified. For Criterion C9, the ranking was changed so that only Alternative VI was considered not to limit the future use of the land. Profit expectations, C11, were decreased to more realistic values.

The results of the first analysis also helped the experts to see how Alternative III could be improved. Groundwater risks in horizontal isolation could be eliminated with little additional cost by combining it with groundwater pumping (Alternative I). This led to the creation of a new Alternative VII. Table 3 displays the revised criteria measurements and the new alternative.

Table 4 presents the results of the second SMAA-O analysis based on the revised data. The rank acceptability indices are also illustrated in Figure 2. The new Alternative VII obtains almost two-thirds of the acceptability for the first rank. Alternative VI has preserved its high acceptability and is the most serious competitor to VII. The acceptability of Alternative IV has decreased significantly, but is still a potential candidate. Alternatives III and V can in practice be eliminated due to their very low acceptability.

Next, the participants wanted to add their preferences into the analysis. The criteria were ranked into an order of importance in the spirit of MACBETH [1]. The experts made pair-wise comparisons between alternatives by answering the question: "Which one of these two criteria would you rather raise from its worst value to its best value?" The participants completed this task surprisingly quickly and were consistently in agreement. The obtained ranking for the criteria was:

TABLE 3. Revised Criteria Measurements and a New Alternative VII. C3 was not Included in the Analyses.

Alternative	C1	C2	C3	C4	C5	C6	C7	C8	C9	C10	C11	C12	C13
I	4	0	1	3	1	3	1	6	3	0.017	0.003	1	3
II	4	2	1	4	1	4	1	4	3	0	0.003	4	1
III	3	2	1	3	0	2	0	3	2	0.25	0.083	2	3
IV	2	1	1	1	0	1	0	2	2	0.42	0.083	3	2
V	1	1	1	2	1	1	0	1	2	3.3–6.7	0.083	6	4
VI	1	0	1	2	0	1	0	5	1	5–10	0.33	5	1
VII	2	0	1	1	0	1	0	2	2	0.27	0.083	1	2

TABLE 4. Results of the Second SMAA-O Analysis after Revising Criteria Values and Adding a New Combined Alternative.

Alternative	a_i^h	p_i^c	b_i^1	b_i^2	b_i^3	b_i^4	b_i^5	b_i^6	b_i^7
VII	83	99	61	34	5	0	0	0	0
VI	61	91	34	20	29	11	4	1	0
IV	50	19	4	43	48	4	0	0	0
III	23	2	0	1	12	46	36	5	0
V	19	23	0	1	5	33	44	12	5
I	8	2	0	0	1	5	13	62	19
II	2	8	0	0	0	1	3	20	75

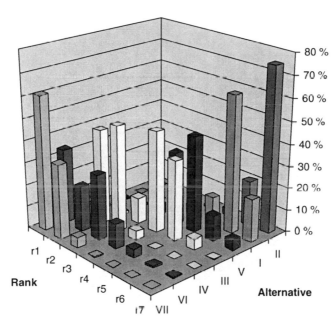

Figure 2. Rank Acceptability Indices from the Second SMAA-O Analysis after Revising Criteria Values and Adding a New Combined Alternative.

C10 > C13 > C12 > C1 > C2 > C6 > C4 > C9 ? C7 > C5 ? C11 > C8

The question mark indicates unspecified ranking between criteria.

Table 5 and Figure 3 present the results of the third SMAA-O analysis after this preference information was included as weight constraints. Alternative VII is the only widely acceptable alternative. Although Alternative IV still receives an acceptability of 6% and could be chosen with the applied model, its confidence factor is very low. This implies that when uncertainty is taken into account, IV is not likely to be the best alternative under any

TABLE 5. Results of the Third SMAA-O Analysis after Adding Preference Information from Eight Experts.

Alternative	a_i^h	p_i^c	b_i^1	b_i^2	b_i^3	b_i^4	b_i^5	b_i^6	b_i^7
VII	97	95	93	7	0	0	0	0	0
IV	60	12	6	86	7	0	0	0	0
VI	29	1	1	6	47	13	15	16	2
III	26	0	0	0	27	47	21	5	0
I	21	0	0	1	15	34	42	9	0
II	9	6	0	0	4	6	18	46	26
V	2	0	0	0	0	1	4	24	71

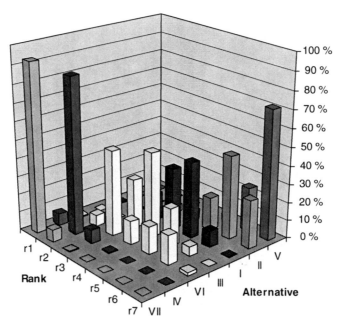

Figure 3. Rank Acceptability Indices from the Third SMAA-O Analysis after Including Preference Information

preferences. Alternative VI is unacceptable due to its high cost, cost being the most important criterion.

In addition, in this phase four new experts, previously unable to attend the meetings, wanted to provide their preference information to the analysis. Interesting enough, these experts did not agree with the ranking defined by the other participants. Instead, their ranking for the criteria was

C10 ? C6 > C1 > C2 ? C4 > C9 > C13 ? C12 > C5 > C7 > C11 > C8.

Moreover, these new experts did not agree with all criteria values approved by the first eight experts. They considered Alternative II worst on

TABLE 6. Results of the Fourth SMAA-O Analysis Using Preference Information from Four Additional Experts.

Alternative	a_i^h	p_i^c	b_i^1	b_i^2	b_i^3	b_i^4	b_i^5	b_i^6	b_i^7
VII	97	95	92	8	0	0	0	0	0
IV	58	14	6	77	16	0	0	0	0
VI	39	3	2	15	63	10	6	4	1
III	21	0	0	0	15	33	45	7	0
V	20	0	0	0	4	49	32	14	0
I	10	0	0	0	1	7	17	75	0
II	0	0	0	0	0	0	0	1	99

Criterion C1 (changing the value from 4 to 5) and gave it the shared last rank on Criterion C13 (changing the value from 1 to 4). They also decreased the rank of Alternative IV on Criterion C13 from 2 to 3 and finally raised the ranks of Alternatives V and VI on Criterion C12 to 5 and 4, respectively. The SMAA-O analysis was repeated using this modified data. The results are presented in Table 6.

Still, the opinions of the new experts did not change the final result of the analysis. There are no significant differences between Tables 5 and 6, although the ranking of the criteria and some criteria values were different in this analysis. Both analyses indicate that VII remains clearly the most acceptable alternative. In fact, the second analysis without preference information (Table 4) already gave a hint of this. The wide acceptability of Alternative VII without preference information means that it most probably will receive high acceptability when different preferences are introduced.

7. Conclusion

The SMAA-O multicriteria method was applied to a real-life problem of selecting among different options for remediating a landfill. Seven of the 13 criteria used in evaluating alternatives were measured on ordinal scales. The result of the first analysis did not satisfy the DMs, who represented various interest groups. Therefore they formed a new alternative by combining existing alternatives. In addition to a basic SMAA analysis without preference information, the DMs ranked the importance of each criterion. Using two different importance rankings, repeated SMAA-O analyses consistently identified the new combined alternative as the most acceptable solution to the problem.

A significant advantage of the SMAA-O method in this application was that it allowed explicit representation of different kinds of uncertain and imprecise information; in particular, both cardinal and ordinal information about criteria measurements and preferences. Many other multicriteria methods require either promoting ordinal information to cardinal, or demoting cardinal information to ordinal. The former situation requires unjustified increasing of the accuracy of the information. The latter situation requires losing some of the accuracy.

References

1. Bana e Costa C.A., Vansnick J-C. 1994. MACBETH an interactive path towards the construction of cardinal value functions. *International Transactions in Operations Research* 4(1):489–500.
2. Bana e Costa C.A. 1986. A multicriteria decision aid methodology to deal with conflicting situations on the weights. *European Journal of Operational Research* 26:22–34.
3. Butler J., Jia J., Dyer J. 1997. Simulation techniques for the sensitivity analysis of multicriteria decision models. *European Journal of Operational Research* 103:531–546.
4. Charnetski J.R., Soland R.M. 1978. Multiple-attribute decision making with partial information: the Comparative hypervolume criterion. *Naval Research Logistics Quarterly* 25:279–288.
5. Durbach I. 2006. A simulation-based test of stochastic multicriteria acceptability analysis using achievement functions. *European Journal of Operational Research* 170(3):923–934.
6. Hokkanen J., Lahdelma R., Miettinen K., Salminen P. 1998. Determining the implementation order of a general plan by multicriteria method. *Journal of Multi-Criteria Decision Analysis* 7:273–284.
7. Hokkanen J., Lahdelma R., Salminen P. 1999. A multiple criteria decision model for analyzing and choosing among different development patterns for the Helsinki cargo harbor. *Socio-Economic Planning Sciences* 33:1–23.
8. Hwang C-L., Lin. M-J. 1987. *Group Decision Making Under Multiple Criteria*. Springer, Berlin, Heidelberg.
9. Jiménez A., Ríos-Insua S., Mateos A. 2003. A decision support system for multiattribute utility evaluation based on imprecise assignments. *Decision Support Systems* 36(1):65–79.
10. Lahdelma R., Hokkanen J., Salminen P. 1998. SMAA - Stochastic multiobjective acceptability analysis. *European Journal of Operational Research* 106(1):137–143.
11. Lahdelma R., Miettinen K., Salminen P. 2003. Ordinal criteria in stochastic multicriteria acceptability analysis (SMAA). *European Journal of Operational Research* 147(1):117–127.
12. Lahdelma R., Miettinen K., Salminen P. 2005. Reference point approach for multiple decision makers. *European Journal of Operational Research* 164(3):785–791.
13. Lahdelma R., Salminen P. 2001. SMAA-2: Stochastic multicriteria acceptability analysis for group decision making. *Operations Research* 49(3):444–454.
14. Lahdelma R., Salminen P. 2002. Pseudo-criteria versus linear utility function in Stochastic Multicriteria Acceptability Analysis. *European Journal of Operational Research* 141(2):454–469.
15. Lahdelma R., Salminen P. 2006. Stochastic multicriteria acceptability analysis using the data envelopment model. *European Journal of Operational Research* 170(1):241–252.

16. Lahdelma R., Salminen P., Hokkanen J. 2000. Using multicriteria methods in environmental planning and management. *Environmental Management* 26(6):595–605.
17. Lahdelma R., Salminen P., Simonen A., Hokkanen J. 2001. Choosing a reparation method for a landfill using the SMAA-O multicriteria method. In Köksalan, Zionts (eds.) Multiple criteria decision making in the new millennium, *Lecture Notes in Economics and Mathematical Systems* 507:380–389.
18. Maystre L.Y., Picted J., Simos J. 1994. *Méthodes Multicritéres ELECTRE.* Presses Polytechniques et Universitaires Romandes, Lausanne, Switzerland.
19. Rietveld P., Ouwersloot H. 1992. Ordinal data in multicriteria decision making, a stochastic dominance approach to siting nuclear power plants. *European Journal of Operational Research* 56:249–262.
20. Rogers M., Bruen M., Maystre L-Y. 1999. *Electre and Decision Support – Methods and Applications in Engineering and Infrastructure Investment.* Kluwer, Dordrecht.
21. SMAA-TRI: A parameter stability analysis methods for ELECTRE TRI. In I. Linkov, G.A. Kiker, R.J. Wenning (eds.) *Environmental Security in Harbors and Coastal Areas,* NATO Security through Science Series – C: Environmental Security, Springer 2007, 217–232.
22. Tervonen T., Figueira J. 2006. A Survey on Stochastic Multicriteria Acceptability Analysis Methods. Research Report 1, The Institute of Systems Engineering and Computers (INESC Coimbra), Coimbra, Portugal, 27 p.
23. Tervonen T., Figueira J., Lahdelma R., Salminen P. 2004. An inverse approach for ELECTRE III. Research Report 20, The Institute of Systems Engineering and Computers (INESC Coimbra), Coimbra, Portugal, 17 p.
24. Tervonen T., Hakonen H., Lahdelma R. 2008: Elevator planning using Stochastic Multicriteria Acceptability Analysis. *Omega* 36(3):352–362.
25. Tervonen T., Lahdelma R. 2007: Implementing Stochastic Multicriteria Acceptability Analysis. *European Journal of Operational Research* 178(2):500–513.
26. Vincke Ph. 1992. *Multicriteria Decision-Aid.* Wiley, New York.
27. Yu W. 1992. Aide multicritère à la décision dans le cadre de la problématique du tri : Concepts, méthodes et applications. Ph.D. thesis, Université Paris-Dauphine.

ORDINAL MEASUREMENTS WITH INTERVAL CONSTRAINTS IN THE EIA PROCESS FOR SITING A WASTE STORAGE AREA

R. LAHDELMA

University of Turku, Department of Information Technology
Joukahaisenkatu 3–5
FIN-20520 Turku, Finland
risto.lahdelma@cs.utu.fi

P. SALMINEN

University of Jyväskylä, School of Business and Economics
P.O. Box 35
FIN- 40014 Finland

Abstract: We describe application of multicriteria decision analysis (MCDA) in the Environmental Impact Assessment (EIA) procedure for siting a waste storage area. The world's largest biofuel-based combined heat and power (CHP) plant was built in Pictarsaari, Finland. The plant produces certain byproducts, such as flue dust and bottom ash, which need to be processed and stored to prevent releases to the environment. Different storage sites will have different impacts on people, nature, and the economy. Cardinal measurement of the different impacts was considered too costly. Instead, ordinal measurement of all criteria was applied; i.e., experts ranked the alternative storage sites according to each criterion. In addition to pure ordinal information, the experts were able to state that some ordinal intervals were more significant than others. To properly treat this kind of information, we extended the Ordinal Stochastic Multicriteria Acceptability Analysis (SMAA-O) MCDA method and used it to evaluate the acceptability of the different sites. Three of the alternatives were found acceptable subject to some valuations and one of them was acceptable subject to a wide range of different valuations. This alternative was then chosen for implementation.

1. Introduction

We consider the problem of siting a storage area for the byproducts of a biofuel-based combined heat and power (CHP) plant. Legislation requires application of the environmental impact assessment (EIA) procedure for this kind of problem. The EIA procedure involves considering different alternative actions; evaluating their effects on nature, the built environment, humans, and society; finding possibilities to decrease harmful effects; and gathering the opinions of different interest groups. The EIA process thus produces a large amount of information to be considered when comparing the different alternatives and deciding which one to implement. Choosing the best alternative can therefore be a difficult problem, because it involves considering several alternatives, multiple criteria, and the opinions of different interest groups.

Comparing alternatives requires a systematic planning process. Multicriteria decision analysis (MCDA) methods can be extremely useful in different phases of EIA processes. MCDA can be used early in the process to describe the preconditions that allow each alternative to be chosen, or later to suggest which alternative to choose [14]. During several past EIA procedures in which we have participated, there has been a clear need for descriptive MCDA, which does not dictate the choice, but leaves the final decision to decision makers (DMs). Application of MCDA methods generally requires the following phases:

1. Defining the problem and identifying the stakeholders (DMs and affected interest groups)

2. Identifying the alternatives and their possible impacts

3. Formulating the decision criteria

4. Measuring the criteria and collecting preference information

5. Using the MCDA method

6. Interpreting the results and forming recommendations

These phases are not necessarily performed in strict sequence. Some phases may be executed in parallel and later phases may reveal the need to return to earlier phases. For example, forming and measuring the criteria can reveal that some alternatives should be rejected as infeasible and also inspire the development of new alternatives.

A characteristic for this application was that the different criteria were measured ordinally; i.e., the experts ranked the alternatives with respect to each criterion. The experts were also able to express that some ordinal intervals were more significant than others. To properly account for such measurements, we developed an extended variant of the SMAA-O MCDA method.

2. The Environmental Impact Assessment Procedure

During the application, the EIA procedure was in Finland constrained both by EU directive [2] and national legislation [12]. EIA is a procedure that identifies, predicts, evaluates, and mitigates the biophysical, social, and other relevant effects of development proposals prior to major decisions being taken and commitments made [6]. The goal of the EIA procedure is to produce information about environmental effects for the planning and decision-making phases as well as to increase public participation and keep citizens informed. The EIA procedure includes:

- Describing the actual project and alternative choices for accomplishing it
- Evaluating the current state of the environment
- Evaluating project impacts
- Studying the possibilities to decrease harmful impacts
- Comparing the differences between the alternatives
- Making suggestions for a possible follow-up program

The procedure can be divided into different phases. First, the assessment program is defined. The assessment program describes how the actual assessments will be carried out. The program will be available for public inspection, and all interest groups can state whether the program covers their points of view or not. Based on the comments, the authorities may approve the program, or request modifications. In the second phase, the actual assessment is conducted according the approved program. At the end of this phase, the assessment report is compiled and published.

Based on the information collected during the EIA procedure, the DMs (those responsible for the project) choose the best alternative and apply for a permit to implement it. The DMs should try to choose a solution that is widely acceptable to different interest groups because the authorities may reject a poorly justified alternative. Also, different interest groups are eager to file complaints against a harmful alternative or an unsound decision process.

A good decision process has the following properties:

1. All feasible alternatives are considered.

2. All possible impacts are taken into account.

3. Both short- and long-term impacts are considered.

4. New alternatives and effects observed during the planning are included in the analysis.

5. The process must not be too expensive or time consuming.

6. The uncertainty/inaccuracy of the measurements is taken into account.

7. All participants should understand the impacts of different decision alternatives.

8. The process should be public, transparent, and rely on well organized interaction between the participants.

3. Case Description

3.1. BACKGROUND

Alholmens Kraft Ltd. was building the world's largest biofuel-based CHP plant in Pietarsaari on the western coast of Finland. The cogeneration plant can use multiple fuels to produce electric power, district heating, and industrial steam. As byproducts, the power plant generates bottom ash, flue dust, and various desulfuration products [16]. The amount of these byproducts is 125,000 t per year. The majority of these byproducts are recycled. Ash can be utilized, for example, in building roads, production of prefabricated construction units, and landscaping [13]. Before recycling, these byproducts must be stored so that environmental impacts can be controlled.

The goal of the EIA process is to evaluate different locations for processing and storing power plant byproducts. The required area is about 30 ha. By legislation, building such a large storage site requires application of the EIA procedure. This area is sufficient for at least 25 years of operation. Future development in recycling technology will most likely increase the operation time of the storage area.

3.2. THE TREATMENT AREA

The volume of compressed flue dust is about 65,000 m^3 per year. The bottom ash requires about 7,000 m^3 per year. The placement area for the byproducts will consist of:

- Placement area of flue dust   21 ha
- Placement area of bottom ash   4 ha
- Maintenance and storing field   0.2 ha
- Water treatment area   1 ha
- Traffic areas   0.5 ha
- Totally about   30 ha

An exclusion area of 50–100 m in width is formed around the functional area. The exclusion area acts as a visual obstruction and damps the wind in the functional area. This requires a dense tree stand, preferably pine. The total area thus needed is 50–70 ha. The flue dust and bottom ash are stored separately. The placement area for bottom ash is built in two to three phases while the placement area for flue dust is built in three to four phases. The phased building decreases the amount of leachate.

Surface soil is removed from the storage area down to the mineral soil layer. In soft soil areas, carrying capacity is strengthened by gravel. The bottom of the storage area consists of a 0.5-m layer of compressed ash. The surface of this layer is modified so that leachate is channeled away from the storage area. Therefore, a 0.5-m-thick water-conducting drainage layer is built on top of the ash mixture. This layer is covered by a filter cloth and a 0.3-m-thick gravel layer for traffic. Leachate is channeled away by a single route. Cutoff drains will prevent other waters from reaching the area. Rainwater is collected from the storage area by underdrains into circumference drains. From there the water is channeled into a balancing reservoir. One side of the balancing reservoir is built into an infiltration polder, through which the water is filtered into an outlet ditch.

Storage is carried out as layered embankment filling. The thickness of a layer is about 1.5 m. To minimize the absorbed rainwater in the storage area, the surface slopes towards the edges. Each storage area is landscaped after the planned filling level is reached. Landscaping is done by covering the area with 0.2 m of humus materials and planting grass and bushes. After that, the area is left for natural reforestation.

4. Problem Formulation

4.1. CITIZEN PARTICIPATION AND TRANSPARENCY OF THE PLANNING PROCESS

One purpose of the EIA is to inform concerned citizens about impacts to their environment. Therefore, the EIA planning process should be open and transparent. Different means are used to collect the opinions of citizens, interest groups, authorities, municipalities, and those who are responsible for the project. Identifying important impacts for the different interest groups is essential for reaching the objectives of the EIA process.

All citizens whose living conditions are affected are allowed to participate in the EIA process. According to the law, citizens can state their opinion about:

- The need to study certain impacts of the project in the assessment program
- The sufficiency of these studies when the program is reported

Alternative ways to hear the opinions of different interest groups include:

1. Supervisory group work
2. Follow-up group work
3. Public meetings
4. Inquiries
5. Key-person interviews

In this case, the first three techniques were applied. The supervisory group consisted of the representatives of the town of Pietarsaari, the municipalities of Pedersöre and Luoto, the Southwest Finland Regional Environment Centre, Oy Alholmens Kraft Ab/PVO-Engineering Oy, and UPM Kymmene Oyj.

The follow-up group consisted of interest groups that may be affected by the project. In addition to the supervisory group, the follow-up group included representatives of several different associations, such as forestry associations and a local lung injury association.

4.2. CHOOSING THE ALTERNATIVES

Several alternative locations for the storage site were studied in the neighborhood of Pietarsaari. The starting points of the evaluation were the requirements stated by the waste law, feasible transportation connections, current land use, and environmental circumstances. The supervisory group formed the alternatives that were included in the assessment program. The initial locations were chosen so that the main road network could reach them. These locations were also chosen so that they are not close to valuable natural conservation targets, important groundwater areas, cultural landscapes, residences, or other objects that might be disturbed.

Initially, five alternative locations were suggested to be evaluated in the EIA program: Luoto I, Luoto II, Söderängsmossen, Hjortermossen, and Lepplax. Much additional information about the alternatives was obtained in the statement phase of the evaluation program. Based on these statements, the alternatives Luoto I and Luoto II were removed from the final evaluation. A new area, Bussimossen, was suggested as a feasible alternative. So, the final set of alternatives consisted of the following four locations:

- Söderängsmossen (S)
- Hjortermossen (H)
- Lepplax (L)
- Bussimossen (B)

According to the EIA legislation, the actual project should be compared also to a so-called zero-alternative, meaning that the project is not carried out. However, the zero-alternative is infeasible, because it can appear only if the power plant is not built. It is therefore not necessary to evaluate this kind of an alternative in the EIA process. If none of the included four alternatives can be chosen, new locations must be sought further from Pietarsaari, and in this case, a new EIA is required.

4.3. CHOOSING THE CRITERIA

The criteria were identified in phases. First, a group of experts identified several impacts caused by the project. Then the supervisory group and the follow-up group stated impacts that they felt were important. Then, a public meeting was arranged where 140 citizens participated. Based on the information obtained from the above-mentioned groups, the final set of criteria was formed. The experts measured the criteria as described below. Criteria that had no significance or had the same values for all alternatives were removed from further consideration, such as—for example—releases from traffic and effects on services in the region. The included criteria were the following:

- (C1) Quality of the groundwater
- (C2) Quality of the surface water
- (C3) Effects on environmental habitats and fragmentation
- (C4) Traffic noise
- (C5) Road safety
- (C6) Scenery
- (C7) Business
- (C8) Recreation
- (C9) Dust and other noise than that of traffic
- (C10) Compatibility with current land use
- (C11) Compatibility with planned land use

5. MCDA Methodology

To compare the different alternatives measured in terms of multiple criteria, we applied a variant of the SMAA method. We applied this method because we had applied it successfully in EIA applications and because it was possible to extend the method to handle ordinal criteria measurements with interval constraints.

The SMAA family of methods, including the original SMAA (SMAA-1) [4, 7], SMAA-3 [3], and SMAA-2 [5, 8] was developed for discrete multicriteria problems, where criteria data is uncertain or inaccurate and where it for some reason is difficult to obtain accurate weight information from DMs. These methods are based on exploring the weight space through Monte Carlo simulation in order to describe the valuations that would make each alternative the preferred one, or that would give a certain rank for an alternative. The SMAA and SMAA-2 methods apply assumed partial value/utility functions and stochastic criteria while SMAA-3 uses a double-threshold model as in the ELECTRE III decision aid (see, for example, [11, 17]). For a recent survey of SMAA methods and applications, see Tervonen and Figueira [15].

In the current problem, the earlier versions of SMAA could not be used since the criteria data is expressed here as criterion-wise ranks. This type of problem led to the development of the SMAA-O method [10]. For a real-life application of SMAA-O, see [9]. To properly treat the interval constraints that the experts provided in this application, we had to further extend SMAA-O, as described below.

As for the earlier versions of SMAA, the results obtained from analysis are descriptive. The DMs are given *rank acceptability indices* for each alternative, describing the variety of preferences that support an alternative for the best rank or any particular rank. This information can be used for classifying the alternatives as more or less acceptable or not acceptable at all. SMAA also computes *central weights,* describing the most typical preferences that make an alternative preferred. It is also possible to measure with *confidence factors* whether the problem data is accurate enough for decision making.

Consider an MCDA problem with m alternatives measured in terms of n different criteria. SMAA-O is based on an assumed additive utility function:

$$u(x_i, w) = \sum_{j=1}^{n} w_j u_j(x_{ij}), \qquad w \in W, \tag{1}$$

where the vector of weights w represents the preferences of the DM, x_i is the ith alternative, and $x_{ij} \in \{1,..., j^{\max}\}$ is the rank of the ith alternative with respect to the jth criterion. The partial utility function $u_j(.)$ is a mapping from ordinal criteria values to a linear scale. However, neither the weight vector nor the mapping needs to be known precisely. Instead, the different alternatives are analyzed by simultaneous simulation of different weight vectors and mapping functions. An appropriate weight distribution $f(w)$ is used in the simulation to model the DMs' unknown or partly known preferences. Absent weight information is represented by a uniform weight distribution in the set of non-negative normalized weights $W = \{w \in R^n: w \geq 0 \text{ and } \Sigma_j w_j = 1\}$.

Similarly, all the different monotonically decreasing mappings $u_j(x_{ij})$ → [0,1] are simulated to model the unknown ordinal-to-cardinal scale mappings. For criterion j with $j^{\max}$ different ranks, we generate a mapping where the best rank (1) corresponds to cardinal value 1, the worst rank ($j^{\max}$) corresponds to cardinal value 0, and the intervals between subsequent cardinal values follow a normalized uniform joint distribution (see David [1], Section 5.4). Generation is implemented by drawing at each iteration $j^{\max}$-2 distinct random numbers from the uniform distribution in the interval [0,1] and sorting the numbers together with '1' and '0' into descending order to get $1 = \gamma_{j1} > \gamma_{j2} > ... > \gamma_{j,jmax} = 0$. These numbers are then used as a sample of stochastic cardinal criteria values ξ_{ij} such that for each alternative i, ξ_{ij} is set equal to γ_{jr}, where r equals x_{ij}. Thus, the partial utility functions obtain random values

$$u_j(x_{ij}) = \xi_{ij}, \tag{2}$$

where $\xi \in X$ follow a joint distribution $f(\xi)$.

It is possible to modify the above process to consider additional constraints, such as *ordinal interval constraints*. An ordinal interval constraint states that for two different ranks, r and s, the interval between ranks r and $r + 1$ is more significant than between s and $s + 1$. This kind of information is represented by cardinal values whose intervals satisfy corresponding inequality constraints; i.e., $\{\gamma_{j,r}\text{-}\gamma_{j,r+1} > \gamma_{j,s}\text{-}\gamma_{j,s+1}\}$. For example, the experts may consider the best alternative much better than the second best, but the second best only a little better than the third. This would result into the constraint $\{\gamma_{j,1}\text{-}\gamma_{j,2} > \gamma_{j,2}\text{-}\gamma_{j,3}\}$. This kind of interval constraint can be easily considered by rejecting during the simulation those ordinal-to-cardinal mappings that do not satisfy the constraints.

Hereafter the computation is similar to that in the SMAA-2 method. The rank of each alternative is determined as an integer from the best rank ($=1$) to the worst rank ($=m$) by means of a ranking function

$$rank(\xi_i, w) = 1 + \sum_{k=1}^{m} \rho(u(\xi_k, w) > u(\xi_i, w)) = 1 + \sum_{k=1}^{m} \rho(\xi_k w > \xi_i w), \tag{3}$$

where $\rho(\text{true}) = 1$ and $\rho(\text{false}) = 0$. The SMAA-O method is based on analyzing the so-called sets of *favorable rank weights* $W_i^r(\xi)$ defined as

$$W_i^r(\xi) = \{w \in W : rank(\xi_i, w) = r\}, \tag{4}$$

The favorable rank weights are simply the (stochastic) set of weights that result in rank r for an alternative. All descriptive measures are then computed as properties of $W_i^r(\xi)$.

The rank acceptability index b_i^r is the expected volume of $W_i^r(\xi)$. The rank acceptability index is a measure of the variety of different valuations granting alternative x_i rank r, and is computed as a multidimensional integral over the criteria distributions and the favorable rank weights using

$$b_i^r = \int_X f(\xi) \int_{W_i^r(\xi)} f(w)\, dw\, d\xi. \tag{5}$$

The rank acceptability indices can be examined graphically in order to compare how different varieties of weights support each rank for each alternative.

Candidates for most acceptable alternatives should be those with high (clearly non-zero) acceptabilities for the first rank. When seeking compromises, alternatives with large acceptabilities for the worst ranks should be avoided. However, the acceptability indices should not be used for ranking acceptable alternatives, because the magnitude and mutual order of the acceptability indices depends on the chosen (assumed) distributions and scaling of criteria.

The central weight vector is the expected center of gravity of $W_i^1(\xi)$ (favorable first-rank weights), and the confidence factor is the probability for the alternative to obtain the first rank if the central weight vector is chosen. We did not apply central weight vectors and confidence factors in this application. For a more complete description of these measures, refer to [8].

6. Measuring the Criteria

Cardinal measurement of the different impacts was considered too costly. Instead, ordinal measurement of the criteria was applied. This means that the experts ranked in consensus the different alternative storage sites under each criterion. The same rank was given to alternatives when the experts considered them about equally good, or disagreed about which one was better. The experts were also able to state that some ordinal intervals were more significant than others. Such augmented ordinal measurements are more precise than pure ordinal measurements, but not as precise as cardinal measurements.

Below we present the (pure) ordinal measurements of the criteria, which are also summarized in Table 1.

- (C1) Groundwater

  The placement area is built so that no harmful emissions will reach the groundwater in any of the alternatives. None of the alternatives reside in a classified groundwater area. However, in Lepplax, groundwater is refreshed and therefore it is considered to be the worst choice from this point of view. The ranking is (1) Hjortermossen, Bussimossen, Söderängsmossen; (2) Lepplax.

TABLE 1. The Criterionwise Ranks of the Alternatives. S = Söderängsmossen, H = Hjortermossen, L = Lepplax, B = Bussimossen.

Alt.	C1	C2	C3	C4	C5	C6	C7	C8	C9	C10	C11
S	1	2	3	1	1	1	1	3	2	1	1
H	1	1	2	3	3	3	2	3	3	4	3
L	2	1	1	2	2	2	3	1	4	2	1
B	1	1	2	2	2	1	2	2	1	3	2

- (C2) Surface water

  No large differences appear between the alternatives for this criterion. In Söderängsmossen the leachate may affect the Gubbträsk water area. The ranking of the alternatives for this criterion is thus (1) Hjortermossen, Lepplax, Bussimossen; (2) Söderängsmossen.

- (C3) Effects on environmental habitats and fragmentation

  The least amount of habitat fragmentation and other harmful effects to the polymorphism of nature will occur in Lepplax. The reason for this is that it is a quarry area where the natural environment has already degraded. Hjortermossen and Bussimossen are forest areas without significant natural values. The common alder wild wood and flying squirrel populations close to Söderängsmossen are vulnerable to disturbance. The ranking under this criterion is (1) Lepplax; (2) Hjortermossen, Bussimossen; (3) Söderängsmossen.

- (C4) Traffic noise

  Söderängsmossen is the best alternative, because it is closest to the power plant and no one would be disturbed by the noise. The choice of Lepplax or Bussimossen would cause small noise effects while the route to Hjortermossen adjoins several facilities likely to be disturbed by the noise; for example, a day nursery. The ranking is (1) Söderängsmossen; (2) Lepplax, Bussimossen; (3) Hjortermossen.

- (C5) Road safety

  Analysis of the alternatives under this criterion generated the same ranking as the traffic noise criterion, for the same reason: (1) Söderängsmossen; (2) Lepplax, Bussimossen; (3) Hjortermossen.

- (C6) Landscape

  All alternatives are mostly sheltered by forests. No harm to the landscape is caused when Söderängsmossen or Bussimossen is chosen. The Lepplax alternative may be seen from the village and from the main road. Hjortermossen was considered to be the worst under this criterion since its visibility is the highest. The ranking thus is (1) Söderängsmossen, Bussimossen; (2) Lepplax; (3) Hjortermossen.

- (C7) Business

  Söderängsmossen, Hjortermossen, and Bussimossen are located in forested areas. Any economic loss will be compensated in the purchase price. However, the region of Söderängsmossen has the best location and infrastructure for the business of refining byproducts. In the Lepplax alternative, the quarry will be closed. The ranking is (1) Söderängsmossen; (2) Hjortermossen, Bussimossen; (3) Lepplax.

- (C8) Recreation

  The quarry area of Lepplax has no recreational use. The recreational use of Bussimossen was estimated to be smaller than in the areas of Söderängsmossen and Hjortermossen. The ranking is (1) Lepplax; (2) Bussimossen; (3) Söderängsmossen, Hjortermossen.

- (C9) Dust and noise

  Dust and noise effects have the least potential to disturb in the Bussimossen alternative. In Söderängsmossen, a waste treatment facility already exists; in Hjortermossen, a recreation route exists. Lepplax is located closest to residences and the village, and therefore it is the worst alternative. The ranking is (1) Bussimossen; (2) Söderängsmossen; (3) Hjortermossen; (4) Lepplax.

- (C10) Compatibility with the current land use

  This criterion was evaluated based on infrastructure and the location of residences and businesses. The ranking is (1) Söderängsmossen; (2) Lepplax; (3) Bussimossen; (4) Hjortermossen.

- (C11) Compatibility with the planned land use

  Söderängsmossen and Lepplax fit best with the planned land use. Bussimossen will be a neutral alternative while the regional plan defines a landscape protection area close to Hjortermossen. The ranking is (1) Söderängsmossen, Lepplax; (2) Bussimossen; (3) Hjortermossen.

7. Results

It is easy to see which individual criteria favor each alternative in Table 1. Söderängsmossen is the best choice when groundwater (C1), traffic noise (C4), road safety (C5), landscape (C6), business (C7), and compatibility with the current (C10) and planned land use (C11) are considered. Hjortermossen obtains only shared first ranks on groundwater (C1) and surface water effects (C2). Lepplax is favored in terms of surface water effects (C2), effects on environmental habitats and fragmentation (C3), recreational use (C8), and compatibility with the planned land use (C11). The choice of Bussimossen is supported by groundwater (C1) and surface water effects (C2), landscape (C6), and dust and noise (C9).

The choice between the alternatives depends on how the different criteria are valued. This application included numerous interest groups, with very different opinions about the importance of different criteria. Therefore, it was not possible to reach consensus on the weighting of different criteria.

In this situation it was meaningful to search for a widely acceptable compromise alternative to suit many different preferences. For this reason, we applied SMAA without weight information.

Table 2 and Figure 1 present the acceptability indices for each alternative and rank. The analysis was implemented using 200,000 iterations in a Monte Carlo simulation. Based on the results, it is possible to conclude that Hjortermossen should not be chosen since its acceptability for the first rank is zero and for the second rank almost zero. The acceptability of the Söderängsmossen is the greatest; more than half (51.5%) of the different possible weights make it the most preferred alternative. However, Bussimossen with 33.3% and Lepplax with 15.2% first-rank acceptability can also be considered acceptable under suitable preferences.

TABLE 2. The Rank Acceptability Indices (%) of the Alternatives with Pure Ordinal Criteria.

Alternative	Rank 1	Rank 2	Rank 3	Rank 4
S	51.5	26.8	18.5	3.2
H	0	0.7	12.2	87.1
L	15.2	23.8	51.3	9.7
B	33.3	48.8	17.9	0

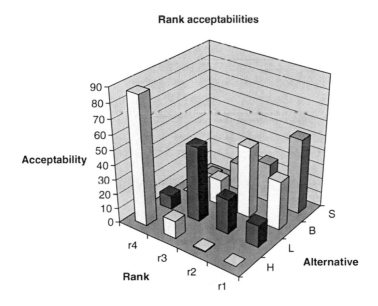

Figure 1. Rank Acceptability Indices (%) of the Alternatives with Pure Ordinal Criteria.

The initial results are not conclusive, because three of the four alternatives are at least in principle acceptable. Note that the initial results were obtained using extremely vague information: absent weight information and only ordinal measurements of criteria. More conclusive results require more accurate information. It is possible make either the preference (weight) information or the criteria information (or both) more accurate. Because of the high controversy in the valuations, it was considered easier to make the criteria measurements more accurate.

Instead of introducing cardinal measurements for the criteria, it was decided to strengthen the ordinal measurements by introducing ordinal interval constraints. This kind of additional information is very easy to produce compared to true cardinal measurements. The experts were able to specify ordinal interval constraints in consensus for the following criteria.

- **(C3)** Effects on environmental habitats and fragmentation: the interval 1–2 is greater (more significant) than 2–3.
- **(C4)** Traffic noise: interval 1–2 is smaller (less significant) than 2–3.
- **(C6)** Landscape: interval 1–2 is smaller than 2–3.
- **(C8)** Recreation: interval 1–2 is greater than 2–3.
- **(C9)** Dust and noise: interval 2–3 is smaller than 1–2, which is smaller than 3–4.

Table 3 and Figure 2 present the rank acceptability indices for the alternatives when these additional constraints are taken into account.

The most significant change is that the Lepplax and Bussimossen alternatives reverse their order according to their first rank acceptabilities. The first rank acceptability of Söderängsmossen increases slightly. Overall, the changes do not make the choice much easier, because all alternatives except Hjortermossen are still acceptable. However, because Söderängsmossen has confirmed its position as the most widely acceptable alternative, this indicates that it is a fairly robust and safe choice. For this reason, the DMs decided to select Söderängsmossen as the location for the storage area.

TABLE 3. The Rank Acceptability Indices (%) with Ordinal Interval Constraints.

Alternative	Rank 1	Rank 2	Rank 3	Rank 4
S	53.2	26.2	18.2	2.4
H	0	0.4	9.9	89.7
L	25.1	24.4	42.5	7.9
B	21.7	48.9	29.4	0

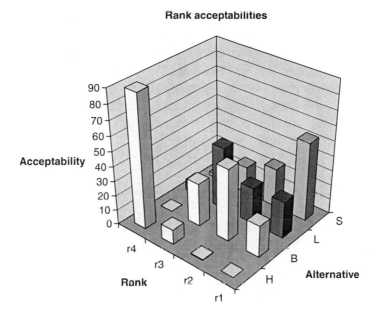

Rank acceptabilities

Figure 2. Rank Acceptability Indices of the Alternatives with Ordinal Interval Constraints.

8. Conclusion

We have described the application of MCDA in a real-life EIA process for locating a storage facility for the byproducts of a biofuel CHP plant in Pietarsaari, Finland. The SMAA-O method was used for evaluating the alternative locations. Due to high controversy in the valuations of different interest groups, the analysis was performed without weight information. Initially, the alternatives were measured with respect to different criteria only on ordinal scales; i.e., experts ranked the alternatives criterion-wise. Later, the experts provided more accurate measurements by stating that some ordinal intervals were more significant than others.

From the four alternative locations, three were found acceptable. The choice between these acceptable alternatives depends on the valuation of the criteria. The Söderängsmossen alternative received in both analyses more than half of the acceptability for the first rank. This means that it is a robust and safe choice. This alternative was later selected for implementation.

The use of ordinal criteria in multicriteria analysis can often save a significant amount of work compared to making cardinal measurements. Experts can often provide such rankings in consensus based on qualitative reasoning. Of course, such information is by nature less accurate than cardinal measurements, and therefore it may not always be sufficient for conclusive results. Ordinal interval constraints, as in this application, constitute a promising approach to providing better accuracy than pure ordinal information, but still with much less effort than with cardinal measurements. Ordinal interval constraints can also be provided by experts using qualitative arguments. Our experiences indicate that experts may be able to easily reach consensus about interval constraints. When consensus cannot be reached for some interval constraints, it is always possible to resort to pure ordinal measures.

Acknowledgements

This research is supported, in part, by the Academy of Finland. We thank Paavo Ristola Consulting Engineers Ltd for its cooperation.

References

1. David, H.A. 1970. Order Statistics. Wiley, New York.
2. EU. 1997. COUNCIL DIRECTIVE 97/11/EC of 3 March 1997 amending Directive 85/337/EEC on the assessment of the effects of certain public and private projects on the environment. The Council of the European Union.
3. Hokkanen, J., Lahdelma, R., Miettinen, K., Salminen, P. 1998. Determining the implementation order of a general plan by using a multicriteria method. *Journal of Multi-Criteria Decision Analysis* 7:273–284.
4. Hokkanen, J., Lahdelma, R., Salminen, P. 1999. A multiple criteria decision model for analyzing and choosing among different development patterns for the Helsinki cargo harbor. *Socio-Economic Planning Sciences* 33:1–23.
5. Hokkanen, J., Lahdelma, R., Salminen, P. 2000. Multicriteria decision support in a technology competition for cleaning polluted soil in Helsinki. *Journal of Environmental Management* 60:339–348.
6. IAIA. 1999. *Principles of EIA Best Practice*. IAIA - International Association for Impact Assessment.
7. Lahdelma, R., Hokkanen, J., Salminen, P. 1998. SMAA - Stochastic Multiobjective Acceptability Analysis. *European Journal of Operational Research* 106(1):137–143.
8. Lahdelma, R., Salminen, P. 2001. SMAA-2: stochastic multicriteria acceptability analysis for group decision making. *Operations Research* 49(3):444–454.
9. Lahdelma, R., Salminen, P., Hokkanen J. 2002. Locating a waste treatment facility by using stochastic multicriteria acceptability analysis with ordinal criteria. *European Journal of Operational Research* 142:345–356.
10. Lahdelma, R., Miettinen, K., Salminen, P. 2003. Ordinal criteria in Stochastic Multicriteria Acceptability Analysis (SMAA). *European Journal of Operational Research* 147:117–127.
11. Maystre, L.Y., Picted, J., Simos, J. 1994. *Méthodes Multicritéres ELECTRE*. Presses Polytechniques et Universitaires Romandes, Lausanne, Switzerland.
12. Ministry of Environment. 1994. *The Act on Environmental Impact Assessment Procedure*. Helsinki, Ministry of Environment.
13. Niutanen, V., Korhonen J. 2002. Management of old landfills by utilizing forest and energy industry waste flows. *Journal of Environmental Management* 65:39–47.

14. Roy, B. 1996. *Multicriteria Methodology for Decision Aiding*. Kluwer, Dordrecht.
15. Tervonen, T., Figueira, J. 2007. A survey on Stochastic Multicriteria Acceptability Analysis methods. *Journal of Multi-Criteria Decision Analysis* (in press), DOI: 10.1002/mcda.407.
16. Vainikka, P., Helynen, S., Hillebrand, K., Nickull, S., Nylund, M., Roppo, J. Yrjas, P. 2004. Alholmens Kraft: optimised multifuel CHP with high performance and low emissions at Pietarsaari pulp and paper mills. Proceedings: PulPaper Conference, June 1–3, Helsinki, Finland.
17. Vincke, Ph. 1992. *Multicriteria Decision-Aid*. Wiley, New York.

BALANCING SOCIOECONOMIC AND ENVIRONMENTAL RISKS AND BENEFITS UNDER MULTIPLE STRESSOR CONDITIONS

A Case Study of the Powder River Basin in Wyoming

S. SHALHEVET

SustainEcon
126 Thorndike Street
Brookline, MA 02446, USA
sarit.shalhevet@gmail.com

Abstract: U.S. national energy policy calls for increasing natural gas production in order to increase environmental security, provide energy security, and increase national security by reducing dependence on imported energy. One of the major sources of natural gas production in the U.S. is the coalbed methane (CBM) gas development project in the Powder River Basin. However, this project is highly controversial and is considered one of the greatest environmental threats to Wyoming. The threat is caused by a large variety of chemical and non-chemical stressors, including physical stressors such as excavation, increased water volume, wind and water erosion, noise, dust, road construction, and increased traffic.

This paper presents a model for multiple stressor analysis of the ecosystem and the derived human values. The objective is to recommend an optimal management policy that maximizes economic profits while minimizing social costs and environmental damage. The adverse effects of the physical and chemical stressors in the Powder River Basin project are assessed under different feasible management alternatives for each of the 18 sub-watersheds involved in the project. The social, economic and environmental impacts on sustainable development are incorporated into a single score model using the economic valuation approach, which assigns a monetary value to environmental damage in order to compare different types of impacts.

The cost-benefit analysis shows that for the project as a whole, the total benefits are lower than the value of the environmental damage. However, an examination by area shows that for some sub-watersheds, the benefits outweigh the costs. Management alternatives that involve improved water treatment methods cost more than the value of their environmental benefits. The recommendation is to continue the project only in those sub-watersheds where the benefits outweigh the environmental costs, while maintaining the current water treatment methods.

I. Linkov et al. (eds.), Real-Time and Deliberative Decision Making.   415
© Springer Science + Business Media B.V. 2008

1. Introduction

The U.S. national energy policy calls for increasing natural gas production in order to increase environmental security, provide energy security, and increase national security by reducing the dependence on imported energy. Development of coalbed methane (CBM) as a natural gas resource in the U.S. began in the mid 1980s and has been increasing substantially since the mid 1990s [18]. The Powder River Basin, located in the states of Wyoming, Montana, and North Dakota, is the largest source of natural gas production in the U.S., supplying 431.3 million tons of coal in 2006, which is 35% of the country's total annual coal production. While the project will contribute to cleaner energy sources and may increase national security, it has many negative socioeconomic and environmental impacts as well.

Many of the project's impacts are a result of the method of CBM production, which is based on withdrawing water from wells to recover the methane gas from the coal beds. Currently, most of the water is disposed of through untreated surface discharge and land application disposal of the produced water. In several wells, however, other water management methods are used, including infiltration impoundments to dispose of the water, and a combination of passive and active water treatment methods. The water management alternative chosen for each well has a major impact on economic profits and well as the magnitude of social and environmental stressors involved.

This research focused on the impact of a proposed project to drill an additional 39,367 CBM wells and 3,200 conventional (non-CBM) oil/gas wells over the next 10 years. Including the ones existing or permitted, this will amount to a total of 51,391 wells by the end of 2011. The overall life of the project will be about 20 years, and will also include construction of well pads, roads, water, and gas pipes [18]. The project includes 18 sub-watersheds; the type of stressors and the degree of their impact vary by sub-watershed, and management policies can be determined separately for each one.

The objective was to develop management recommendations for the Powder River Basin project that will maximize the economic profits while minimizing the social costs and environmental damages caused by the multiple stressors associated with the project. This was achieved by constructing a model for optimal decision making that evaluates the benefits to companies and society against the environmental damage, including the effect on agriculture, on biodiversity, and on the landscape, under different water treatment methods. This model is based on measuring the relevant economic, social, and environmental stressors on a comparable scale by making an economic evaluation of the social and environmental impacts. The impact of each stressor is then incorporated into a model that takes into account the private and public profits from each part of the project.

2. Main Stressors and Their Effects

2.1. THE EFFECT OF MULTIPLE STRESSORS ON THE PUBLIC

Some of the effects of the project are positive. Extraction of CBM will provide the public with an cleaner source of energy, lowering the level of greenhouse emissions; it may reduce dependence on imported energy, thus increasing national security [1]. The companies involved in the project will profit and increase the government's income from federal taxes.

The effect on the local economy is mixed. The number of jobs and level of wages in the area will rise, increasing the overall income in the area by about $570 million for the total duration of the project [18]. However, as it will cause a temporary increase of up to 7% in the total population [18], this also places additional stress on the local public services, increasing the pressure on firemen, schools, and other services. Additionally, the temporary nature of the project makes the area more vulnerable to "boom and bust" economic cycles [1]. The local farmers and ranchers, however, bear the main brunt of the project: income from crops has decreased [16], and drilling makes some areas too dangerous for livestock. One rancher was quoted as saying that his 1,110-acre ranch in Wyoming has been "torn to hell" by the gas companies [7], and he has had to cut back his livestock operation by two-thirds.

The landscape has already been changed by multiple stressors. Before the project began, the area consisted of prairie land and other areas of undisturbed natural vegetation. The project, with its wells, roads, and facilities, scars the landscape, turning large parts into an unattractive industrial area. The U.S. Bureau of Land Management uses five classes of visual resource management (VRM) to classify the degree of visual sensitivity of each area according to how much aesthetic degradation of that area will affect the public. Approximately 3,000 wells are in areas classified as having a relatively high visual sensitivity [18].

Noise is a major stressor on the local public. Noise is one of the main factors influencing quality of life, and people often rate this as the most important environmental factor, even more important than air pollution and water quality [6]. Nearby residents complain that the noise from the compressor stations impairs their quality of life [12]. According to the U.S. Environmental Protection Agency, the level of acceptable environmental noise is up to 55 dBA [18], which is the noise level of normal conversation [6]. Residences within 1,500 ft of construction would experience temporary noise levels above 55 dBA during daylight hours, and within 2,000 ft the level would still be close to the maximum: 53–54 dBA. The anticipated noise level in rural areas is approximately 45–50 dBA during the day and 35–40 dBA during the night. While the public health would not be at risk from noise beyond 1,600 ft from project, it may still impact the quality of life at nearby residences [18].

Additional stressors are caused by the construction of more than 10,000 miles of new roads for the project [18], increasing traffic congestion, accidents, noise and air pollution. The total value of these impacts for the U.S. is about $730 billion [10]. and the impact of road construction on the valuation of the project is likely to be significant as well. Finally, multiple stressors affect public health; one of the major stressors is the increased air pollution caused by dust from the project [18].

2.2. THE EFFECT OF MULTIPLE STRESSORS ON THE ENVIRONMENT

The greatest effects of the project's stressors are on the local water resources. Pumping out water can deplete groundwater resources [1] and cause dewatering of local and regional aquifers [13]. Discharge of CBM produced water to surface drainage can cause several negative effects on water quality and quantity. The effect on water quality occurs because CBM produced water has increased salinity (measured by electrical conductivity, or EC) and a high level of sodicity (measured by the sodium adsorption ratio, SAR) [19]. Salinity is the most important factor in determining the quality water for irrigated agriculture [15], and sodicity compounds this problem by slowing the rate of water infiltration into the soil [9]. The environmental impact statement identified two limits of SAR and EC for each watershed: a least restrictive proposed limit (LRPL) and a most restrictive proposed limit (MRPL) for irrigated agriculture. In seven of the 18 sub-watersheds, the SAR levels will be higher than the LRPL due to the project, and in three of these seven sub-watersheds the level will be even higher than the MRPL.

The increased quantity of water due to the project may have some positive effects, including more water for livestock range [9], which is the major source of income from agriculture [16]. However, unlike natural water flows, which normally depend on the season, CBM produced water is discharged year-round with small fluctuations, increasing the frequency and magnitude of the flows. The increased flows can cause water erosion and increase sedimentation downstream, which may cause local flooding that will damage the nearby fields [18].

The main effects on the soils include wind erosion, water erosion, and slope hazards. Excavation for the project could cause slope steepening; combined with soil compaction and loss of vegetation, this could reduce the soil's resistance to water and wind erosion, and cause loss of organic matter in the soil. Most of the damage is concentrated in a few watersheds. The long-term surface disturbances associated with the project may be 85,000–95,000 acres [18].

The project, including the wells, roads, and other facilities, causes some permanent loss of vegetation. There are also some indirect effects caused by the changes of water flows mentioned above and by increased soil salinity.

A total of about 175,000 acres of vegetation will be disturbed by the project, mainly shortgrass prairie, mixed grass prairie, and sagebrush shrublands. About 3,500 acres of wetlands and riparian areas (1.5% of total wetlands) would be disturbed by the project; these disturbances would consist mainly of habitat loss, road impacts, and impacts from the changes in water quantity and quality [18].

The project may result in loss of terrestrial wildlife because of increased hunting and accidents in the area, change in habitat that may affect the predation rate, and habitat fragmentation caused by road construction and other human activities. The project would disturb about 160,000 acres of pronghorn range (about 2% of total range area), 130,000 acres of mule deer range (also about 2%), and 10,000 acres of white-tailed deer range (about 1%), as well as ranges of elk and moose [18].

Birds may be driven away by noise and other disturbances from the project. The project is expected to cause disturbances for the bald eagle, the mountain plover, and the sharp-tailed grouse. It will also disturb one species categorized as a "sensitive species," the greater sage grouse, with 15% of sage grouse leks in the 0.25-mile buffer surrounding oil and gas developments, and 81% of the leks in the 2.0-mile buffer [18].

Finally, aquatic life will be affected by the timing and quantity of stream flows, sedimentation, and spills of fuel and drilling fluid. The project causes high concentrations of selenium in the reservoirs, which are toxic to fish. The high water salinity level increases the accumulation of dissolved metals in the sediments of the reservoirs, which may be toxic for diving ducks that ingest sediments [18].

2.3. SUMMARY OF MAIN EFFECTS

The main benefits associated with the project include:

- Environmentally friendly: Supplies a cleaner source of energy (low-sulfur coal)
- Energy security: Reduces the dependence on foreign energy
- Economically profitable: Increases local income

The main stressors associated with the project include:

- Water withdrawal reduces groundwater supply.
- Water discharge into streams increases water erosion and flooding.
- Discharged water has high salinity and sodicity, causing negative impact on water quality in streams.
- Construction, roads and drilling disturb and endanger livestock and wildlife, and cause vegetation loss.

- Aesthetic degradation from facilities and drilling.
- Noise from compressor stations disturbs residents and drives away wildlife.
- Dust from the project increases air pollution.
- Excavation causes slope steepening, causing reduced soil resistance to wind and water erosion.

3. Methodology

The methodology is based on previous models combining financial and environmental aspects [5], and on data from the project's environmental impact statement [18]. A model for multiple stressor analysis of the ecosystem and the derived human values was constructed in order to recommend an optimal management policy that maximizes economic profits while minimizing social costs and environmental damage. The adverse effects of the physical and chemical stressors were assessed under the feasible water management alternatives for each of the 18 sub-watersheds involved in the project. The social, economic, and environmental impacts on sustainable development were incorporated into a single score model using the economic valuation approach, which assigns a monetary value to environmental damages in order to compare different types of impacts. The net value of the project is the sum of the company profits and public income (taxes), minus the economic value of the impact of multiple stressors. The model results show the total net value of each treatment policy for each watershed, considering both the company profits and the costs of the stressors resulting from the project.

The social value to the public includes use and non-use (existence) values. Use value can be measured by the cost incurred from use or the income derived from it. For example, the use value of forested areas includes the value for the area's residents, usually measured by the effect on the property value, and the value for visitors, usually measured by tourist expenditures. The use value of wildlife can include the value of hunting and wildlife photography, which is often measured by expenditures on travel and equipment used [3]. Existence value is defined as the value derived from reserving the option of future use or from just knowing it is there [14]. It is commonly measured by surveys of willingness to pay to preserve the existence of that option, for example, to preserve the region's natural forest scenery or the local wildlife.

Five major indicators of the project's impact were evaluated. These included two financial indicators, the companies' profits and the income from taxes (assumed to be a public benefit); and three stressors, the impacts on agriculture, on biodiversity and on the landscape. The stressors' impact on agriculture was measured through their impact on the income from agricultural land, which is mostly rangeland. The stressors' impact on biodiversity was measured by multiplying the resulting reduction in wildlife by the existence value of wildlife preservation. The stressors' impact on the landscape was measured by multiplying the resulting decrease in forested area by the existence value of forest preservation.

The project's profits were calculated using financial data (costs, taxes, and price) from Goerold [4] and physical data (number of wells and amount of gas and water produced) from the environmental impact statement [18]. The company profits are revenues minus production costs, treatment costs, and taxes. Income to society is the total income from taxes (state and federal) and other royalties. Wages should also be counted as social income, but were left out due to inadequate data. The profit to society ("social profits") was the income minus the environmental costs (as taxes can compensate for the value of environmental damage). The total profit from the project was the profit to the companies plus the profit to society. The data was calculated annually for 16 years and discounted to net present value at an 8% interest rate.

3.1. MULTI-STRESSOR EVALUATION

The environmental costs were calculated as the sum of the cost of damage to soils and water, wildlife disturbance, and landscape degradation.

The cost of damage to soils and water or each watershed was based on the resulting changes in income from crops and livestock. Saline water is dumped onto agricultural areas in the basin, and may sodify the soil, causing a reduction of water uptake into the soil, increasing water runoff, and reducing crop yields [12]. The main crops in the region include alfalfa, wheat, and oats. Field crops and forage crops are generally tolerant or moderately tolerant of salinity; wheat is highly tolerant of salinity [15]. The effect on livestock range is controversial, ranging from positive effects (supplying more water in an arid area) to negative effects due to disturbed areas and ditches [7].

Most of the region's agricultural income is from livestock rather than crops. Between 1997 and 2001, the total number of cattle remained stable, but the number of sheep and lambs decreased by 20%. Both the total cultivated area and the yield per acre decreased between 1997 and 2001 in all the project's counties; the total income from crops under fixed prices decreased significantly. These changes are probably partly caused by the beginning of the project, as production in other parts of the region was not affected. The loss of income from agriculture at each watershed within the county was estimated based on USDA data on crops and livestock in each county, the

agricultural acres in that watershed disturbed by the project for crops, and
rangeland disturbed for livestock [16, 17].

The valuation of wildlife disturbance was based on the existence value of
wildlife preservation in the Powder River Basin for Wyoming residents, which
was estimated based on willingness-to-pay surveys [2]; the value by watershed
was calculated according to the reduction in wildlife effectiveness in each
area. As pronghorn antelopes have a high preservation value [8], and the
pronghorn are among the animals most affected by the project [18], the reduction of pronghorn effectiveness served as a measure of the level of reduction
in wildlife effectiveness, and the value by watershed was calculated according
to the percent of reduction in pronghorn effectiveness in each watershed.

The valuation of landscape degradation was based on the existence value
of forested areas, using surveys of the willingness to pay for preservation of
forested areas [11]. The value per watershed was calculated according to the
acres of forested areas damaged in each watershed [18].

4. Results and Discussion

The results for the project as a whole are presented in Figure 1, in terms of
total net present value of the project for 16 years. The figure presents the
impact of different water management alternatives on the companies' profits,
on the total economic valuation of the impacts of the relevant stressors, and
on total net value of the project. Under the current water treatment strategy,
the total companies' profits from the project are positive, but improved water

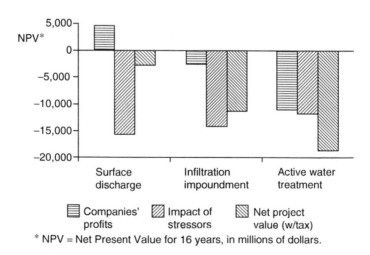

Figure 1. Economic Valuation by Water Management Alternative.

treatment methods result in a total net economic loss. In any case, the economic valuation of the stressors is higher than the companies' profits, resulting in a total negative net value from the project under every alternative.

Although improved water treatments do reduce the impact of the stressors, the negative impact of improved treatments on the companies' profits is higher than their contribution to the valuation of the multiple stressors. Progressing from surface discharge of the CBM water to infiltration impoundments reduces the net value of the project, which includes economic profits as well as the economic valuation of the stressors. Further progression from infiltration impoundments to active water treatments results in a greater loss in the project's net value.

The breakdown of the valuation of the multiple stressors by type is presented in Figure 2. The figure shows the net present value for 16 years of each stressor, assuming the continuation of the current water treatment methods. The figure shows that the project's environmental costs are mostly caused by the loss of scenery, as measured by the willingness of the public to pay for forest conservation. This is consistent with previous research, which shows that forest conservation has a high social value. Furthermore, the impact on forest scenery on the public is higher than the value of the companies' profits from the project.

The breakdown of the total net value of the project by sub-watershed is presented in Figure 3. There are significant differences in the impact on different sub-watersheds. The net value of six sub-watersheds is positive, and 86% of the total value of the damage is concentrated in four sub-watersheds. The net value of the project is negative—that is, the economic value of the stressors is higher than the profits from the project—for the watersheds located at the Upper Tongue River, Clear Creek, Little Powder River, and Middle Powder

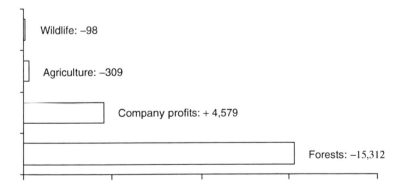

*NPV = Net Present Value for 16 years, in millions of dollar.

Figure 2. Economic Valuation of Stressor Impacts (Under the Current Water Treatment Method).

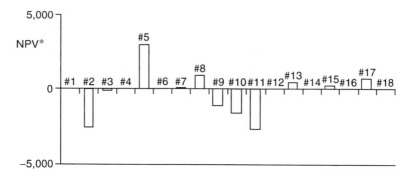

* NPV = Net Present Value for 16 years, in millions of dollars.

Figure 3. Net Value by Sub-Watershed Number (Under the Current Water Treatment Method).

River. The net value of the project is positive—that is, the benefits from the project are higher than the value of the damages—for the watersheds located at the Upper Powder River, Salt Creek, Crazy Woman Creek, Antelope Creek, Upper Cheyenne River, and the Upper Belle Fourche River.

5. Conclusions and Recommendations

The Powder River Basin project has many regional and national economic advantages, including increased employment and income from taxes, the important environmental benefit of providing low-sulfur coal, and other potential advantages, including lowering energy prices and reducing reliance on foreign energy sources. But the project is also associated with many social and environmental stressors which cause large-scale, irreversible damages.

The results of this research show that the project's total benefits are lower than the value of its social and environmental stressors; but for some sub-watersheds, the benefits outweigh the value of the stressors. The improved water treatment methods cost more than the value of their benefits. Therefore, the recommendation is to continue the project only in the sub-watersheds where the benefits outweigh the environmental costs, while maintaining the current water treatment methods.

Further research is needed to include a more extensive examination of the environmental damages, and estimation of the cost of damage remediation, as well as an economic valuation of the greenhouse gas emissions associated with each alternative, which were not taken into account in this project. Furthermore, comparison with other energy projects is needed in order to judge the impact of the project as a whole as compared with other energy production options.

References

1. Blend, J., 2002. *Important Economic Issues to Address with Coal-Bed Methane.* Resource Protection and Planning Bureau, Montana Department of Environmental Quality, October 22.
2. Cooper, A., n.d. *The Natural Capital of the Southern Lake Michigan Coastal Zone.* Graduate paper, University of Illinois at Chicago, IL.
3. Fausold, C. J., and Lilieholm, R. J., 1996. The Economic Value of Open Space. *Land Lines* 8, 5.
4. Goerold, W. T., 2002. *Revised Powder River Basin Coalbed Methane Financial Model.* Report PRB-CBM-FM), Lookout Mountain Analysis.
5. Haruvy, N., Shalhevet, S., and Yaron, D., 2001. Effect of urban development on water quality and environmental concerns. *International Water & Irrigation* 21(2):24–32.
6. LaMuth, J., 1998. *Community Development: Noise.* Report CDFS-190-198. The Ohio State University Fact Sheet.
7. Laurence, C., 2003. Cowboys fight to run drillers off the range. *Sunday Telegraph,* 2 March.
8. Loft, E. R., 1998. Economic Contribution of Deer, Pronghorn Antelope and Sage Grouse Hunting to Northeastern California and Implications to the Overall "Value" of Wildlife. *California Wildlife Conservation Bulletin,* California Department of Fish and Game, 11.
9. Munn, L. C., 2000. Water in a Dry Land: A Problematic Gift. *Frontline Report,* Fall. Available at: http://www.wyomingoutdoorcouncil.org/news/newsletter/docs/2000d/cbmh2o.php
10. Murphy, J. J., and Delucchi, M. A., 1998. A review of the literature on the social cost of motor vehicle use in the United States. *Journal of Transportation and Statistics* January:15–42.
11. Pearce, D., 2003. *Renewable Resources: The Tropical Forest.* Publication no. B48 2002–3. A lecture given on Nov. 21, 2002, at the University College London. Available at: http://www.ucl.ac.uk/~uctpa15/B48_forest.pdf
12. Power River Basin Resource Council Website. Available at: http://www.powderriverbasin.org
13. Regele, S., and Stark, J., 2000. *Coal-Bed Methane Gas Development in Montana, Some Biological Issues.* Montana Department of Environmental Quality, Industrial and Energy Minerals Bureau. Available at: http://ipec.utulsa.edu/Conf2001/kurz_89.pdf
14. Sagoff, M., 1988. *The Economy of the Earth.* Cambridge studies in Philosophy and Public Policy, Cambridge University Press, Cambridge.
15. Shalhevet, J., 1994. Using water of marginal quality for crop production: major issues. *Agricultural Water Management* 25:233–269.
16. U.S. Department of Agriculture (USDA), 2003. *Wyoming Agricultural Statistics Service.* Wyoming Office of USDA's National Agricultural Statistics Service.
17. U.S. Department of Agriculture (USDA), 2003. *Agricultural Statistics 2003.* National Agricultural Statistics Service, Washington, DC.
18. U.S. Department of the Interior, 2003. Final Environmental Impact Statement and Proposed Plan Amendment for the Powder River Basin Oil and Gas Project. Report WY-070-02-065, Bureau of Land Management, Wyoming State Office, Wyoming.

RISK ASSESSMENT OF RADIONUCLIDE CONTAMINATION IN POTABLE WATER

A.A. BAYRAMOV

Institute of Physics
National Academy of Science of Azerbaijan
G. Javide av.33
Baku Az1143, Azerbaijan
bayramov_azad@mail.ru

S.M. BAYRAMOVA

Institute of Geology
National Academy of Science of Azerbaijan
G. Javide av.29
Baku AZ1143, Azerbaijan

Abstract: A risk assessment analysis procedure for determining the radiation safety of potable water was considered. The determination of safety was based on measuring specific alpha and beta activity. If one or both activity values is exceeded. it is necessary to carry out radionuclide analysis. Using this method, water samples taken from the transboundary with Georgia Kura River in Azerbaijan were analyzed. It was established that water in the Kura River contains safe levels of radiation. This procedure can be used for assessing radiation risk in any water intended for household use.

1. Introduction

As is well known, to be safe for human consumption, drinking water must be free from microorganisms capable of causing disease. It must not contain minerals, organic substances, or radioactive nuclides at concentrations that could produce adverse effects. [1, 2]. The International Committee on Radiological Protection (ICRP) makes recommendations to regulatory bodies for radiation standards. ICRP advocates defining a justification for radioactive practices, which then justifies the level of exposure allowed by the radiation standard [3]. In 1979, the U.S. established national interim primary drinking water regulations. The maximum contaminant levels (MCLs) for some radiation elements in potable water according to these regulations are shown in Table 1 [4, 5].

I. Linkov et al. (eds.), Real-Time and Deliberative Decision Making.
© Springer Science + Business Media B.V. 2008

TABLE 1. National Interim Primary Drinking Water Regulations.

	Contaminant	MCL*
Alpha Emitters	Radium-226	5 pCi/L
	Radium-228	5 pCi/L
	Gross alpha activity (excluding radon and uranium)	15 pCi/L
Beta and Photon Emitters	Tritium	20,000 pCi/L
	Strontium-90	8 pCi/L

1.1. DEFINITIONS

1.1.1. Maximum Contaminant Level

The highest level of a specific compound that is allowed in treated drinking water. MCLs are established by regulation and are set as close to the maximum contaminant level goal (MCLG) as possible using the best available technology to remove the contaminant.

1.1.2. Maximum Contaminant Level Goal

The level of a specific compound in drinking water below which no adverse health effects are expected to occur. MCLGs are derived using either observed or predicted health endpoints and are nonenforceable public health goals.

1.2. RADIOACTIVE POLLUTION AFTER CHERNOBYL

Radioactive pollution of water can happen, for example in regions of a disposition of accelerators [6] or nuclear reactors.

According to a report on water quality impacts (radionuclide contamination of rivers, lakes, and groundwater in Ukraine) after the Chernobyl accident [7], today the Dnieper and Prypyat rivers remain the main source of radionuclide transfer. Over the post-accident period about $150 \cdot 10^{12}$ Bq of ^{90}Sr and $120 \cdot 10^{12}$ Bq of ^{137}Cs have been transferred by the Prypyat river to the Kyiv reservoir. Further radionuclide accumulation in bottom sediments is observed in static water bodies (ponds, lakes, and reservoirs). Since autumn 1986, radioactive river and reservoir contamination has been caused by radioactive surface runoff from catchment areas and backwaters. Contaminated runoff has also infiltrated groundwater.

The distinguishing feature of this period is the prevalence of ^{90}Sr and ^{137}Cs in radionuclides; ^{90}Sr has a gradually increasing share in the radioactive contamination of water. Estimates indicate that ^{90}Sr will continue to play a leading role. The dose from this radionuclide may exceed 2 to 35 times the dose from ^{137}Cs; at the same time it will constitute no more than 0.1 mSv annually.

In our paper, a risk assessment procedure estimating the radiation safety of potable water is considered. The estimation is based on measuring specific alpha and beta activity. If one or both values for these activities is exceeded, it is necessary to carry out a radionuclide analysis. Using this procedure, water samples from the transboundary with Georgia's Kura River were analyzed in Azerbaijan. It was established that water in the Kura contains safe levels of radiation. This procedure can be used for the radiation analysis of any water intended for household use.

2. Procedures and Methods

Potable water meets applicable hygienic specifications and is intended to sustain human life and be suitable for the production of food, drinks, and other products. Determination of potable water conformance to radiation safety criteria is the result of measuring the value of a specific activity and error at a confidence probability $(P = 0.95)$. Absolute error of measure consists of statistical Δ_s and systematic Δ_c. Total error Δ is: $\Delta = \Delta_s + \Delta_c$. Systematic error Δ_s is:

$$\Delta_s = \sqrt{\Delta_1^2 + \Delta_2^2},$$

Where: Δ_1 are the errors of attested metrological characteristics of measuring tools identified in the verification certificate, and Δ_2 is a truncation error of sampling preparation. If Δ_2 is not indicated in the method, it is equal to 10%. For preliminary estimation of potable water conformance to radiation safety criteria, the measured value of a specific total alpha (A_α) and beta (A_β) activity and its absolute errors Δ_α and Δ_β are used. For groundwater at A_α and A_β, they should be measured at the same time.

For potable water of underground water supply sources, simultaneously with measuring A_α and A_β, it is necessary to determine radon content: to measure a specific activity of radon A_{Rn} and absolute error Δ_{Rn}. Potable water satisfies the requirements of radiation standards if the following requirements are simultaneously satisfied [8]:

$$A_\alpha + \Delta_\alpha \leq 0,1 \, \text{Bq}/\text{kg} \tag{1}$$

$$A_\beta + \Delta_\beta \leq 1,0 \, \text{Bq}/\text{kg} \tag{2}$$

$$A_{Rn} + \Delta_{Rn} \leq 60 \, \text{Bq} / \text{kg} \tag{3}$$

If (1) and (2) requirements are not satisfied, it is necessary to execute a radionuclide analysis. A full radionuclide analysis is recommended to make a determination of the correspondence between total activity and the sum of radionuclide activity by this test.

$$A_\alpha + \Sigma K_i A_i \leq 0.2 \tag{4}$$

Where A_i is a specific activity of radionuclide measured in water; K_i are the coefficients describing the nonconformity of energy spectrums of the standard of comparison and the real sample (Table 2); 0.2 is the coefficient accounting for presence in a water sample of other alpha-emitting nuclides with activity of $\leq 5\%$ from value SL^{water} (SL represents a safe level of radionuclide concentration in potable water). This determination was not fulfilled during the analysis.

If requirement (4) is satisfied, it is adopted until all main alpha-emitting nuclides in the sample have been defined and subsequent measuring is not necessary.

Water satisfies the requirements of a radiation safety if

$$\sum \frac{A_i}{SL_i} + \sqrt{\sum \left(\frac{\Delta A_i}{SL_i} \right)^2} \leq 1 \tag{5}$$

Where: A_i is a specific activity of radionuclide measured in water (including ^{222}Rn); and ΔA_i is an absolute error of measuring a specific radionuclide activity.

The establishment of local control levels for the given water source is recommended at the realization of option (5) for the further monitoring of potable water. These levels are determined based on indications that the

TABLE 2. Coefficients K_i at Using the Standard of Comparison With $E_\alpha \approx$ 5,15 MeV and Low Level of Discrimination of the Alpha-Radiometry ≈ 3 МэВ.

Alpha-emitting nuclides	Energy of alpha-radiation (MeV)	K_i
^{232}Th	4.01	0.60
^{238}U	4.195	0.65
^{230}Th	4.685	0.85
^{234}U; ^{226}Ra	4.77; 4.78	0.90
$^{239+240}Pu$; ^{210}Po	5.155÷5.168; 5.5	1.00
^{228}Th; ^{241}Am; ^{238}Pu	5.42; 5.486; 5.5	1.10
^{224}Ra; ^{223}Ra	5.68; 5.61	1.15

common alpha activity and/or beta activity will not exceed the level of dose 0.1 Sv/year.

Noncompliance with option (5) indicates it is necessary to further research the water to determine the annual inflow of radionuclides:

1. Measuring should characterize water quality throughout the year. For groundwater, no fewer than four water samples should be taken (one per season); for surface water, no fewer than 12 water samples should be taken (one per month).

2. Analyses should indicate the quality of the water really consumed by the population. Depending on the availability of water treatment or mixing at different water supply points, radiation control may be carried out before water is introduced to a supply line, and for some radionuclides (gaseous or with a brief half-life; for example, ^{222}Rn) radiation control is carried out as part of water supply distribution.

When the stable presence of radionuclides above a level of interference is detected in a water source, it is necessary to elicit sanitary and epidemiologic expertise about the possibility of further use of the water source and the necessity of implementing protective actions.

3. Experimental Results and Discussions

Experimental research was carried out on the water samples taken from the Kura River in Azerbaijan. As is known, water from the Kura River plays an important role in water supply and is used as a potable water source in some regions of Azerbaijan, including Baku. A series of sanitary and epidemiological investigations has shown that water in the Kura River is saturated by harmful and hazardous chemical agents. Chemical concentration varies depending on weather conditions. Therefore, a radiation risk assessment analysis of Kura River water samples was conducted.

Specific total alpha (A_α) and beta (A_b) activity and the absolute errors of their determination Δ_α and Δ_β in water samples have been measured:

$$A_\alpha + \Delta_\alpha = 0.16\,\text{Bq}/\text{kg}$$

$$A_\beta + \Delta_\beta = 0.15\,\text{Bq}/\text{kg}$$

From requirements (1) and (2) it is clear that the level of total activity is exceeded; i.e., it is necessary to carry out the radionuclide analysis. To choose the radionuclides subject to determination in this sample, we use Table 3.

TABLE 3. Consecution the Radionuclide Analysis in Dependence on the Measured Levels Total Alpha-and-Beta activity.

№	Measured level of total alpha-and beta-activity (Bq/kg)	Tested radionuclides	Comments
1	$A_\alpha + \Delta_\alpha \leq 0.10$ $A_\beta + \Delta_\beta \leq 1.0$	Radionuclide composition not tested	
2	$0.1 < A_\alpha + \Delta_\alpha \leq 0.20$ $A_\beta + \Delta_\beta \leq 1.0$	Brief analysis $^{210}Po, \,^{210}Pb$	Check up of requirement of (5) and so.
3	$0.2 < A_\alpha + \Delta_\alpha \leq 0.40$ $A_\beta + \Delta_\beta \leq 1.0$	Broadened analysis $^{210}Po, \,^{210}Pb, \,^{226}Ra, \,^{228}Ra$	Check up of requirement of (5) and so.
4	$A_\alpha + \Delta_\alpha > 0.40$ $A_\beta + \Delta_\beta \leq 1.0$	Complete analysis $^{210}Po, \,^{210}Pb, \,^{226}Ra, \,^{228}Ra,$ $^{238}U, \,^{234}U$	If requirements (4) are not satisfied it is necessary additional determination of $^{232}Th, \,^{230}Th,$ $^{228}Th, \,^{239+240}Pu, \, 238Pu, \,^{241}Am$ in nearby regions; check up of requirement of (5) and so.
5	$A_\alpha + \Delta_\alpha > 1.0$ any value $A_\beta + \Delta_\beta$	$^{137}Cs, \,^{90}Cr, \,^{40}K$	

Comparing the measured samples with data of Table 3, we conclude:

$$0.1 < A_\alpha + \Delta_\alpha = 016 \leq 0.20$$

Thus, we fulfill the radionuclide analysis (^{210}Po and ^{210}Pb are detected in the water sample). Further analysis of water samples has shown:

$$\text{a specific activity of } ^{210}Po - 0.003 \pm 0.001 \, Bq / kg$$

$$\text{a specific activity of } ^{210}Pb - 0.040 \pm 0.015 \, Bq / kg$$

Check up of requirement (5). The SL determination for ^{210}Po K ^{210}Pb uses Table 4.

From (5) we have

$$\sum \frac{A_i}{SL_i} + \sqrt{\sum \left(\frac{\Delta A_i}{SL_i} \right)^2} = \left(\frac{0.003}{0.12} + \frac{0.04}{0.20} \right) + \sqrt{\left(\frac{0,001}{0.12} \right)^2 + \left(\frac{0.015}{0.20} \right)^2} = 0.30 < 1 \tag{6}$$

Thus. from the risk assessment analysis of the water samples taken, it is concluded that the water contains safe levels of radiation and is suitable for household use. Further, radionuclide analysis is not required.

TABLE 4. Safety Level of Radionuclides Concentration in Potable Water.

Radioactive nuclide	Half-life ($T_{1/2}$)	SL (Bk/kg)
$^3H(\beta)$	12.1 years	7.7 + 3
$^{14}C(\beta)$	5.73 years	2.4 + 2
$^{60}Co(\beta.\gamma)$	5.27 years	4.1 + 1
$^{89}Sr(\beta)$	50.5 days	5.3 + 1
$^{90}Sr(\beta)$	29.1 years	5.0
$^{129}I(\beta)$	1.57 years	1.3
$^{131}I(\beta.\gamma)$	8.04 days	6.3
$^{134}Cs(\beta.\gamma)$	2.06 years	7.3
$^{137}Cs(\beta.\gamma)$	30.0 years	1.1 + 1
$^{210}Pb(\beta)$	22.3 years	2.0 −1
$^{210}Po(\alpha)$	138 days	1.2 − 1
$^{224}Ra(\alpha)$	3.66 days	2.1
$^{226}Ra(\alpha)$	1.6 years	5.0 − 1
$^{228}Ra(\beta)$	5.75 years	2.0 − 1
$^{228}Th(\alpha)$	1.91 years	1.9
$^{230}Th(\alpha)$	7.70 years	6.6 − 1
$^{232}Th(\alpha)$	1.4 years	6.0 − 1
$^{234}U(\alpha)$	2.44 years	2.9
$^{238}U(\alpha)$	4.47 years	3.1
$^{238}Pu(\alpha)$	87.7 years	6.0 − 1
$^{239}Pu(\alpha)$	2.41 years	5.6 − 1
$^{240}Pu(\alpha)$	6.54 years	5.6 − 1
$^{241}Pu(\alpha)$	4.32 years	6.9 − 1
$^{222}Rn(\alpha)$	3.82 days	60

4. Conclusion

A risk assessment analysis procedure for determining the radiation safety of potable water was considered. The determination of safety was based on measuring specific alpha and beta activity. If one or both activity values is exceeded. it is necessary to carry out radionuclide analysis. Using this method, water samples taken from the transboundary with Georgia Kura River in Azerbaijan were analyzed. It was established that water in the Kura River contains safe levels of radiation. This procedure can be used for assessing radiation risk in any water intended for household use.

References

1. Pontius. F. W. and Clark. S. W. Chapter 1. (1999). *Water Quality & Treatment.* 5th ed. AWWA, Denver, CO.
2. Opflow. (2000). American Water Works Association.

3. Radiation standards. Preston. (1987). RERF TR 9–87.
4. Drinking Water and Health. Safe Drinking Water Committee. (1977). National Academy of Sciences. Washington, DC.
5. Water Quality and Treatment. (1999). *A Handbook of Community Water Supplies.* 5th ed. American Water Works Association. New York, McGraw-Hill.
6. Environmental Monitoring of Ionizing Radiation. (2006). Jefferson Lab EH&S Manual - Rev. 8.6.
7. Official statement by the Ministry of Emergencies of the Ukraine on important issues concerning the nuclear accident at Chernobyl. Available at: http//*www.chernobyl.info*
8. Radiation test of water. Methodical comments. (2000). Ministry of Health of Russia. Moscow.

ASSESSMENT OF GROUNDWATER CONTAMINATION RISK FROM NONCHEMICAL STRESSORS IN WASTEWATER

N. HARUVY

Netanya Academic College
1 University Street
Netanya, Israel 42365
navaharu@netvision.net.il

S. SHALHEVET

SustainEcon
126 Thorndike Street
Brookline, MA 02446, USA
sarit.shalhevet@gmail.com

Abstract: Chemical risk analysis approaches have been applied extensively to examine the effect of chemical stressors in wastewater on the risk of groundwater contamination. We have developed multidisciplinary water management models that examine the risk to groundwater caused by chemical stressors such as nitrogen and chlorides. These models examine the impact under a variety of scenarios, taking into account hydrological, technological, economic, and regional planning aspects.

Nonchemical stressors are a major concern in wastewater, but methods of risk assessment for these stressors are less developed. Pharmaceuticals are a well known source of contamination of urban wastewater, and their negative impact on the environment and on human health has been well documented. Microbial agents are considered a major biological stressor and their existence in wastewater has caused epidemic outbreaks in several cases.

This paper analyzes the application of our water management models to evaluate the effect of nonchemical stressors in wastewater on the risk of groundwater contamination. We examine different scenarios of concentrations of pharmaceuticals and microbial agents in wastewater, and show how chemical risk assessment methods can be applied to decision making for optimal wastewater treatment, based on economic considerations as well as environmental security and human health considerations.

I. Linkov et al. (eds.), Real-Time and Deliberative Decision Making.   435

1. Introduction

The risk to groundwater from chemical stressors such as nitrogen and chlorides in wastewater has been researched extensively. Much has been published about the negative environmental impact arising from the use of these chemicals [2, 11, 12], and several models have been developed for risk assessment and optimization of chemical use [3, 4]. We have developed a water management model that examines the risk to groundwater caused by chemical stressors and applied it to case studies in Israel [5, 6].

In addition to chemical stressors, there are many types of nonchemical stressors in wastewater that can cause damage to groundwater resources, including stressors such as pharmaceuticals, personal care products (PPCPs), and microbial agents (pathogens of different types). However, research on the risk to groundwater from nonchemical stressors is still in its beginning stages, and the nature and extent of the damage is not as well known.

We are now developing methods of application of the water management model to evaluate the effect of nonchemical stressors in wastewater on the risk of groundwater contamination. Our analysis is based on the water management model mentioned above [5, 6], which examines the risk to groundwater caused by chemical stressors. This research examines the application of the model to evaluate the impacts of nonchemical stressors in wastewater on the risk of groundwater contamination, using three examples of nonchemical stressor types in wastewater: pharmaceuticals, personal care products (PPCPs), and microbial agents.

Pharmaceuticals have been found in high concentrations in surface water and wastewater; some pharmaceuticals are not significantly absorbed in the subsoil, and may reach the groundwater. Sampling results in France [8] show that caffeine is the most common drug. It is commonly encountered in surface waters and may form up to 95% of the total drug concentration in surface water; it has been detected in groundwater as well. Paracetamol, a non-opioid analgesic found in brand names such as Tylenol, was found at high concentrations—up to 11 $\propto$g/L (wastewater effluent) and 211 ng/L (drinking water). Ibuprofen, a nonsteroidal anti-inflammatory analgesic found in brand names such as Advil and Moltrin, was found at 219 ng/L in wastewater, and up to 4.5 ng/L in groundwater. The total drug concentration varies from 1,028 to 15,800 ng/L (wastewater) and from 7.7 to 300 ng/L (aquifer water). Research in the UK [1] detected and quantified ibuprofen, paracetamol, and salbutamol in surface water samples; ibuprofen was consistently found at the highest concentrations (up to 3 $\propto$g/L). It should be noted, however, that the concentrations of drugs in groundwater are low relative to the quantity allowed for human consumption. For example, the therapeutic dose for paracetamol, is about 1,200 mg/day. By comparison, the highest concentration in groundwater was found to be 211 ng/L, meaning that a person would need to drink $5,687 m^3$ of water to reach the therapeutic dose. The therapeutic dose for ibuprofen is about 600–1,200 mg/day.

Personal care products (PPCPs) are still an emerging research subject, and the hazards to the environment and human health of PPCPs in the water supply are still poorly understood. Some initial research has shown that this is indeed a matter for concern. For example, in Baltimore, 10 of 18 major PPCPs were detected in the treated sewage effluent, signifying incomplete removal for the majority of the PPCPs during the wastewater treatment process [13].

Microbial agents are considered a major biological stressor and their existence in wastewater has caused epidemic outbreaks in several cases. Human wastewater is a source of pathogenic microorganisms. Bacteria and protozoa show poor survival outside a human host, but viruses and helminths can remain infectious for months to years.

2. Multidisciplinary Water Management Model

Haruvy et al. [5, 6] developed a water management model that examines the risk to groundwater caused by chemical stressors, such as nitrogen and chlorides, in wastewater and in other water supply sources. The model incorporates regional planning, economic, hydrological, and technological considerations to plan the optimal combination and timing of water supply from different sources, including desalination and wastewater treatment, and assesses the risks and benefits from different levels of treatment under a variety of scenarios. The model was applied to a case study of several hydrological cells in several regions in Israel.

The outputs of the model include regional planning, hydrological, and technological information, and an economic model of water management. The regional planning outputs include area allocations, demand for water, and water sources. The hydrological model, which is guided by the local water management policy, is one of the major components of the multidisciplinary water management model; its outputs include pumped water, chloride and water balances, forecasts of levels, groundwater salinity, and maximum permitted salinity. The technological information outputs concern desalination and other treatment processes. The economic model supplies information on the required quantity of water, and total costs of water supply and treatment.

The objective of this paper is to examine the application of the water management model described above for the risk assessment of nonchemical

stressors in wastewater on groundwater contamination. The following
sections describe the adaptations required for including nonchemical stressor
assessments in the model.

3. Adaptation of the Model for Risk Assessment of Nonchemical Stressors

3.1. OVERALL STRUCTURE

The optimization model compares the economic valuation of the damage
from groundwater contamination with the cost of water treatment. It comprises five submodels (Figure 1) including a regional planning model, hydrological model, risk exposure model, technological model, and economic
model. The inputs to the regional planning, hydrological, risk exposure,
and technological models are local area data (regional planning, hydrologi-

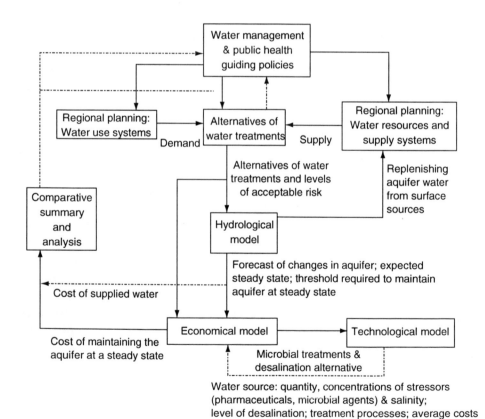

Figure 1. Diagram of the Model's Basic Functions.

cal, technological, economic, demographic, and lifestyle data). The outputs from these models are fed as inputs into the economic model. The potential water sources include groundwater, the National Water Carrier, and wastewater from different sources. The water consumers and sources of wastewater supply include the domestic, industrial, and agricultural sectors.

The guiding policy includes water management policy as well as public health management policy. The local water policy is the driving force of the model, and includes the following principles. The first principle is based on setting a predetermined threshold of water quality for drinking and irrigation; or, alternatively, a predetermined threshold of water quality in groundwater (measured in the chemical risk assessment model by the concentration of chlorides). The second principle is pumping at an appropriate level to achieve stable groundwater levels and salinity and meet pumping capacity limitations. The third principle is setting limitations on the intrusion of seawater into the aquifer. The fourth is that for each hydrological cell, the water supply is equal to the demand of water. The public health management policy includes predetermined public health policies, which currently consists of national microbial regulations. Different levels of microbial regulations cause different health and environmental impacts; a higher level of treatment reduces the damage from wastewater, but increases the costs.

Representative major stressors were selected based on the research cited in the introduction. From the large number of possible pharmaceuticals we selected caffeine, ibuprofen, and paracetamol as major and representative pharmaceuticals for the model. Personal care products were not used in the model due to the limited amount of data presently available. Selection of representative microbial agents for the model is particularly complicated. Biological parameters have to indicate all potential pathogenic organisms, including viruses, bacteria, and parasites of different origins, but testing for all potential pathogenic organisms is too expensive and time consuming. The biological indicator parameters most commonly used for evaluating microbial water quality are total coliforms and E. coli; although total coliforms are not likely to cause illness in themselves, their presence indicates that pathogens may be present.

The model scenarios include:

- Initial concentrations of $1 \propto g/L$ for each of the following pharmaceuticals: caffeine, ibuprofen and paracetamol
- Initial concentrations of $10 \propto g/L$ for each of these three pharmaceuticals
- Initial concentrations of $50 \propto g/L$ for each of these three pharmaceuticals
- Initial concentration of 1,000 fecal coliform/100 mL, based on the WHO 1989 guidelines

■ Restrictions for agricultural reuse of wastewater in the range of 1–10 fecal coliforms/100 mL, based on existing microbial guidelines in Israel

3.2. REGIONAL PLANNING MODEL

The inputs for each hydrological cell include the following information:

■ Physical characteristics—built area, agricultural area, industrial production by type (e.g., hospitals, cosmetic production) and water availability

■ Demographic characteristics—population size and level of consumption

The outputs include estimates for the consumption and disposal of the main nonchemical stressors.

3.3. HYDROLOGICAL MODEL

The hydrological model describes the area as a geographical layout of hydrological cells. Each hydrological cell contains two layers. The upper layer is the unsaturated zone, and its depth is the distance between the ground surface and the top surface of the groundwater. The lower layer represents the saturated zone (groundwater) and its depth is the distance between the top surface of the groundwater and the base of the aquifer. The model assumes that only part of the water that leaches downward to the unsaturated zone actually reaches the aquifer. However, the whole mass of pharmaceuticals and PPCPs is assumed to penetrate the unsaturated zone and reach the aquifer eventually. The rate of penetration is identical within each hydrological cell, but may differ among the cells within the area. The model is based on solution of a system of simultaneous equations, described in Haruvy et al. [5], and on input data including physiochemical characteristics (pharmaceuticals, distributed water), hydraulic conditions, groundwater scenarios, surface water scenarios, and treatment methods. The output includes the predictions of the annual concentrations of each stressor in groundwater for the next two decades, based on initial concentrations, additional contamination, and rate of penetration into the aquifer water.

3.4. RISK EXPOSURE MODEL

The risk exposure model depends on the local customs. For example, the degree to which it is customary to wash fruit and vegetables before use differs between countries, and influences the level of risk caused by identical microbial regulations.

3.5. TECHNOLOGICAL MODEL

The basic principles of the technological model for chemical risk assessment are described in Haruvy [2]. That model provides information about the relevant treatment methods, such as desalination, and their costs, based on the following parameters [5]:

- Input data: water quantity, raw water quality, absorption and water-holding capacity of the soil, depth of drilling, lengths of the pipelines, height of the desalination plant, and volume of the raw water reservoir
- Technological parameters: size of the desalination plant, quality of the water fed to the plant, and location of the plant
- Economic data: returns on investment, plant availability, and cost components; i.e., costs of energy, membranes, chemicals, manpower, maintenance, and overheads

The adaptation of the model to nonchemical risk assessment includes the addition of following parameters:

- Input data: the data on pharmaceuticals includes dosing information, where relevant, and application scenarios. The data on microbial agents includes residence time, condition and type of piping material, water temperature, and disinfectant residuals.
- Technological parameters: technological aspects of different testing and treatment options.
- Economic data: costs of treatment components by type of treatment technology used.

3.6 ECONOMIC MODEL

The economic model evaluates the costs of the water supply to the region and the economic value of the damage incurred from the different stressors under the various scenarios. For example, the model takes into account, for each level of microbial regulation, the economic valuation of the environmental and health damage caused by the risk associated with a specific concentration of fecal coliform/mL, and the cost of treatment. The inputs to this model comprise output data from the other models.

The inputs for the economic model are:

- The water requirements of the various sectors, and the policy guidelines regarding the threshold of pharmaceutical concentrations of caffeine, ibuprofen and paracetamol, and the concentration of microbial agents, as calculated by the regional planning model

- Forecasts of concentrations by stressor for each year, as calculated by the hydrological model, based on the results of the various scenarios

- Average cost of each water source and treatment method, as calculated by the technological model

The economic evaluation is based on a tradeoff between cost and risk reduction, measured as the economic valuation of the environmental and health damages incurred; at higher levels of water treatment, the risk is reduced, but economic costs increase. The optimal level of treatment is the level at which the economic value of the damage equals the cost of wastewater treatment.

The risks of pharmaceutical and microbial groundwater contamination include damages to the environment, including air, soil, and water resources, as well as increased illness prevalence and mortality rates. In order to make decisions based on the tradeoff between damage and treatment costs, we need to translate these noneconomic damages into monetary terms.

Several methods have been developed for the economic evaluation of environmental risks. One of these methods is calculating the cost of remediation of the damages. A second method is estimating the willingness to pay for the prevention of environmental damage, which is done through surveys of the relevant public. A third method is hedonistic pricing, which examines the value of the land and housing in areas with a specific environmental damage, compared to equivalent areas without this damage.

Different methods are used for the economic evaluation of health risks. The World Health Organization (WHO) has adopted Disability Adjusted Life Years (DALYs) as the most appropriate metric for expressing the burden of a disease [7]. DALYs attempt to measure the time lost because of disability or death from a disease compared with a long life free of disability in the absence of the disease [9]. The economic evaluation of the DALYs associated with a specific health risk can be estimated using the cost of medical care and loss of income. The additional cost of life insurance policies is another proxy used to evaluate the cost of health damages.

Previous research found that irrigation of salad crops in Israel under the WHO guidelines would involve a risk of disease of $10^{-6}-10^{-8}$ per person per year or one case per 1 million to 100 million person-years of exposure. The additional cost of treatment to avoid one case of mild enteric disease under the Israeli wastewater reuse guidelines (instead of WHO guidelines) could be $500,000 to $1,000,000 [9].

3.7. COUNTRY-SPECIFIC MODIFICATIONS

The model was initially developed based on Israeli data. The optimal level of treatment is likely to be different in each country. Therefore, application of the model to other countries should be based on the specific data in each country. Local influencing factors include, for example:

- Geographical factors (climate, type of soil).

- Existing levels of nonchemical stressors.

- Local customs; for example, the degree in which it is customary to wash fresh fruit and vegetables before use, which influences the risk level at equal concentrations of pathogens.

- Economic variables such as the gross national product per capita, which influence the ability to finance water treatment.

Recommendations for microbial regulations vary among countries depending on the optimal level of treatment in each country. Therefore, a model has to separately take into account relevant data for each country.

The cost of treatment should be taken into account as a function of the GNP, since the ability to pay for treatment differs among countries.

4. Conclusions

The risks from nonchemical stressors should be examined in the context of a model that takes into account the impact of the available alternatives, including cost of treatment to remove pharmaceuticals; cost of non-treatment (economic value of health costs), environmental damages (soil, aquatic ecosystem), hydrological aspects (length of time for penetration into groundwater); regional planning aspects (local water resources), and technological aspects (the different treatment options).

The model described here provides a skeleton for combining and quantifying the different risks from chemical and nonchemical stressors. The model can be applied to decision making for optimal wastewater treatment, based on economic considerations as well as environmental security and human health considerations.

References

1. Bound, J. P., and Voulvoulis, N., 2006. Predicted and measured concentrations for selected pharmaceuticals in UK rivers: Implications for risk assessment. *Water Research* 40 (15):2885–2892.
2. Haruvy, N., 1997. Agricultural reuse of wastewater: nation-wide cost-benefit analysis. *Agriculture, Ecosystems and Environment* 66:113–119.
3. Haruvy, N., Shalhevet, S., and Ravina, I., 2004. Irrigation with treated wastewater in Israel - financial and managerial analysis. *Journal of Financial Management and Analysis* 16 (2):65–73.

4. Haruvy, N., and Shalhevet, S., 2005. Land use and water management in Israel—economic and environmental analysis of sustainable reuse of wastewater in agriculture. *Proceedings of the 45th Congress of the European Regional Science Association*, Free University of Amsterdam.
5. Haruvy, N., Shalhevet, S., and Bachmat, Y. (in press). Risk management of transboundary water resources: sustainable water management of Jordan River Basin Area, *International Journal of Risk Assessment and Management*. Special Issue: Sustainable Management of Water Resources in Transboundary River Basins: Risk Assessment and Modeling.
6. Haruvy, N., Shalhevet, S., and Bachmat, Y. 2008. A model for integrated water resources management in water-scarce regions: irrigation with wastewater combined with desalination processes. *International Journal of Water*. Special Issue: Integrated Water Resources Management in Arid and Semi-Arid Areas. 4(1/2): 25–40.
7. Murray, C. J. L., and Acharya, A. K., 1997. Understanding DALYs. *Journal of Health Economics* 16 (6):703–730.
8. Rebiet, M., Togola, A., Brissaud, F., Seidel, J. L., Budzinski, H., and Elbaz-Poulichet, F., 2006. Consequences of treated water recycling as regards pharmaceuticals and drugs in surface and ground waters of a medium-sized Mediterranean catchment. *Environmental Science & Technology* 40 (17):5282–5288.
9. Shuval, H., 2006. Evaluating the World Health Organization's 2006 health guidelines for wastewater reuse in agriculture. *The 36th Conference of the Israel Society for Ecology and Environmental Quality Sciences (ISEEQS)*, Haifa, Israel. June 27–28, 2006.
10. Wajsman, D., and Rudén, C., 2006. Identification and evaluation of computer models for predicting environmental concentrations of pharmaceuticals and veterinary products in the Nordic environment. *Journal of Exposure Science & Environmental Epidemiology* 16 (1):85–97.
11. Yaron, D., Bachmat, Y., Wallach, R., Mayers, S., and Haruvy, N., 1999. Not "the age of wastewater followed by the age of desalination" but combined wastewater and desalination. *Water and Irrigation* 393:5–13.
12. Yaron, D., Haruvy, N., and Mishali, D., 2000. Economic considerations in the use of wastewater for irrigation. *Water and Irrigation* 400:19–23.
13. Yu, J. T., Bouwer, E. J., and Coelhan, M., 2006. Occurrence and biodegradability studies of selected pharmaceuticals and personal care products in sewage effluent. *Agricultural Water Management* 86 (1/2):72–80.

WASTEWATER REUSE

Risk Assessment, Decision Making, and Environmental Security: A Technical Report

M.K. ZAIDI

Idaho State University, College of Engineering
Pocatello, ID 83209, USA
zaidmoha@isu.edu

Abstract: Wastewater reuse—risk assessment, decision making, and environmental security—a NATO funded workshop, was conducted in Istanbul, Turkey, from October 12 to 16, 2006. Forty-eight research papers were presented by the participants. To address selected regional problems, three task groups conducted a series of breakout sessions to explore specific needs and formulate recommendations for decision makers. Local situations were scrutinized to improve the environmental security and stability of the region.

1. Introduction

The search for fresh water for domestic, agricultural, and industrial use is a problem for many countries throughout the world. According to the World Health Organization (WHO), 1.1 billion people around the world lack access to "improved water supply" [1] and more than 2.4 billion lack access to "improved sanitation" [2].

The United Nations designated 2005 to 2015 as the international decade for action, "Water for Life." In April 2005, its Commission on Sustainable Development identified water as one of its three important issues. On World Environment Day in June 2005, the UN Secretary General said that by 2030, more than 60% of the world's population will live in cities, and the growth will impose huge problems, including clean water supplies. This implies that concern about the critical condition of water resources is widely shared by the international community. The Middle East in particular has witnessed significant degradation of water resources both in quality and quantity over the past decades, which is one of the major threats to future sustainable development and political stability in the region.

Wastewater, if well treated, can be an important source of water and nutrients for irrigation in developing countries, particularly—but not restricted

I. Linkov et al. (eds.), Real-Time and Deliberative Decision Making.   445

to—those located in arid and semi-arid areas. The use of wastewater is widespread and represents around 10% of the total irrigated surface worldwide, although varying widely at local levels [2]. While the use of wastewater has positive effects for farmers, it may also have negative effects on human health and the environment. The negative effects impact not only farmers but also a wide range of people. The global burden of human disease caused by sewage pollution of coastal waters alone is estimated at 4 million lost 'man-years' every year, which equals an economic loss of approximately US$16 billion a year; losses may even be higher for discharge to fresh waters, but have yet to be fully estimated [3].

There is no complete global data on the extent to which wastewater is used to irrigate land, mostly due to a lack of heterogeneous data and the fear that countries have about disclosing information; economic penalties can be imposed if produce is found to have been irrigated with low-quality water. Nonetheless, the global figure commonly cited is that at least 20 million hectares in 50 countries (around 10% of irrigated land) are irrigated with raw or partially treated wastewater [4]. As wastewater reuse is becoming a necessity due to shortage in freshwater supply, it is important for governments to put in place wise but feasible management practices. In order to implement sustainable reuse of wastewater and to contribute to food security, reuse projects need to be planned and constructed for the long term, based on local needs [5].

During the NATO workshop, an array of government diplomats, security specialists, and social and physical scientists from the Middle East, North Africa, Europe, and North America reviewed the actions of past and current resources in the Mediterranean. Their focus was environmental security, environmental consequences, and challenges for the future. This workshop provided a multilateral forum for continued cooperation, information exchange, and dialogue among the environmental, developmental, foreign, and security communities within the Mediterranean region. It may provide a basis for further cooperation and partnership, including other more advanced conferences and publications, on assessing the condition of the entire region and the subsequent impacts and linkages to environmental security.

2. Observations

We made three task groups—risk assessment, decision making, and environmental security—to study regional problems in those areas. Our task teams recommend that each situation needs to be very carefully studied. Recommendations were presented and discussed with the decision makers and others involved.

2.1. WORKING GROUP 1: RISK ASSESSMENT OF CHEMICAL CONTAMINANTS

Risk assessment is the structured analytical method that uses various techniques to reach the most probable figure of the risk. In the context of wastewater reuse the scope of risk assessment consists of quantifying negative effects on human health, the environment, and socioeconomic conditions. The goal is to provide decision makers with simple information to support decision making. Risk management includes:

1. Hazard identification

2. Analysis of risk control measures

3. Making control decisions

4. Risk control implementation

5. Supervision and review

Risk is ranked according to severity and frequency and was assessed by monetary evaluation of Willingness to Pay (WTP) and Willingness to Accept Compensation (WTA). Economic methodologies to evaluate WTP and WTA (Contingent Valuation and Choice Experiments) incorporate WTP or WTA in cost-benefit analysis (CBA). A case study from Cyprus, Egypt, Greece, Italy, Jordan, and USA helped to understand the issue in different countries.

2.2. WORKING GROUP 2: REUSE OF RECLAIMED WASTEWATER IN NORTHERN JORDAN

Whether water reuse will be appropriate depends upon careful economic considerations, potential uses for the reclaimed water, stringency of waste discharge requirements, and public policy, where the desire to conserve rather than develop available water resources may override economic, aesthetic, and public health considerations. In addition, the varied interests of many stakeholders, including those representing the environment, must be considered. A number of factors affect the implementation of water reuse. Historically, the impetus for water reuse has risen from three prime motivating factors:

1. Availability of high-quality effluent

2. Increasing cost of freshwater development

3. Desirability of establishing comprehensive water resources planning and management, including water conservation, water reuse, and environmental protection

Water reclamation and reuse can serve several objectives. To achieve these objectives one has to follow a decision-making process while planning, designing, implementing, and maintaining a reclaimed water project. Seven steps were identified:

1. Objectives and criteria
2. Develop Strategies
3. Data Collection
4. Scenario Analysis
5. Implementation
6. Feedback
7. Optimization

It was recommended that wastewater should be treated to the minimum allowable treatment level for irrigating industrial crops. It was also recommended to consider using reclaimed water near its source to minimize system loss and investment needed to bring the reclaimed water to the irrigators. The reclaimed water should be used for growing native trees and fruit crops as they adapt to local climate in Jordan more successfully than imported species of crops and trees. The team recommended conducting further studies with the goal of optimizing Jordan's import and export to enhance overall gross domestic product of the country and eradicate poverty.

2.3. WORKING GROUP 3: ENVIRONMENTAL SECURITY ISSUES ARISING FROM THE REUSE OF WASTEWATER

The U.S. Army Environmental Policy Institute defines environmental security as relative public safety from environmental dangers caused by natural or human processes due to ignorance, accident, mismanagement, or design and originating within or across national borders [6].

Environmental security issues arise from the reuse of wastewater and its disposal as it affects humans and wildlife through the life-supporting media of soil, water, and air. The objective is to minimize adverse impact subject to economic constraints. The problem is that the impact mechanisms are very complex and not always well understood; and, there are economic consequences. The means to solve these problems include:

1. To understand processes, including quantitative relationships between cause and effect
2. To develop strategies and policies to limit contaminants (regulations, awareness, education)
3. To develop technologies to reduce outputs
4. To develop cleanup technologies to mitigate legacy problems

The state of practice in environmental security is:

1. A lot of the science is already in existence.
2. Some more work is needed to quantify cause and effect, especially in the long term.
3. Limits on the user side.
4. Reduction on the disposal side (treatment).
5. Treatment technologies evolve continually.

3. Conclusion

The strategies laid out by the team could be applied for decision-making processes in other countries for using treated wastewater for irrigation after analyzing local data. In addition, for such application to other countries, due consideration should be given to socioeconomic, cultural, political, and other externalities. It was mentioned that each situation needs to be very carefully studied.

Acknowledgement

The task teams worked very hard to get things worked out and we had very nice presentations from each team. We are very thankful to all team members—Paul West, Orhan Gunduz, Maria Helena Marecos do Monte, Mohamed Tawfic Ahmed, Nizar Al Halasha, Azad Bayramov, Phoebe Kondouri, Mihaela Lazarescu, Sureyya Meric, Baqar Zaidi, Jobaid Kabir, Saul Arlosoroff, Joop DeSchutter, Kallali Hamadi, Kubik Jiri, Nsheiwat Zein, M. Tahir Rashid, Yehuda Kleiner, John Letey, Yoooi Inbar, Hussein Abdel Shafy, Tarek Abu Elmaaty, Zahide Acar, Mu'taz Al Alawi, Gul Asiye Aycik, Alper Baba, Vijacheslav Charsky, Ozan Deniz, Gürdal Gülbin, Renat Khayadarov, Mohamed Abdel Geleel, Islam Mustafaev, Azamat Kalyevich Tynybekov—for their solid contribution in helping me to put this report together.

References

1. World Health Organization. 1996. Water and Sanitation. WHO Information Fact Sheet No. 112. Geneva. Available at: http://www.who.int/inf-fs/en/fact112.html
2. Rose JB. 2007. Water Reclamation, Reuse and Public Health. Water Science and Technology, 55(1/2): 275–282.

3. United Nations. 2002. World Summit on Sustainable Development Report, Johannesburg, S. Africa, 8/26-9/4/2002. A/CONF.199/20. New York. 2004. Available at http://ods-dds ny.un.org/doc/UNDOC/GEN/N02/636/93/ PDF/N0263693.pdf? OpenElement
4. United Nations. 2003. Water for People, Water for Life. The UN World Water Development Report. Barcelona: UNESCO.
5. Blanca Jiménez. 2006. "Irrigation in Developing Countries Using Wastewater". International Review for Environmental Strategies, 6(2): 229–250
6. Glenn, J. C., Gordon, T. J., Perelet, R., Landholm, M. 1998. Defining Environmental Security: Implications for the U.S. Army. Report #AEPI-IFP-1298. Army Environmental Policy Institute, Atlanta, Georgia.

LIST OF PARTICIPANTS

Abdel-Shafy, Hussein	Water Research & Pollution Control Department, National Research Centre, Cairo, Egypt	husseinshafy@yahoo.com
Bana e Costa, Carlos	Instituto Superior Tecnico Technical University of Lisbon, DEG, Tagus Park, Av. Cavaco Silvia, Porto Salvo 2780-990, Portugal	carlosbana@netcabo.pt
Bana e Costa, Joao	Bana Consulting Rua Professor Carga 33 Lisboa 1600, Portugal	joao@bana-consulting.pt
Barroso, Victor	IST-Alameda ISR Torre Norte Piso 7 Portugal	Tel: +351218418286 Fax: +35121841829 vab@isr.ist.utl.pt
Barzilai, Jonathan	Dalhousie University Department of Industrial Engineering Halifax B3J2X4, Canada	barzilai@scientificmetrics.com
Bastos, José	CNPCE Palácio Bensaude, 151 Lisbon 1600-153, Portugal	Tel: +351 21 72146 00 Fax: +351 21 727 05 66 jbastos@cnpce.gov.pt
Bayramov, Azad	Institute of Physics National Academy of Sciences G. Javide 33 Baku AZ 1143, Azerbaijan	Tel: +994124394057 Fax: +994124470456 bayramov_azad@mail.ru
Clarckson, J. Rutgers	Arcadis 10352 Plaza Americana Drive Baton Rouge, LA 70816, USA	jclarkson@arcadis-us.com
Cofrancesco, Al	U.S. Army Corps of Engineers 3909 Halls Ferry Road Vicksburg, MS 39180, USA	cofrana@wes.army.mil
Costa, Joao Carlos	ISQ Tagus Park Porto Salvo 2740-120, Portugal	jccosta@isq.pt
Cormier, Susan	U.S Environmental Protection Agency 26 W. M. L. King Drive Cincinnati, OH 45227, USA	Tel: 513-569-7995 Fax: 513-569-7438 cormier.susan@epa.gov

Dias, Susete	IST-Alameda Portugal	Tel: +351218419074 Fax: +351218419062 susetedias@ist.utl.pta
Fernandes, G.	Instituto Superior Tecnico Technical University of Lisbon DEG, Tagus Park, Av. Cavaco Silva Porto Salvo 2780-990, Portugal	
Fernandez Barberis, Gabriela	University San Pablo – CEU Julián Romea 23 Madrid 28003, Spain	Tel: 659638056 Fax: +34915548496 gabriela.ferbar@gmail.com
Ferguson, Elizabeth	U.S. Army Corps of Engineers 3909 Halls Ferry Road Vicksburg, MS 39180, USA	Elizabeth.A.Ferguson@erdc. usace.army.mil
Figovsky, Oleg	Israel Research Center Polymate 3a Shimkin Street Haifa 34750, Israel	Tel: 972-4-8248072 Fax: 972-4-8248050 figovsky@netvision.net.il
Figueira, Jose Rui	Instituto Superior Tecnico Technical University of Lisbon DEG, Tagus Park, Av. Cavaco Silva Porto Salvo 2780-990, Portugal	Tel: 351-21-423-35-07 Fax: 351-21-423-35-68 figueira@ist.utl.pt
Ganoulis, Jacques	Aristotle University of Thessaloniki Department of Civil Engineering Thessaloniki 54124, Greece	Tel: +30-2310-99-5682 Fax: +30-2310-99-568a1 igaanouli@civil.auth.gr
Geleel, Mohamed Abdel	National Centre for Nuclear Safety & Radiation Control 3 Ahmed El Zomor Street Nasr City 11762, Egypt	Tel: 202-7575223 Fax: 202-2740238 magelee12000@yahoo.com
Grebenkov, Alexandre	Joint Institute for Power and Nuclear Research – Sosny Minsk 220109, Belarus	Tel: +375 17 299-4542 Fax: +375 17 299-4355 greb@sosny.bas-net.by
Guinto, Darlene	HydroPlan Portland, Oregon, USA	dguinto@q.com
Haraza, Mahmoud Shafy	Atomic Energy Authority Ahmed El Zomor St. P.O. Box 755 Nasr City, Cairo 11672, Egypt	shafymahmoud@yahoo.com
Haruvy, Nava	Netanya Academic College One University Road Netanya 42100, Israel	Tel: +972-8-946-3189 Fax: +972-8-936-5345 navaharu@netvision.net.il
Kapustka, Larry	Golder Associates Ltd. 1000, 940-6th Avenue, SW Calgary T2P 3T1, Canada	Tel: 403-260-2265 Fax: 403-299-5606 Larry_Kapustka@golder.com

Kiker, Gregory	University of Florida Department of Agricultural & Biological Engineering P.O. Box 110570 Gainesville, FL 32611-0570 USA	Tel: +1 352-392-1864 gkiker@ufl.edu
Kozine, Igor	Obninsk State University (IATE) Studgorodok 1 Obninask 249040, Russia	igor.kozine@risoe.dk
Krymsky, Victor	Ufa State Aviation Technical University 12 K. Marx Street Ufa 450000, Russia	Tel: +7 347 272 99 09 Fax: +7 347 272 99 09 kvg@mail.rb.ru
Lahdelma, Risto	University of Turku Joukahaisenkatu 3-5 Turku 20014, Finland	Tel: +358405031030 Fax: +358 2 333 8600 risto.lahdelma@cs.utu.fi
Letnik, David	Israel Engineering Academy Haifa, Israel	daile@012.net.il
Levner, Eugene	Holon Institute of Technology 52, Golomb St. Holon 58102, Israel	Tel: 972546246757 Fax: 972 3 502 6733 elevner@ull.es
Linkov, Igor	U.S. Army Engineer Research and Development Center 83 Winchester Street, Suite 1 Brookline, MA 02446, USA	Tel: +1 617-233-9869 ilinkov@yahoo.com
MacDonnell, Margaret	Argonne National Laboratory 9700 S. Cass Ave., Building 900 Argonne, IL 60439, USA	Tel: 001-630-252-3243 Fax: 001-630-252-4336 macdonell@anl.gov
Magar, Victor	ENVIRON International Corporation 333 W Wacker Drive, Ste. 2700 Chicago, IL 60606, USA	Tel: 312-288-3840 vmagar@environcorp.com
Maia Martins, Joao	Marinha Portuguesa	maia.martins@marinha.pt
Martinho, Adriano	PO Civil Emergency Planning Council Estrada da Luz 153 Lisboa 1600-153, Portugal	Tel: +351 217 214 603 Fax: +351 217 270 522 amartinho@cnpce.gov.pt
Merad, Myriam	INERIS National Institute for Industrial Risks & Environment Protection BP 2 – Parc Alata Verneuil en Halatte 60550 France	Tel: +33 (0)3 44 55 69 25 Fax: +33 (0)3 44 55 69 55 myriam.merad@ineris.fr

Narmine Salah, Mahmoud	The National Center for Nuclear Safety and Radiation Control Atomic Energy Authority 3 Ahmed El-zomor St., Nasr City P.O. Box 7551 Cairo, Egypt	natonarmine@yahoo.com
Pais, Isabel	National Council for Civil Emergency Planning Estrada da Luz 151 Lisboa 1600-153, Portugal	Tel: +351 963058682 Fax: + 351 21 7270522 ipais@cnpce.gov.pt
Pinhol, Anabela	Merck R.Alfredo da Silva 3-C Lisbon, Portugal	Tel: +351213613549 Fax: +351213605549 Anabela.Pinhol@merck.pt
Pité, Antonio	Tagus Park Porto Salvo 2740-120, Portugal	AGPite@isq.pt
Plum, Martin	Idaho National Laboratory US Department of Energy P.O. Box 1625 Idaho Falls, ID 83415-3878, USA	Martin.Plum@inl.gov
Ramadan, Abou Bakr	Atomic Energy Authority Nasr City P.O. Box 7551 Cairo, Egypt	Tel: 202 27 48 787 Fax: 202 27 40 238 ramadan58@yahoo.com
Reed, Russell V.	HydroPlan Portland, Oregon, USA	russellvreed@msn.com
Ribeiro, Lidia	Fuzzy, Engenharia de Sistemas e Decisão Lda Praceta Gastão Ferreira, 4B Paço de Arcos 2770-073, Portugal	Tel: 351 919423864 Fax: 351 214406062 lidia.m.ribeiro@gmail.com
Sá Marques, Francisco	Fuzzy, Engenharia de Sistemas e Decisão Lda Praceta Gastão Ferreira, 4B Paço de Arcos 2770-073, Portugal	Tel: 917837560 Fax: 214406061 mota.sa@netcabo.pt
Sanderson, Hans	National Environmental Research Institute, Aarhus University 399 Fr.borgvej Roskilde DK-4000 Denmark	Tel: +45-4630-1822 Fax: +45-4630-1114 hasa@dmu.dk
Santos Alves dos, António Manuel	Ordem Dos Engenheiros Avª. Ivens, Nº.1 Bloco B - 2º Esqº. Alfragide – Amadora 2610-268 Portugal	Tel: 217949119/8 Fax: 217941003 cpett@dgtt.pt

Shai, Hanan	Staregy Leadership 4/21 Habessor Street Herzlia 46328, Israel	hanansy@zahav.net.il
Shalhevet, Sarit	Shalhevet Consulting 126 Thorndike Street Brookline, MA 02246, USA	Tel: 617 879-0577 Fax: 617 879-0577 sarit.shalhevet@gmail.com
Shamir, Eitan	IDF Staff & Command College Israel	eitan.shamir@skynet.be
Sheppard, Ben	Simfore 16 Brickwood Road Croydon CR0 6UL, UK	Tel: +447900225534 Fax: +448450523460 ben.sheppard@simfore.com
Slavin, David	Institute of Biomedical Engineering Imperial College London London, UK	David.Slavin@pfizer.com
Soares, Carlos Guedes	Technical University Lisbon Avenue Ravisco Pais Lisbon 1048-001, Portugal	guedess@mar.ist.utl.pt
Soares, Manuel	Technical University Lisbon Avenue Ravisco Pais Lisbon 1048-001, Portugal Soares, Mário Merck 870-54 Ruan Rel Ramiro Villa Nova DeGalo 4400, Portugal	mario.soares@merck.pt
Steevens, Jeffery	U.S. Army ERDC 3909 Halls Ferry Road Vicksburg, MS 39180, USA	Tel: 601-634-4199 Fax: 601-634-2263 Jeffery.A.Steevens@erdc.usace. army.mil
Strimling, David	DecisionAiding 20/10 Roalovan Ra'Ana 43072 Israel	david.strimling@gmail.com
Sullivan, Terry	Environmental Research and Technology Division Brookhaven National Laboratory Bldg 830 34 N. Railroad Ave Upton, NY 11973, USA	Tel: 631 344-2840 Fax: 631 344-4486 TSullivan@bnl.gov
Tervonen, Tommi	Centre for Management Studies Instituto Superior Técnico, Taguspark Porto Salvo 2780-990, Portugal	Tel: +351 96 529 1326 Fax: +351 214 233 568 tommi.tervonen@it.utu.fi

Thomaz, Joao

Instituto Superior Tecnico
Technical University of Lisbon
DEG, Tagus Park, Av. Cavaco
Silva
Porto Salvo 2780-990, Portugal

Tel: +351214233507
Fax: +351214233568
joaothomaz@netcabo.pt

Trindade, Elói

Tagus Park
Porto Salvo 2740-120, Portugal

DETrindade@isq.pt

Waller, Robert

Canadian Museum of Nature
Box 3443, Station "D"
Ottawa K1P 6P4, Canada

Tel: 613-566-4797
Fax: 613-364-4027
rwaller@mus-nature.ca

Wenning, Richard J.

ENVIRON International
Corporation
6001 Shellmound Street, Ste. 700
Emeryville, CA 94608, USA

Tel: 510-420-2556
Fax: 510-658-1191
rjwenning@environcorp.com

Yatsalo, Boris

Obninsk State University (IATE)
Studgorodok 1
Obninsk 249040, Russia

Tel: +7 48439 78560
Fax: +7 48439 70822
yatsalo@prana.obninsk.org

Zaidi, Mohammed

Idaho State University
121 North 18th Avenue
Pocatello, ID 83201-3345, USA

Tel: 208-282-2902
Fax: 208-282-4538
zaidmoha@isu.edu

Printed in the United States
132753LV00002B/13-21/P